Thermodynamics and Energy Engineering

Edited by Petrică Vizureanu

Published in London, United Kingdom

IntechOpen

Supporting open minds since 2005

Thermodynamics and Energy Engineering
http://dx.doi.org/10.5772/intechopen.79004
Edited by Petrică Vizureanu

Contributors
Murat Kunelbayev, Yedilkhan Amirgaliyev, Aliya Kalizhanova, Ainur Kozbakova, Salauat Daulbayev, Timur Merembayev, Didar Yedilkhan, Massimo Guarnieri, Piergiorgio Alotto, Federico Moro, Lindiwe Khotseng, Rabea Q. Nafil, Munaf S. Majeed, Jack Denur, Jonathan Deseure, Tianlong Deng, Samadiy Murodjon, Xiaoping Yu, Mingli Li, Ji Duo, Mugur Balan, Ancuta M Magurean, Octavian Pop, Adrian G Pocola, Alexandru Serban, Irina Petreanu, Mirela Dragan, Silviu Laurentiu Badea, Majdi Hazami, Farah Mehdaoui, AmenAllah Guizani, Hichem Tahouthi, Marco Noro, Renato Lazzarin, Paritosh C. Kulkarni

First published in London, United Kingdom, 2020 by IntechOpen
IntechOpen is the global imprint of INTECHOPEN LIMITED, registered in England and Wales, registration number: 11086078, 7th floor, 10 Lower Thames Street, London, EC3R 6AF, United Kingdom
Printed in Croatia

British Library Cataloguing-in-Publication Data
A catalogue record for this book is available from the British Library

Additional hard and PDF copies can be obtained from orders@intechopen.com

Thermodynamics and Energy Engineering
Edited by Petrică Vizureanu
p. cm.
Print ISBN 978-1-83880-568-5
Online ISBN 978-1-83880-569-2
eBook (PDF) ISBN 978-1-83880-570-8

For EU product safety concerns:
IN TECH d.o.o., Prolaz Marije Krucifikse Kozulić 3, 51000 Rijeka, Croatia, info@intechopen.com or visit our website at intechopen.com.

Meet the editor

Petrică Vizureanu was born on October 17, 1967, in Bârlad, Romania. He obtained an MSc and PhD in Heating Equipment from The "Gheorghe Asachi" Technical University from Iasi in 1992 and 1999, respectively. Dr. Vizureanu is currently full Professor and Scientific Supervisor in Materials Engineering (since 2009) at the same university. His research interests include expert systems for heating system programming, computer assisted design for heating equipment, heating equipment for materials processing, heat transfer, biomaterials, and geopolymers. Dr. Vizureanu has published more than 140 papers in international journals and conferences (proceedings) and thirty-three books.

Contents

Preface

The theoretical approach of this book is to develop a primary survey of basic thermodynamic concepts, allowing one to predict states of a fuel cell system, including its potential, temperature, pressure, volume, and moles. The specific topics explored include enthalpy, entropy, specific heat, Gibbs free energy, net output voltage irreversible losses in fuel cells, and fuel cell efficiency. Thermodynamics is the study of energy change from one state to another. The predictions that can be made using thermodynamic equations are essential for understanding fuel cell performance, as a fuel cell is an electrochemical device that converts the chemical energy of a fuel and an oxidant gas (air) into electrical energy.

Any system producing energy obeys the laws of thermodynamics. The amount of work/heat produced depends on thermodynamic values for reversible reactions, whereas for irreversible reactions overpotential is required to complete the work.

Herein is a review of modeling techniques for three types of fuel cells that are gaining industrial importance, namely, polymer electrolyte membrane (PEMFC), direct methanol (DMFC), and solid oxide (SOFC) fuel cells (FCs). The models presented are both multi-dimensional, suitable for investigating distributions, gradients, and inhomogeneities inside the cells, and zero-dimensional, which allows for fast analyses of overall performance and can be easily interfaced with or embedded in other numerical tools. The thermal dependence is considered in all models. Some special numerical approaches for facing specific problems are also presented.

In addition, the book presents research on the F-diagram method for the solar system via a thermosyphon circulation-diagram method based on the correlation of a number of simulations and computed non-dimensional variables. Modeling conditions varies in corresponding ranges of thermosyphon with double circuit circulation of the solar system. By means of the F-diagram method, environmental monthly temperature values with correction index have been computed, showing that the monthly average daily heating degree and direct solar radiation decrease according to weather conditions.

Starting from the equation of Einstein, the next chapter proposes a simple and fundamental presentation of fission and fusion principles, together with some of their applications, including in nuclear reactors and nuclear propulsion vessels and submarines. Fission and fusion are chosen as the most important forms of nuclear energy directly related with the equation of Einstein.

Renewable energies, which are by nature variable, are subject to both daily and/or seasonal intermittencies. Electrochemical devices have been successful in proving their applicability in terms of energy storage (power to gas). Controlling in real time, predicting the performance is the advantage that electrochemical generators can offer.

Regarding how to build simple models of PEMFC for fast computation, Artificial Intelligence and mathematic tools can make smart grids smarter. The electrochemical modeling can be coupled successfully with an AI approach if these models can be quickly computed with good numerical stability.

This book contains twelve chapters over two parts: "Theoretical Models" and "Applications."

The second section of the book includes an overview and general considerations about improving heat-engine performance via high-temperature recharge. Moreover, it investigates the efficiencies of heat-engine operation employing various numbers of heat reservoirs. It discusses operation with the work output of the heat engines sequestered, as well as with it being totally frictionally dissipated. We consider mainly heat engines whose efficiencies depend on ratios of a higher and lower temperature or on simple functions of such ratios, but also provide brief comments concerning more general cases.

Opportunity and perspective of energy storage in Phase Change Material (PCM) wall used in buildings' application are described further and the heat transfer phenomena and effect of it integrated inside a building on its indoor thermal comfort is investigated. A numerical investigation using specific software was also achieved to solve the energy and the exergy mathematic relations to evaluate the PCM wall performances by determining the melting phase proprieties during the charging and the discharging process. PCM is discussed as an application for water desalination to store solar energy. A possible solution to this problem is harnessing solar energy to engender thermal energy for solar distillation. Thus, solar distillation is one of the potential solutions to asses both the ever-increasing demands for clean water and for finding eco-friendly techniques to yield water. This analysis was undertaken to discover the possible utilization of PCM on solar distillation in a double slope solar still.

Another important subject covered in the book is lithium recovery from brines including seawater, salt lake brine, underground water, and geothermal water. This is very important due to the fact that the lithium market is expected to grow from 184,000 TPA of lithium carbonate to 534,000 TPA by 2025. To ensure the growing consumption of lithium, it is necessary to increase the production of lithium from different resources. Natural lithium resources mainly associate within granite pegmatite type deposits, salt lake brines, seawater, and geothermal water. Among these, the reserves of lithium resource in salt lake brine, seawater, and geothermal water are 70%–80% of the total, and are excellent raw materials for lithium extraction. Compared to the minerals, the extraction of lithium from water resources is promising because this aqueous lithium recovery is more abundant, more environmentally friendly, and cost-effective.

Last two chapters of the book examine fuel cells as alternative energy sources. Thus, they are expected to be a replacement for thermal engines and rechargeable batteries within the next few years as they are emission-free and not subject to Carnot restrictions. Fuel cells can be manufactured in different sizes depending on the amount of energy required. Herein, you will find a demonstration of the principle of work involved in fuel cells and their structural components as well as ideas to enhance output power.

Furthermore, the reader will find a classification and an overview of fuel cells, including working principles, equations of the governing reactions, and main applications. A brief exposure of thermodynamically and electrochemical theory describe the functioning of fuel cells. Also, the book details PEMFC assembly, starting with the schematic presentation of the main components, the role of each component, specific materials and their requested properties, and the way of assembling the components into the device. Conclusions are presented, challenges related to reliability and cost are addressed, and targets for future development of PEMFC for mobile and stationary applications are discussed.

In this volume, scientists will find information starting from a theoretical approach of fuel cells to their use as alternative energy sources for stationary, mobile, and automotive applications.

Petrică Vizureanu
The "Gheorghe Asachi" Technical University Iaşi,
Romania

Section 1

Theoretical Models

Chapter 1

Fuel Cell Thermodynamics

Lindiwe Khotseng

Abstract

Thermodynamics is the study of energy change from one state to another. The predictions that can be made using thermodynamic equations are essential for understanding fuel cell performance, as a fuel cell is an electrochemical device that converts the chemical energy of a fuel and an oxidant gas into electrical energy. When a fuel cell is operating, some of the input is used to create electrical energy, but another portion is converted into thermal energy, depending on the type of fuel cell. Based on the first and second laws of thermodynamics, one can write down thermodynamic potentials to specify how energy can be transferred from one form to another. This chapter examines how electrical energy and thermal energy are transferred in the hydrogen fuel cell system. It also defines how reversible fuel cell voltages, which are the maximum fuel cell performances, are affected by departures from the standard state. Basic thermodynamic concepts allow one to predict states of the fuel cell system, including the potential, temperature, pressure, volume and moles of a fuel cell. The specific topics explored in this chapter include enthalpy, entropy, specific heat, Gibbs free energy, net output voltage irreversible losses in fuel cells and fuel cell efficiency.

Keywords: enthalpy, entropy, Gibbs free energy, specific heat, fuel cell efficiency, hydrogen fuel cell

1. Introduction

Fuel cells are electrochemical devices that convert chemical energy into work in the form of electric energy and heat. Any system producing energy obeys the laws of thermodynamics. The amount of work/heat produced depends on thermodynamic values for reversible reactions, while for irreversible reactions overpotential is required to complete the work. Hydrogen and oxygen are used to illustrate the simplest case. A general thermodynamic analysis of hydrogen fuel cells of the reversible work for the reversible reaction is performed. The concepts enthalpy, specific heat, entropy and Gibbs free energy are related to the reacting systems in fuel cells. Gibbs free energy is the thermodynamic potential that measures the reversible work by a thermodynamic system at constant pressure and temperature. Change in enthalpy and change in entropy are significant in particular to fuel cells; they indicate spontaneity of the adsorption process and increased randomness of adsorbate molecules on the solid surface, respectively. Specific heat is a measure of the amount of heat energy required to increase the temperature of a substance by 1°C. The fuel cell performance is examined through the reversible voltage, and the actual output voltage is after overpotential. The efficiency of a fuel cell is the useful energy output which is the electrical energy produced, and the energy input is the enthalpy of hydrogen.

2. The thermodynamics of a fuel cell

A fuel cell, also known as a galvanic or voltaic cell, is a well-known example of a device that works by changing chemical energy into electrical energy, which is exhibited in terms of cell potential and electrical current output. The maximum possible electrical energy output and the corresponding electrical potential difference between the cathode and anode are achieved when the fuel cell is operated under thermodynamically reversible conditions, as shown in **Figure 1**, a fuel cell system to which fuel and oxidant streams enter and product stream exits. Unfortunately, it is inevitable that some of the energy will be dissipated as heat.

The overall electrochemical reactions occurring inside the fuel cell system boundary are described as follows:

$$H_2\ (\text{fuel}) + O_2\ (\text{oxidant}) \rightarrow W + Q + H_2O\ (\text{product}) \tag{1}$$

where W is the rate of electrical work done by the system and Q is the rate of heat transferred into the system from the surroundings at constant pressure and temperature.

Electrical work is, in general, described by the relation:

$$W = EI\Delta t \tag{2}$$

where E is the cell voltage and I is the current. In a fuel cell reaction, electrons are transferred from the anode to the cathode, generating a current.

The amount of electricity ($I\Delta t$) transferred when the reaction occurs is given by nF, where n is the number of electrons transferred and F is Faraday's constant = 96,493 coulombs. The electrical work can hence be calculated as:

$$W = -nFE_{cell} \tag{3}$$

The Gibbs free energy is the maximum amount of work done on the system:

$$W_{el} = \Delta G \tag{4}$$

$$\Delta G = -nFE_{cell} \tag{5}$$

Hence the maximum cell potential or the reversible cell potential becomes:

$$E_{rev} = -\frac{\Delta G}{nF} \tag{6}$$

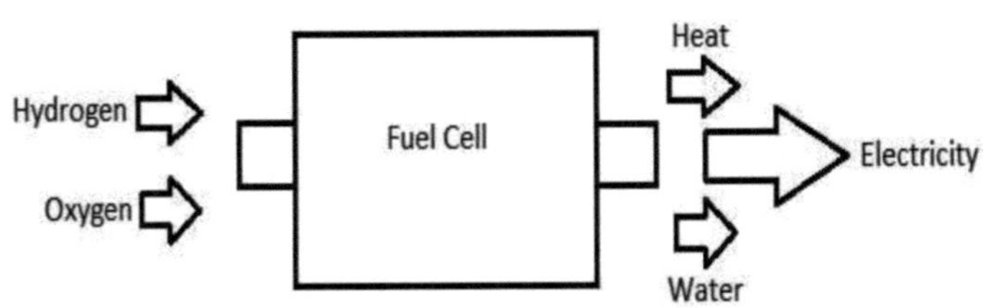

Figure 1.
Simple H_2/O_2 fuel cell diagram.

where E° is also called the reversible voltage, because it is the maximum possible voltage without any irreversible losses. This is the maximum possible voltage of an electrochemical cell, since it is attained assuming a reversible process. If we are looking at the redox reaction on a per-mole-of-fuel basis, the absolute Gibbs function is equivalent to the molar specific value. All fuel cell losses are associated with deviation from this maximum. Since F and n are constants for a particular global redox reaction, the functional dependence of the maximum possible voltage of an electrochemical cell is related strictly to the dependencies of the Gibbs free energy, namely, temperature and pressure of the reactants and products. If all the potential chemical energy for a reaction went into electrical work and there was no heat transfer, there would be no entropy change; dG = dH. In this case, we can show that:

$$E_{rev} = -\frac{\Delta H}{nF} \tag{7}$$

For a generic reaction:

For a general reaction or process of A and B giving products C and D:

$$aA + bB \rightarrow cC + dD \tag{8}$$

$$\Delta G_f = \Delta G^o{}_f + RT\ \ln\left[\frac{aC^c aD^d}{aA^a aB^b}\right] \tag{9}$$

where a's are the thermodynamic activity coefficients for the reacting species. To convert to voltage, we can divide it by nF:

$$E\,(T,P) = \underbrace{E^o}_{I} - \underbrace{\frac{RT}{nF}\ \ln\left[\frac{aC^c aD^d}{aA^a aB^b}\right]}_{II} \tag{10}$$

where I is the standard voltage evaluated at 1 atm pressure for all components and II accounts for the thermodynamic activity dependence on the Nernst voltage.

i. For an ideal gas, a = P_i/P^o, where P_i is the partial pressure of the species of interest and P° is the reference pressure, 1 atm.

ii. For water vapour, the partial pressure of the vapour cannot exceed the saturation pressure, P_{sat}, which is a function of temperature. Thus, the reference pressure is set to P_{sat}, and a = P_v/P_{sat}, which is the relative humidity, RH. This can normally be considered to be 1.0 in the immediate molecular region of the water-generating electrode. This is a reasonable assumption because water generation is always at the catalyst surface and the activity of water here is 1.0. Also, the reaction itself is not limited by the product water concentration at this surface:

$$E\,(T,P) = E^o - \frac{RT}{nF}\ \ln\left[\frac{\left(\frac{P_C}{P^o}\right)^c\left(\frac{P_D}{P^o}\right)^d}{\left(\frac{P_A}{P^o}\right)^a\left(\frac{P_B}{P^o}\right)^b}\right] \tag{11}$$

where the partial pressures are evaluated at the particular electrode where the reaction involving the species occurs. Using this expression, we can solve for the expected maximum (Nernst) voltage for a given fuel cell reaction. Two important points are as follows:

i. The Nernst equation is a result of the equilibrium established at the electrode surfaces. A particular gradient can exist between the concentration of a species in the channel of a fuel cell and the electrode, especially under high-current-density conditions, which cannot be considered a true thermodynamic equilibrium situation anyway.

ii. Only species directly involved in the electrochemical reaction of Eq. (8) are represented directly in the activity terms of Eq. (10). Species not participating in the electrochemical charge transfer reaction only indirectly alter the voltage through the species mole fractions of the participating species.

For the H_2/O_2 fuel cell potential, the open-circuit voltage is the maximum operating voltage (when no current is flowing) and is determined by the chemical thermodynamics of the overall cell reaction. The Nernst equation provides a relationship between the standard potential (E^o) for the cell reaction and the open-circuit voltage, where it can be determined at the partial pressures of reactants and products at temperature (T):

$$E\,(T,P) = E^o - \frac{RT}{2F}\ \ln\left[\frac{\left(\frac{yH_2OPcathode}{Psat}\right)}{\left(\frac{yH_2Panode}{P^o}\right)\left(\frac{yO_2Pcathode}{P^o}\right)^{1/2}}\right] \qquad (12)$$

To understand how the reversible voltage varies with temperature and pressure, respectively, we have previously shown that the Gibbs free energy is related to the reversible cell voltage by Eq. (5):

$$\Delta G = -nFE_{cell}$$

At constant pressure, the above relationship produces a Maxwell relation that links the change in open cell voltage with temperature T (a measurable quantity) to the change in entropy S [1]:

$$\left(\frac{\partial E}{\partial T}\right)_Q = \left(-\frac{\partial S}{\partial Q}\right)_T \qquad (13)$$

$$\left(\frac{\partial E}{\partial T}\right)_P = \frac{\Delta S}{nF} \qquad (14)$$

$$\Delta S = nF\left(\frac{\partial E}{\partial T}\right)_P \qquad (15)$$

$$E_{rev} = E^o{}_{rev} + \frac{\Delta S}{nF}\,(T - T^o) \qquad (16)$$

At constant temperature, Eq. (5) produces an equation that links voltage with pressure, p, to the change in volume:

$$\left(\frac{\partial E}{\partial P}\right)_T = -\frac{\partial v}{nF} \qquad (17)$$

If the volume change of the reaction is negative (if fewer moles of gas are generated by the reaction than consumed, for instance), then the cell voltage will increase with increasing pressure.

Usually, only gas species produce an appreciable volume change. Assuming that the ideal gas law applies, we can write Eq. (17) as:

$$\left(\frac{\partial E}{\partial P}\right)_T = -\frac{\Delta n_g RT}{nFp} \quad (18)$$

where Δn_g represents the change in the total number of moles of gas upon reaction. Pressure, like temperature, turns out to have a minimal effect on reversible voltage:

$$E_{rev} = E^o{}_{rev} - \frac{(\Delta n)RT}{nF} \ln P \quad (19)$$

2.1 Enthalpy of reaction for a fuel cell

Enthalpy is the thermodynamic quantity that states the total heat content of the system, which is the sum of all internal process in a closed system [2]. For homogeneous systems, enthalpy is solely based on the size of the system as it is an extensive property. Enthalpy (H) is the sum of the internal energy of the system (U) and the product of pressure (P) and volume (V) of the system. The change in enthalpy in a system is equal to the heat gained or lost in the system:

$$H = U + PV \quad (20)$$

The enthalpy change (ΔH) for a reaction in a fuel cell indicates the full amount of heat released by the reaction at a constant pressure; hence, enthalpy is simply equal to the heat released:

$$H = Q_p \quad (21)$$

At constant pressure and volume, the thermodynamic properties of the cell are related to the behaviour of its potential and are defined from the Gibbs-Helmholtz equation [3]:

$$\Delta H = \Delta G - T\left(\frac{\partial \Delta G}{\partial T}\right)_P \quad (22)$$

In accordance with Eq. (5), in terms of electrochemical processes, change in enthalpy can be written as:

$$\Delta H = nF\left[T\left(\frac{\partial Ecell}{T}\right)p - E_{cell}\right] \quad (23)$$

The overall reaction in Eq. (1) is the same as the reaction of hydrogen combustion. Combustion is an exothermic process, which means that there is energy released in the process [4, 5]:

$$H_2 + ½\, O_2 \rightarrow H_2O \qquad \Delta H < 0 \quad (24)$$

From the table of enthalpies of formation ($h^o{}_f$) and absolute entropies of formation ($s^o{}_f$) obtained from the basic thermodynamic data (see **Table 1**), the heat of formation of both liquid and vapour water can be calculated using the equation above to form:

The heat (or enthalpy) of a chemical reaction is the difference between the heat of formation of products and reactants. This means [6]:

$$\begin{aligned}\Delta H_f &= \sum \text{products} - \sum \text{reactants} \\ &= (h_f)_{H_2O} - \left[(h_f)_{H_2} + 1/2\,(h_f)_{O_2}\right]\end{aligned} \tag{25}$$

$\Delta H_{f,\ H_2O(l)}$ = HHV = −285.8 kJ/mol and $\Delta H_{f,\ H_2O\ (g)}$ = LHV = −241.8 kJ/mol.

The enthalpy of the hydrogen combustion reaction (Eq. (25)) is also called hydrogen's heating value. The 285.83 kJ/mol is known as hydrogen's higher heating value (HHV), which means that 1 mol of hydrogen is fully combusted with ½ mol of oxygen and cooled down to 25°C. If hydrogen is combusted with sufficient excess oxygen and cooled down to 25°C, the value will become 241.82 kJ/mol, which is known as hydrogen's lower heating value (LHV) [7]. The difference between the LHV and HHV of 44.01 kJ/mol is equal to the molar latent heat of water vaporisation at 25°C.

In the heating value for reactions involving water as a product, there is a choice in the calculation of thermodynamic voltages between a high heating value (HHV) and a low heating value (LHV), defined as follows for a given reaction:

- High heating value: It is assumed all the product water is in the liquid phase.
- Low heating value: It is assumed all the product water is in the gas phase.

Note that all calculations are based on HHV or LHV and do not necessarily correspond to the actual physical state of the product water at the fuel cell electrode. The terms HHV and LHV are used in combustion calculations as well, where the product water is nearly always in the gas phase. The difference between the two values is proportional to the latent heat of vaporisation of the liquid. The use of the LHV (gas-phase vapour product) will result in a lower calculated thermal voltage, since some energy is used for the latent heat of vaporisation of the liquid. In practice, the LHV is completely appropriate for high-temperature fuel cells, but the HHV is also commonly used. An important point regarding low-temperature fuel cells that is often confusing is that the choice of HHV or LHV is arbitrary and 100°C is not a point of demarcation between the two. Often 100°C is thought of as a natural boundary between the HHV and LHV because it is the phase change temperature of water at 1 atm pressure. The delineation between liquid and gas, however, is more complex and is related to the local vapour pressure and total pressure.

2.2 Entropy (S)

The concept of entropy is one of the thermodynamic parameters that are important to the science of fuel cells to understand. Entropy is defined as the

	h^o_f (kJ/mol)	s^o_f (kJ/mol.K)	ΔG^o_f (kJ/mol)
Hydrogen, H_2	0	0.131	0
Oxygen, O_2	0	0.205	0
Water (liquid), H_2O (l)	−285.8	0.070	−237.2
Water (vapour), H_2O (g)	−241.8	0.189	−228.6

Table 1.
Enthalpies of formation and absolute entropies of formation of fuel cell reactants and products (at 25°C and 1 atm).

measure of the unavailable energy in a closed thermodynamic system that is usually considered to be a measure of the system's disorder, known as the second law of thermodynamics. The total entropy of a system increases over time, as the molecular disorder increases. Therefore, if the system is in equilibrium, the change between the initial state and the final state, the system is going through a reversible change. Since entropy represents the unavailable energy used in the system, a system of zero entropy optimises the work output of the system. The fuel cell generates the amount of electricity and rejects an amount of thermal energy Q to its environment. As there is heat transfer, and it is a real system, there must be an increase in entropy. The amount of heat rejected and the maximum amount of electrical power that a fuel cell will generate can be determined by formulating the entropy changes occurring in the cell:

$$H_2 + O_2 \rightarrow H_2O + Q \tag{26}$$

The entropy of H_2 and O_2 will disappear, but the new entropy of H_2O and heat formation will appear. As long as the process is reversible, which is the assumption for the fuel cell, the entropy appearing in the rejected heat can be written as [8]:

$$\Delta S = \frac{\Delta Qrev}{T} \tag{27}$$

The equation for the change of entropy (ΔS) is equal to the change in enthalpy (ΔQ) divided by the temperature (T) of the system.

As there is no heat transfer in electrical work, the entropy is zero. The entropy in an H_2/O_2 fuel cell can be calculated using the absolute entropy values given in **Table 1**. The second law of thermodynamics requires that in a fuel cell, there will be a net increase in entropy. Therefore, the entropy that shows up in the rejected heat and the product water (liquid water) must be greater than the entropy contained in the reactants (H_2 and O_2) [9, 10]:

Entropy gain ≥ entropy loss

$$\frac{Q}{T} + \sum \text{products} \geq \sum \text{reactants} \tag{28}$$

$$Q \geq T\,[\text{products} - \text{reactants}] \tag{29}$$

To calculate the amount of heat rejected per mole of H_2:

$$\sum S_{reactants} = 0.131\ \frac{kJ}{mol.k} \times 1\ mol\ H_2 + 0.205\ \frac{kJ}{mol.k} \times ½\ mol\ O_2$$

$$= 0.2335\ kJ/mol.K$$

$$\sum S_{products} = 0.070\ \frac{kJ}{mol.k} \times 1\ mol\ H_2O\ (l)$$

$$= 0.070\ kJ/mol.K$$

The minimum of heat released during the reaction:

$$Q_{min} = T\left(\sum S_{reactants} - \sum S_{products}\right) = 298\ k\ (0.0134 - 0.070)$$

$$= 48.72\ kJ\ per\ mole\ of\ H_2$$

As heat capacity effects are generally minor, ΔH and ΔS values are usually assumed to be independent of temperature. A simplified entropy calculation can be with entropy values obtained from **Table 1**:

$$\Delta S_f = (s^o)_{H_2O} - \left[(s^o)_{H_2} + 1/2\,(s^o)_{O_2}\right] \tag{30}$$

$\Delta S_{f,\,H_2O\,(l)}$ = HHV = −0.163 kJ/mol.K and $\Delta S_{f,\,H_2O\,(g)}$ = LHV = −0.045 kJ/mol.K.

2.3 Gibbs free energy

From the second law of thermodynamics, the change in free energy, or maximum useful work, can be obtained when a 'perfect' fuel cell operating irreversibly is dependent upon temperature. Therefore, W_{el}, the electrical power output, is [11]:

$$\Delta G = \Delta\,H - T\,\Delta S \tag{31}$$

where H is the total energy of the system, S is the 'unavailable' energy and G is the 'free' energy, or the energy available to do useful work.

The change in Gibbs free energy varies with both temperature and pressure. It can be shown that for a H_2/O_2 fuel cell:

$$\Delta G_f = \Delta G^o{}_f + RT\,\ln\left[\frac{pH_2O}{pH_2 p\sqrt{O_2}}\right] \tag{32}$$

where $\Delta G^o{}_f$ is the change in Gibbs free energy at standard pressure, which varies with the temperature T of the fuel cell, in Kelvin; p_{H_2}, p_{O_2} and P_{H_2O} are the partial pressure of the hydrogen, oxygen and vapour, respectively; and R is the universal gas constant (8.314 J/(kg.K)). The fact that the value of $\Delta G^o{}_f$ is negative means that the energy is released from the reaction [12–14].

For the H_2/O_2 fuel cell reaction, the change in Gibbs free energy is:

$$\Delta G_f = G_{f,products} - G_{f,reactants} = G_{f,H_2O} - \left[G_{f,H_2} + G_{f,O_2}\right] \tag{33}$$

The maximum possible electrical energy output and the corresponding electrical potential difference between the cathode and anode are achieved when the fuel cell is operated under the thermodynamically reversible condition. This maximum possible cell potential is called 'reversible cell potential', one of the significantly important parameters for FC.

From Eq. (23) we calculated ΔH to be 285.830 kJ/mol for hydrogen's HHV and 241.98 kJ/mol for hydrogen's LHV, while for ΔS in Eq. (28), the entropy of reaction is 0.163 kJ/mol for HHV and for LHV 0.044 kJ/mol.

To calculate Gibbs free energy for H_2/O_2 fuel cell reaction:

$$\Delta G_f\ (l) = HHV = -285.8\ kJ/mol - (273.15\ K)\ (-0.1633\ kJ/mol) = -237.1\ kJ/mol$$

$$\Delta G_f\ (g) = LHV = -241.8\ kJ/mol - (273.15\ K)\ (-0.045\ kJ/mol) = -228.6\ kJ/mol$$

The fact that the value of $\Delta G^o{}_f$ is negative means that the energy is released from the reaction. From Eq. (6), the potential or reversible open cell voltage E^o of any electrochemical device is defined as:

$$E^o = -\frac{\Delta G}{nF}$$

where n is the amount of exchanged electrons and F is Faraday's constant. For the hydrogen oxidation or water formation, n = 2. The free enthalpies ΔG of water formation are either:

$$\Delta G_{f,\,H_2O\ (l)} = -237.13\ kJ/mol \text{ or } \Delta G_{f,H_2O\ (g)} = -228.57\ kJ/mol$$

The corresponding potential is therefore (**Table 2**):

$$E_l{}^o = \frac{-\Delta G_f H_2O(l)}{2F} = -1.23\ V \text{ and } E_g{}^o = \frac{-\Delta G_f H_2O(g)}{2F} = -1.18\ kJ/mol$$

2.4 Specific heat capacity

Another property that is important in thermodynamics and the study of fuel cells is the specific heat. The specific heat of a solid or liquid is usually defined as the heat required to raise unit mass through 1 degree temperature rise. For a gas there are an infinite number of ways in which heat may be added between any two temperatures, and hence a gas could have an infinite number of specific heats. However, only two specific heats for gases are defined, as specific heat at constant volume, c_v, and specific heat at constant pressure, c_p.

For a reversible ideal gas, a non-flow process at constant pressure and at constant volume is given as, respectively:

$$dQ = Cp\ dT \tag{34}$$

and

$$dQ = Cv\ dT \tag{35}$$

For a substance that is modelled as incompressible, the specific heats are assumed to be equal, Cp = Cv [15]:

$$Cp = Cv = \frac{dQ}{dT} \tag{36}$$

For an ideal gas in which a fuel cell is assumed to be, the specific enthalpy depends only on temperature:

$$h_f = dQ = CpdT \tag{37}$$

$$h_T = h_{298.15} + \int_{298.15}^{T} C_p dT \tag{38}$$

where $h_{298.15}$ is the enthalpy at a reference temperature.

The specific heat can also be related to specific entropy at temperature T:

	ΔH (kJ/mol)	ΔG (kJ/mol)	ΔS (kJ/mol.K)	E (V)
$H_2 + 1/2O_2 \rightarrow H_2O$ (l)	−285.8	−237.1	−0.163	1.23
$H_2 + 1/2O_2 \rightarrow H_2O$ (g)	−241.8	−228.6	−0.045	1.18

Table 2.
Enthalpies, entropies and Gibbs free energy of H_2/O_2 fuel cell reaction in (kJ/mol.K) and the resulting theoretical cell potential at 25°C.

$$dS = \frac{dQ}{T} = Cp\,\frac{dT}{T} \tag{39}$$

$$s_T = s_{298.15} + \int_{298.15}^{T} \frac{1}{T} C_P dT \tag{40}$$

The specific enthalpy and entropy for H_2/O_2 fuel cell are given by [16]:

$$\Delta H = \left[h^{o}{}_{f,H_2O} + Cp,\ _{H_2O}\,(T - 298.15)\right]_{H_2O} - ½\left[h^{o}{}_{f,O_2} + Cp,\ _{O_2}(T - 298.15)\right]_{O_2} - \left[h^{o}{}_{f,H_2} + Cp,\ _{H_2}(T - 298.15)\right]_{H_2} \tag{41}$$

$$\Delta S = \left[h^{o}{}_{f,H_2O} + Cp,\ _{H_2O}\,\ln\frac{T}{298.15}\right]_{H_2O} - ½\left[h^{o}{}_{f,O_2} + Cp,\ _{O_2}\ln\frac{T}{298.15}\right]_{O_2} - \left[h^{o}{}_{f,H_2} + Cp,\ _{H_2}\ln\frac{T}{298.15}\right]_{H_2} \tag{42}$$

The values of molar entropy and enthalpy of formation at 298.15 K are given in **Table 1** [17]. Eqs. (38) and (40) can be used to determine the specific heat capacity, Cp, at constant pressure. Specific heat relationships are generally modelled with a high-order polynomial, such as those listed below for hydrogen fuel cell gases, valid in the range of 300–1000 K [18]. It is reported that over a range of temperatures, Cp is not constant, while over the range of 300–350 K, the obtained Cp values are 0.6% accurate [4].

Hydrogen, H_2:

$$Cp\ (T) = 3.057 + 2.677 \times 10^{-3}\ (T) - 5.810 \times 10^{-6}\ (T)^2 + 5.521 \times 10^{-9}\ (T)^3 - 1.812 \times 10^{-12}\ (T)^4$$

Oxygen, O_2:

$$Cp\ (T) = 3.626 - 1.878 \times 10^{-3}\ (T) + 7.055 \times 10^{-6}\ (T)^2 - 6.764 \times 10^{-9}\ (T)^3 + 2.156 \times 10^{-12}\ (T)^4$$

For H_2o:

$$Cp\ (T) = 4.070 - 1.808 \times 10^{-3}\ (T) + 4.152 \times 10^{-6}\ (T)^2 - 2.964 \times 10^{-9}\ (T)^3 + 0.807 \times 10^{-12}\ (T)^4$$

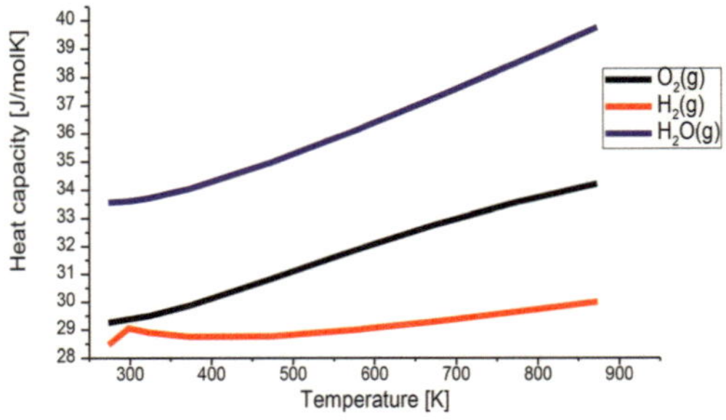

Figure 2.
Specific heat values for hydrogen, oxygen and water as a function of temperature.

The heat capacity values for hydrogen, oxygen and water as a function of temperature are shown in **Figure 2**. The data were obtained from Kabza [4].

3. Fuel cell thermodynamic efficiency

In the process of energy conversion in a fuel cell, the initial chemical energy between the enthalpy of the products and reactants is converted into electrical energy and thermal energy, as stated in the first law of thermodynamics. The efficiency of any energy conversion device is defined as the ratio between useful energy output and energy input [19–22]:

$$\eta = \frac{\text{actual electrical work}}{\text{maximum available work}}$$

$$\eta = \frac{\Delta G}{\Delta H} = \frac{\Delta H - T\Delta S}{\Delta H} \tag{43}$$

The maximum possible thermodynamic efficiency of a fuel cell can be written as [9]:

$$\eta == 1 - \frac{T\Delta S}{\Delta H} \tag{44}$$

In the case of a fuel cell, the useful energy output is the electrical energy produced, and the energy input is the enthalpy of hydrogen, that is, hydrogen's HHV. Assuming that all of the Gibbs free energy can be converted into electrical energy (the reaction is reversible), the maximum theoretical efficiency of a fuel cell is [23, 24] (**Figure 3**):

$$\eta = \frac{\Delta G}{\Delta H} = \frac{237.34}{286.02} \text{ x } 100\% = 83\%$$

For hydrogen's LHV, the fuel cell efficiency would be [26]:

$$\eta = \frac{\Delta G}{\Delta H} = \frac{228.74}{241.98} \times 100\% = 94.5\%$$

The LHV has higher efficiency compared to HHV, because the reversible efficiency of the fuel cell decreases as the operating temperature increases [27].

The expected fuel cell efficiency is not always achieved due to thermodynamic and electrochemical irreversible losses [28].

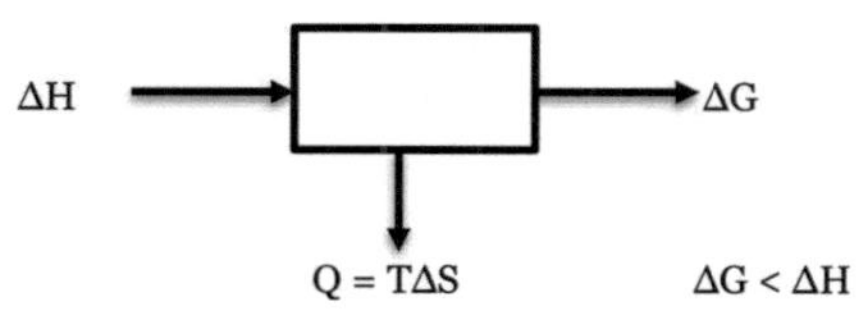

Figure 3.
Energy inputs and outputs for a fuel cell as an energy conversion device [25].

4. Irreversible losses

Other than calculating energy quantities during the conversion of chemical energy to electrical energy, there is also the matter of electron flow through materials in the fuel cell process. The single fuel cell provides a voltage dependent on operating conditions such as temperature, applied load and fuel/oxidant flow rates [29, 30]. If a fuel cell is supplied with reactant gases, but the electric current is not closed, it will not generate any current, and one would expect the cell potential to be at the theoretical cell potential for the given conditions (temperature, pressure and concentration of reactants). In reversible conditions, the energy loss is the heat lost towards the environment, $T\Delta s$, due to negative entropy [12].

However, in practice, the thermodynamic cell potential is decreased from its ideal potential, usually less than 1 V, due to irreversible losses known as overpotential or polarisation [17]. The fuel cell performance overpotential is due to [14, 31]:

i. Activation overpotential: The activation polarisation is related to the charge transfer processes occurring during the electrochemical reactions on electrode surfaces. The losses are caused by the slowness of the reactions taking place on the surface of the electrodes [32]. Activation polarisation depends on the nature of type of electrode, ionic interactions, ion-solvent interactions and the electrode-electrolyte interface [33].

ii. Ohmic overpotential: In most fuel cells, the most important contribution to this resistance is the electrolyte, due to the ionic nature of its conductivity, resistance to the flow of electrons through the electrodes and the contact resistance at the cell terminals.

iii. Mass transport (concentration) overpotential: Concentration polarisation occurs due to a decrease in the concentration of the reactants at the electrode-electrolyte interface. Due to diffusion or convection problems in the electrolyte, the concentration of the reactants is not maintained at the initial level. Reaction product accumulation can also cause a dilution of reactants. The concentration gradient thereby formed causes a drop in electrode activity, and the terminal voltage is reduced.

iv. Fuel crossover overpotential: 'Crossover' is one of the common effects occurring in alcohol fuel cells [22]. Although the electrolyte, a polymer membrane, is not electrically conductive and practically impermeable to reactant gases, some amount of fuel will diffuse from anode to cathode to react with oxygen, resulting in fewer electrons in the generated current of electrons that travel through an external circuit [34]. With this transit the cathode potential decreases, thus reducing the overall efficiency of a fuel cell. It occurs when the intermediates generated by fuel oxidation have higher concentration than oxygen at the cathode. The increase of temperature escalates the crossover effect [35].

The standard measure of performance is the polarisation curve, which represents the cell voltage behaviour against operating current density (**Figure 4**). From the figure, the voltage loss caused by mixed potential and crossover, activation polarisation, ohmic polarisation and mass transport losses is the most significant in the tail of the I-V curve. The maximum fuel cell is then examined through the reversible voltage of the system, which is calculated using thermodynamics and the actual voltage of the system [36]. The final voltage is lower than the thermodynamic voltage and is usually between 0.5 and 1.0 V. Although polarisations cannot be eliminated, material choice and electrode designs can contribute to their minimisation [37]:

$$E = E_{therm} - \eta_{act} - \eta_{ohm} - \eta_{conc} \quad (45)$$

where E is the cell potential, E_{therm} is the thermodynamic potential, η_{act} is the voltage loss due to activation polarisation, η_{ohm} is the voltage loss due to ohmic polarisation and η_{conc} is the voltage loss due to mass transport polarisation and the entropy generation results [38–41]:

$$S = \frac{E}{nFT} \quad (46)$$

In addition, the related heat lost for irreversibility can be calculated as (**Figure 4**):

$$q = T\Delta s - nFE = \Delta H - \Delta G - nFE \quad (47)$$

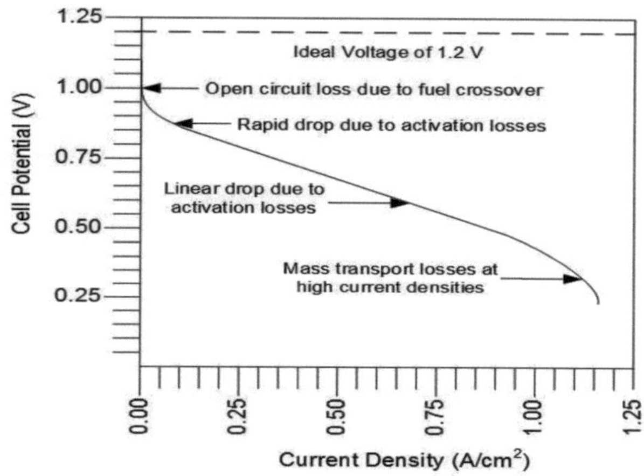

Figure 4.
H_2/O_2 polarisation curve at equilibrium and voltage losses in fuel cell [30].

5. Conclusion

Thermodynamics is used to understand the process of energy conversion in fuel cells. The determination of a fuel cell's performance depends on thermodynamic evaluation. The heat potential of a fuel is given by the enthalpy of the reaction. Not all heat potential of a fuel can be used to perform useful work; the reversible work of a fuel is defined by Gibbs free energy, which is the electrical work. The study of the electrical effects shows that the molar flow of the fuel used is proportional to the electric current and the reversible work is proportional to the reversible voltage. The cell voltage varies with temperature, pressure and reactant/product activities. Irreversible losses cause a difference in the efficiency of reversible and real processes, with efficiency of real processes always less than reversible processes. The losses are due to two major reasons, namely, irreversible kinetic losses and fuel utilisation losses.

Acknowledgements

We greatly appreciate the National Research Foundation and the Tertiary Education Support Programme for financial support as well as the Chemistry Department at the University of the Western Cape.

Author details

Lindiwe Khotseng
Department of Chemistry, University of the Western Cape, Cape Town, South Africa

*Address all correspondence to: lkhotseng@uwc.ac.za

References

[1] Yong-Qiang X, Chun-Hui L, Jin-Chuan F. The electrochemical thermodynamics for chemical reactions in dispersed cells. Journal of Colloid and Interface Science. 1999;**217**:107-110. DOI: 10.1006/jcis.1999.6221

[2] Tran D, Tanner G, Yang K, Woodley R. Thermodynamic effects of nanotechnological augmentation of hydrogen fuel cells. PAM Review. 2017: 76-86. DOI: 10.5130/pamr.v4i0.1443

[3] Spiegel C. PEM Fuel Cell Modeling and Simulation Using Matlab. New York, NY: Elsevier; 2008. 443 p

[4] Kabza A. Fundamentals. In: Fuel Cell Formulary [Internet]. 2015. pp. 30–34. Available from: http://www.pemfc.de/FCF_A4.pdf [Accessed: 2019-09-11]

[5] Winter M, Brodd J. What are batteries, fuel cells, and supercapacitors. Chemical Reviews. 2004;**104**(10): 4245-4270. DOI: 10.1021/cr02030k. Available from: https://doi.org/10.1021/cr02030k

[6] Filho LP. Thermodynamic studies and applications of polymeric membranes to fuel cells and microcapsules [thesis]. Tarragona: Universitat Rovira i Virgili; 2007

[7] Blomen LJMJ, Mugerwa MN. Fuel cell systems. In: Srinivasan S, Dave BB, Murugesamoorthi KA, Parthasarathy A, Appleby AJ, editors. Overview of Fuel Cell Technology. New York, NY: Plenum Press; 1993. pp. 37-72

[8] Bejan A. Advanced Engineering Thermodynamics. Hoboken, NJ: Wiley; 2016. 741 p

[9] Larminie J, Dicks A. Fuel cell systems explained. West Sussex: Wiley; 2003. 401 p

[10] Sieniutycz S. Thermodynamic aspects of power generation in imperfect fuel cells: Part I. International Journal of Ambient Energy. 2010;**31**(4): 195-202. DOI: 10.1080/01430750.2010.9675812

[11] Kakaç S, Pramuanjaroenkij LA, Vasiliev A. Mini-Micro Fuel Cells. In: Barbir F, editor. Fuel Cell Basic Chemistry and Thermodynamics. New York, NY: Academic Press; 2012. pp. 13-26. DOI: 10.1007/978-1-4020-8295-5_2

[12] Offer GJ, Mermelstein J, Brightman E, Nigel PB. Thermodynamics and kinetics of the interaction of carbon and Sulphur with solid oxide fuel cell anodes. Journal of the American Ceramic Society. 2009; **92**(4):763-780. DOI: 10.1111/j.1551-2916.2009.02980.x

[13] Kee RJ, Zhu H, Goodwin DG. Solid oxide fuel cells with hydrocarbon. Proceedings of the Combustion Institute. 2005;**30**(2):2379-2404. DOI: 10.1016/j.proci.2004.08.277

[14] Nether P. Two-dimensional transient model of a cascaded micro-tubular solid oxide fuel cell fed with methane. Journal of Power Sources. 2006;**157**(1):325-334. DOI: 10.1016/j.jpowsour.2005.07.077

[15] Moran MJ, Shapiro HN. Fundamentals of Engineering Thermodynamics. 5th ed. New York, NY: Wiley; 2006. 831 p

[16] EG & G Technical Services Inc. Fuel Cells: A Handbook. 7th ed. Morgantown; 2004. 427 p

[17] Kotz JC, Treichel PM, Townsend JR, Treichel DA. Principles of chemical reactivity: Energy and chemical reactions. In: Chemistry and Chemical Reactivity. 10th ed. Stamford: Cengage Learning; 2015. pp. 180-217

[18] Mench M. Fuel Cell Engines. Hoboken, NJ: Wiley; 2008 515 p

[19] Singhal SC, Kendall K, editors. High-Temperature Solid Oxide Fuel Cells: Fundamentals, Design and Applications. Amsterdam: Elsevier; 2003. 406 p

[20] Li X. Thermodynamic performance of fuel cells and comparison with heat engines. In: Zhao TS, Kreuer K-D, Van Nguyen T, editors. Advances in Fuel Cells. Oxford: Elsevier; 2007, p. 1-46

[21] Haynes C. Clarifying reversible efficiency misconceptions of high temperature fuel cells in relation to reversible heat engines. Journal of Power Sources. 2001;**92**:199-203. DOI: 10.1016/S0378-7753 (00)00541-3

[22] Dos Santos MC, Parreira LS, De Moura Seuza F, Camarge Junior J, Gentil T. Fuel Cells: Hydrogen and Ethanol Technologies. In: Hashmi S, editor. Reference Module in Materials Science and Materials Engineering. Oxford: Elsevier; 2017. p. 1-21. DOI: 10.1016/8978-0-12-803581-8.09263-8

[23] Pavelka M, Marski F. Detailed thermodynamic analysis of polymer electrolyte membrane fuel cells efficiency. International Journal of Hydrogen Energy. 2013;**38**(17): 7102-7113. DOI: 10.1016/j. ijhydene.2013.03.149

[24] Barbir F. Fuel cell basics. In: Kakaç S, Pramuanjaroenkij A, Vasilie L, editors. Mini-Micro Fuel Cells: Fundamentals and Applications. Amsterdam: Springer; 2007. pp. 13-26

[25] Barbir F. PEM Fuel Cells: Theory and Practice. Waltham: Elsevier; 2013. 509 p

[26] Lutz AE, Larson RS, Keller JO. Thermodynamic comparison of fuel cells to the Carnot cycle. International Journal of Hydrogen Energy. 2002; **27**(10):1103-1111. DOI: 10.1016/ S0360-3199(02)00016-2

[27] Reimer U, Lehnert W, Holade Y, Kokoh B. Irreversible losses in fuel cells. In: Hacker V, Mitsushima S, editors. Fuel Cells and Hydrogen: From Fundamentals to Applied Research. Amsterdam: Elsevier; 2018. pp. 15-38

[28] Sieniutycz S. Thermodynamic basis of fuel cell systems. Cybernetics and Physics. 2012;**1**(1):67-72

[29] Harned HS, Owen BB. The Physical Chemistry of Electrolytic Solutions. 3rd ed. New York: Reinhold; 1958

[30] Explanation of the thermodynamics behind fuel cell and electrolyser design [Internet]. 2018. Available from: www.fuelcellstore.com/blog-section/explanation-of-the-thermodynamics-behind-fuel-cell-and-electrolyser-design [Accessed: 2019-08-27]

[31] Pilatowsky I, Romero RJ, Isaza CA, Gamboa SA, Sebastian PJ, Rivera W. Thermodynamics of Fuel Cells. London: Springer; 2011. pp. 25-36. DOI: 10.1007/ 978-1-84996-028-1

[32] Acharya PR. An advanced fuel cell simulator [thesis]. Houston-Dallas-Austin, TX: Texas A&M University; 2004

[33] Zhang J, Tang Y, Song C, Zhang J. Polybenzimidazole-membrane-based PEM fuel cell in the temperature range of 120–200°C. Journal of Power Sources. 2007;**172**(1):163-171. DOI: 10.1016/j. jpowsour.2007.07.047. Available from: https://doi.org/10.1016/j. jpowsour.2007.07.047

[34] Corre GPG, Irvine JTS. High temperature fuel cell technology. In: Stolten D, editor. Hydrogen and Fuel Cells: Fundamentals, Technologies and Applications. Wenham: Wiley-VCH Verlag; 2010. pp. 61-88

[35] Rajput RK. Engineering Thermodynamics. 3rd ed. Laxmi: New Delhi; 2007. 922 p

[36] Srikar VT, Turner KT, Yung TIA, Spearing SM. Structural design considerations for micro-machined solid-oxide fuel cells. Journal of Power Sources. 2004;**125**(1):62-69. DOI: 10.1016/j.jpowsour.2003.07.002

[37] Umberto L. Overview on fuel cells. Renewable and Sustainable Energy Reviews. 2014;**30**:164-169. DOI: 10.1016/j.rser.2013.09.025

[38] Moran MJ, Shapiro HN. Fundamentals of Engineering Thermodynamics. 5th ed. New York, NY: Wiley; 2006. 831 p

[39] Wark K Jr. Advanced Thermodynamics for Engineers. New York, NY: McGraw-Hill; 1995 622 p

[40] Bejan A. Advanced Engineering Thermodynamics. Hoboken, NJ: Wiley; 2016. 741 p

[41] Fuel Cell Technology. Hydrogen Fuel Cell Engines and Related Technologies. 2001. Available from: https://energy.gov/sites/prod/files/2014/03/f9/fcm00r0.pdf [Accessed: 09-09-2019]

Chapter 2

Distributed and Lumped Parameter Models for Fuel Cells

Massimo Guarnieri, Piergiorgio Alotto and Federico Moro

Abstract

The chapter presents a review of modeling techniques for three types of fuel cells that are gaining industrial importance, namely, polymer electrolyte membrane (PEMFC), direct methanol (DMFC), and solid oxide (SOFC) fuel cells (FCs). The models presented are both multidimensional, suitable for investigating distributions, gradients, and inhomogeneities inside the cells, and zero-dimensional, which allows for fast analyses of overall performance and can be easily interfaced with or embedded in other numerical tools, for example, for studying the interaction with static converters needed to control the electric power flow. Thermal dependence is considered in all models. Some special numerical approaches are presented, which allow facing specific problems. An example is the Proper Generalized Decomposition (PDG) that allows overcoming the challenges arising from the extreme aspect ratio of the thin electrolyte separating anode and cathode. The use of numerical modeling as part of identification techniques, particularly by means of stochastic optimization approaches, for extracting the material parameters from multiple in situ measurements is also discussed and examples are given. Merits and demerits of the different models are discussed.

Keywords: fuel cells, PEMFC, SOFC, DMFC, modeling, multiphysics, optimization, identification

1. Introduction

Several types of fuel cells (FCs) are under development for small to mid-size applications, both mobile and stationary. The modeling of FCs is the subject of a vast body of literature, with contributions coming from the fields of chemistry, material science, and engineering. This paper focuses on the authors' experience and provides references for further reading and for the derivation of more models. In particular, the distributed and lumped modeling of direct methanol fuel cells (DMFCs), proton exchange membrane fuel cells (PEMFCs), and solid oxide fuel cells (SOFCs) is addressed. A section is devoted to the numerical optimization of such devices and to the identification of the parameters appearing in their equations by means of stochastic optimization algorithms. The two approaches, that is, distributed and lumped modeling, derive from opposite necessities, namely on one hand, the necessity of studying local details of the physical phenomena and on the other hand the necessity of having computationally efficient tools for large-scale simulations and integration of systems.

2. Modeling challenges

The performance of a fuel cell depends on the materials used, on the sizes of the cell components, on their geometry and arrangement, and on the combination and interactions of all these factors, involving interdisciplinary effects of, notably, interface, mass charge and heat transport, electrochemistry, catalysis, and materials science. Consequently, identifying an optimized stack configuration is a very demanding task that can require long and expensive experimental programs. In such a framework, models are very helpful in exploring possible system behavior and addressing the search for optimal solutions. However, due to the diversity and complexity of such phenomena, which occur at multidimensional level and on a wide range of length and time scales, model analyses have to necessarily be carried out by computer simulations. The numerical modeling of fuel cells, often dubbed CFCD (computational fuel cell dynamics), deals with multidimensional mass transport phenomena, electrochemical kinetics, and transport of charge (electrons and ions), in complex temperature-dependent relationships that are strongly coupled to each other. Namely, they are strongly coupled nonlinear problems and their solutions require advanced iterative algorithms capable of efficiently ensuring converged and accurate solutions. The analyses can be conveniently formulated in terms of electric potentials of the electronic and electrolyte phases, whose equations are strongly nonlinear and coupled to each other at electrochemical kinetics level. Techniques have been developed for efficiently solving these potential equations. Fuel cells present a stratified structure, made of thin layers of different materials where interactions occur. These domains with very different dimensions along coordinate axes give rise to additional computational challenges. Electrochemical activity sites generate current densities that can reach values in the order of 1 A/cm^2 (referred to the cross-sectional area) so that a current of 500 A requires membrane cross sections of 20 $\times$ 20 cm or more, whereas typical thicknesses of gas diffusion layers (GDLs) are in the order of 2–3 $\times$ 10^{-2} cm. These GDLs are separated by proton-exchange membranes (PEMs) which are thin electrolytes, with widths in the order of 1–5 $\times$ 10^{-3} cm.

PEMs allow ion transport while preventing the passage of electrons, a key feature of electrochemical devices in order to force electrons to flow in the external electric circuits. Ion-conductive membranes allow the flow of protons when sufficiently hydrated. In optimal conditions, the proton conductivity can reach values as high as 0.2 S/cm at 100°C, which is a fairly good value for ions, but a poor one if compared with the electronic conductivity of metals, so that in order to reduce the inherent voltage and power losses such electrolytes must be very thin. At the same time, the relative fluctuations of the thickness due to manufacturing must be small, to ensure the membrane performance to be very uniform. The numerical simulation of such a domain, with aspect ratio exceeding 10^3, involves severe size problems: a regular hexahedral tessellation with 10 elements in the thickness direction implies 10^9 nodes, namely a very demanding computational dimension [1]. Commercial CFD codes commonly use un-structured meshes, but this fact does not alter the dimensional complexity of the problem. As a further concern, depending on the transient time scale, a large number of time steps may be needed in order to accurately compute time dynamics. These features raise very challenging problems that can only be faced with parallel computing and multiprocessor computers. Analyzing a fuel cell behavior requires the full characterization of the materials used, that is, the determination of their chemical, physical, thermal, and electrical parameters [2], which are involved in: mass, heat, and charge transport in the electrolyte; thermodynamics and electrokinetics in the catalyst layers; mass, charge, and heat transport in the diffusion layers; and mass, heat, and charge transport in the bipolar plates and their flow channels. In fact, these parameters are needed in the equations (Nernst equation, Butler-Volmer equation, Darcy's equation, Fourier's law, Ohm's law, etc.) constituting FC models [3]. Since these equations are strongly nonlinear and coupled, the models present a strong vulnerability to parameter uncertainties. A number of diagnostic methods, such as cyclic voltammetry, thin-film rotating ring-disc electrode (CV-TF-RRDE), electrochemical impedance spectroscopy (EIS), and broadband electrical spectroscopy (BES), can provide accurate ex situ measurements. Nevertheless, the use of the measurement data thus obtained to model fuel cells in working conditions presents several setbacks. On the other hand, parameter values fully suitable for operative conditions can be achieved by means of *in situ* measurements, but few are available and they are often difficult to carry out, allowing to determine a limited number of parameters. Examples are EIS, neutron radiography, and voltammetric and chronoamperometric approaches in the "driven-cell" mode. To cope with material characterization and parameters extractions, a number of sophisticated numerical techniques that resort to optimization methods for extracting more parameters at once from a large number of experimental data have been developed in recent years. Stochastic methods have proven to be particularly successful to meet such target. Efficient modeling suitable to provide a comprehensive understanding of fuel cell dynamics thus involves: physicochemical model identification, advanced numerical algorithms, materials parameters extraction, and model validation against detailed distribution data. This last aspect arises since data of global nature, such as I-V curves, can result inadequate to capture the material parameters with their correlations.

3. Analytical models

3.1 Proton exchange membrane fuel cells

PEMFCs have been subject to a vast body of studies aimed at modeling and some of the most important issues are described hereafter. In PEMFCs, the hydrogen oxidation reaction (HOR) occurring at the anode catalyst layer (CL) and the oxygen reduction reaction (ORR) at the cathode catalyst layer

$$\begin{array}{lllll} \text{anode}: & H_2 & \rightarrow & 2H^+ + 2e^- \\ \text{cathode}: & \frac{1}{2}O_2 + 2H^+ + 2e^- & \rightarrow & H_2O \end{array} \tag{1}$$

are segregated by the proton exchange membrane (**Figure 1**). According to the Nernst equation, the cell's reversible voltage E varies with temperature T and gas pressures p_{H_2}, p_{O_2} (or equivalently, with concentrations c_{H_2}, c_{O_2}) [4]:

$$\begin{aligned} E &= E^0 + \Delta E_s(T) + \Delta E_c\left(T, p_{H_2}, p_{O_2}\right) \\ &= E^0 + \frac{1}{nF}\int_{T^0}^{T} \Delta\hat{s}(T)\,dT + \frac{T}{f_e}\ln\left[\left(\frac{c_{H_2}}{c^0_{H_2}}\right)\left(\frac{c_{O_2}}{c^0_{O_2}}\right)^{0.5}\right] \end{aligned} \tag{2}$$

where E^0 = 1.229 V is the value in standard temperature and pressure conditions, ΔE_s is the entropic variation due to $\Delta\hat{s}$, and ΔE_c is the term related to the variation of gas pressures and hence of gas concentrations. By introducing the "bulk" (undisturbed) concentrations $\bar{c}_{H_2}, \bar{c}_{O_2}$, ΔE_c can be split into two terms:

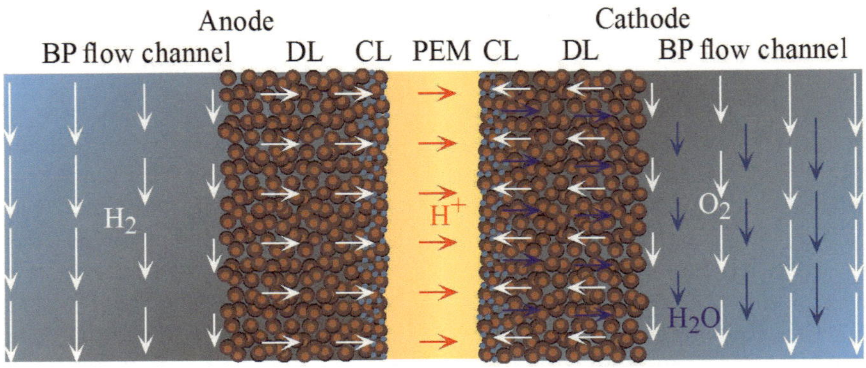

Figure 1.
Sketch of a PEMFC section with anode and cathode flow channels of bipolar plates (BPs), diffusion layer (DLs), catalyst layers (CLs), and proton exchange membrane (PEM). Convective (in BPs) and diffusive (in DLs) fluid flows are sketched (courtesy of Journal of Power Sources).

$$\Delta E_c = \frac{T}{f_e} \ln\left[\left(\frac{\bar{c}_{H_2}}{c^0_{H_2}}\right)\left(\frac{\bar{c}_{O_2}}{c^0_{O_2}}\right)^{0.5}\right] + \frac{T}{f_e} \ln\left[\left(\frac{c_{H_2}}{\bar{c}_{H_2}}\right)\left(\frac{c_{O_2}}{\bar{c}_{O_2}}\right)^{0.5}\right] = \Delta E_{co} + \Delta E_{cl} \tag{3}$$

Here $f_e = nF/R$, where R is the gas constant, F is the Faraday constant, and n is the number of electrons transferred in the reaction. ΔE_{co} arises from the differences between the actual bulk concentrations and the standard-condition values, particularly in a no-load state. ΔE_{cl} springs from the differences between the concentrations at the triple-phase boundaries (TPBs, which are the sites of electrochemical reaction in the catalyst layers—CLs) and the bulk concentrations in the CLs, namely from the concentration gradients which arise to sustain the species molar flows needed in load conditions. In order to allow an accurate modeling over a wide temperature range, ΔE_s can be calculated by integration. Combining the above and excluding the gradient-dependent term ΔE_{cl}, we have:

$$E_{OC} = E^0 + \frac{1}{nF}\int_{T^0}^{T} \Delta\hat{s}(T)\,dT + \frac{T}{f_e}\ln\left[\left(\frac{\bar{c}_{H_2}}{c^0_{H_2}}\right)\left(\frac{\bar{c}_{O_2}}{c^0_{O_2}}\right)^{0.5}\right] \tag{4}$$

The cell open circuit voltage (OCV) is slightly different from E_{OC} because some gas crossover through the membrane occurs in every condition, including no-load, producing small concentration gradients and hence a minimal ΔE_{cl}. On the other hand, when the cell is operated at a steady-state load with an electric current density j, its voltage V differs markedly from E_{OC}, due to much larger concentration gradients, and hence much larger ΔE_{cl}, and to overpotentials η_{kh} (voltage drops ΔV_{kh}, in electrical engineering terms, [3]):

$$V(j) = E_{OC} - \eta_{aa} - \eta_{ac} - \eta_{ca} - \eta_{cc} - \eta_m \tag{5}$$

The first subscript of the overpotentials indicates a, activation losses, or c, concentration losses, whereas the second subscript stands for a, anode, or c, cathode. The single subscript m indicates the membrane (PEM), where ohmic losses occur. All these overpotentials are strong functions of j.

3.1.1 Exchange current density: activation losses

The activation overpotentials or activation voltage drops, ΔV_{aa} and ΔV_{ac}, are an effect of the electrochemical kinetics that appear when the species react at the anodic and cathodic CLs. ΔV_{aa} and ΔV_{ac} increase with the rate of charge density separation $\partial_t \rho_e$ (namely the proton and electron creation at the anode and recombination at the cathode—∂_t represents the partial time derivative), which, in steady-state operation equates the current density at the TPB, j_{TPB}. ΔV_{aa} and ΔV_{ac} are typically modeled with the Butler-Volmer equation [4]. Due to the particular porous structure of the CLs, the area A_{TPB} of the TPBs where j_{TPB} is produced is much larger than the active cell cross-sectional area A (A_{TPB}/A can be larger than 10^3) and, when modeling is devoted to analyzing the cell electric performance, the current density is preferably reported to the cross-sectional area, as:

$$j_0 = \frac{A_{TPB}}{A} j_{0TPB} \tag{6}$$

By using Eq. (6), the Butler-Volmer equation allows to write the current density at the cross-sectional area of each CL as

$$j_t = j_0 \left(\frac{c_R}{\bar{c}_R} e^{\alpha f_e \Delta V_a / T} - \frac{c_P}{\bar{c}_P} e^{-(1-\alpha) f_e \Delta V_a / T} \right) \tag{7}$$

In this equation, the total equivalent current density j_t accounts for the effect of hydrogen crossover on the overpotentials, that can be modeled as an equivalent internal loss current to be added to the cell internal current density and is not available at current collectors, but contributes to the activation overpotentials. j_0 is the exchange current density. The accurate evaluation of its values at the anode and cathode half-reactions is important, because they strongly affect the cell performance and round-trip efficiency. To take into account the effects of the temperature on j_0, an Arrhenius-like dependence with T can be considered [5]. As a consequence of the low temperatures at which PEMFCs operate and of the exponential dependence of j_t on ΔV_a, the activation losses are the major responsible factors for voltage drops at low current densities. In addition, ΔV_{ac} is typically one order of magnitude larger than ΔV_{aa}, so that the cathodic activation losses are the dominant effect at low current densities [3].

3.1.2 Concentration gradients

When the cell operates in load condition, the inflow of reagents and outflow of products are necessary to sustain the electrochemical reactions at the CLs. In turn, these species flows are ensured by convective mass flow in the transport channels of the bipolar plates (BPs) and by the diffusive mass transport in the diffusion layers (DLs, **Figure 1**). These flows are sustained by concentrations and pressure gradients of reagents and products, namely by values c and p at the CLs different from the bulk (and inlet) values $\bar{c}$ and $\bar{p}$ [6]. In order to model the concentration gradients c, Fick's first law $\boldsymbol{N} = -D\, c$ is often used, which invokes the medium diffusivity D and the gas molar flow vector $\boldsymbol{N}$, related in turn to the current density vector $\boldsymbol{j}$ through the Faraday constant F and the charge carriers n as $\boldsymbol{j} = nF\boldsymbol{N}$, for both hydrogen at anode and oxygen at cathode. Since the diffusivity depends on temperature, the effect of the latter on concentration gradients can also have important role in the FC performance.

3.1.3 Concentration losses and limit conditions

Gas concentration and pressure gradients c and p produce the term ΔE_{cl} of the Nernst equation, reducing the electromotive force (emf) with respect to the no-load value E_{OC} and constitute the concentration losses which dominate the cell's performance at high current densities. ΔE_{cl} can be split in the anodic and cathodic concentration voltage drops as:

$$-\Delta E_{cl} = \eta_{ca} + \eta_{cc} = \kappa_{ca} \frac{T}{f_e} \ln\left(\frac{\bar{c}_{H_2}}{c_{H_2}}\right) + \kappa_{cc} \frac{T}{2f_e} \ln\left(\frac{\bar{c}_{O_2}}{c_{O_2}}\right) \quad (8)$$

where κ are the mass transfer coefficients. These two terms provide the anodic and cathodic current density limits j_{La} and j_{Lc}, namely the theoretical values of the current densities which cause the CL concentrations and pressures to vanish at the TPBs and the half-reactions stop, when the fuel cell starvation occurs. Since the smaller limit current density occurs at the cathode, due to the lower diffusivity of oxygen compared to hydrogen, j_{Lc} sets the device's current density limit.

3.1.4 Membrane ohmic losses

The voltage drop in the membrane is proportional to the current and to its thickness and inversely proportional to its conductivity σ which depends on temperature and hydration, which is the ratio $\lambda = c_w/c_{as}$ (with c_{as} = 1970 mol m^{-3}) between water and sulfonic acid concentrations that varies in the range 0–22 for typical Nafion® membranes. The dependence of conductivity on hydration can be expressed as:

$$\sigma(\lambda) = \sigma_0(T)B\lambda \quad (9)$$

The linear dependence on λ via the dimensionless coefficient B is the adaptation of an empirical model [7] aimed at avoiding a negative value of σ at lower λ. The temperature dependence can be expressed with the model of Vogel-Tamman-Fulcher [8]. Although λ varies along the PEM's thickness according to back-diffusion and electro-osmotic drag [9], the average between the PEM boundary values λ_a and λ_c can be used, consistently with a linear profile between λ_a and λ_c. These values depend on the water activities a_{wa} and a_{wc} of the reacting gases at the CLs, and are computed with an empirical polynomial [7], which depends on the water vapor saturation pressure p_{ws} which is also a function of the temperature [3].

3.1.5 Crossover

Fuel crossover consists of hydrogen that, instead of reacting at the anode, migrates through the PEM and reacts with oxygen at the cathode, without producing electric power. This is a major side effect that affects the FC performance and efficiency and depends on two mechanisms, diffusion, and electro-osmotic drag. These two contributions of hydrogen mass flow can be modeled as equivalent current densities so that the resulting equivalent crossover current density is:

$$j_{co} = \frac{nFD_{mH_2}}{d_m} c_{H_2} + n\xi\lambda j \quad (10)$$

where c_{H_2} is the hydrogen concentration at the anodic CL, D_{mH2} is the hydrogen diffusivity (with an Arrhenius-like temperature dependence), d_m is the membrane thickness and ξ is a dimensionless electro-osmotic drag coefficient (giving a maximum value $\nu_w = \xi\lambda = 2.5$ at full hydration $\lambda = 22$, as reported). Crossover hydrogen is the main cause of the difference between the open circuit emf E_{OC} and the observed OCV $V(0)$ [10]. It also causes a loss of stored energy that reduces round-trip efficiency. Also, oxygen crosses the PEM, but, since its diffusivity is much lower than hydrogen's [11], this effect is usually neglected.

Dissipations occurring inside the cell produce thermal gradients which affect the temperature-dependent parameters. The main loss phenomena are Peltier heating (thermodynamic heat generation), losses due to the electrochemical kinetic activity at the anode and cathode CLs, and Joule losses in the PEM, so that, inside the cell, the dissipated power per unit area is:

$$P_{diss} = j\left(-\frac{T\Delta\hat{s}}{nF} + \eta_a(j)\right) + \frac{d_m}{\sigma}j^2 \quad (11)$$

Heat transport inside the cell depends on conduction, diffusion, and convection and interacts with thermal capacity in dynamic conditions [3]. An accurate enough estimation of the mean temperature T inside the cell with respect to the room and gas inlet temperature T_r can be obtained by:

$$T = T_r + k_T P_{diss} \quad (12)$$

with k_T a global thermal exchange coefficient. This expression can be reformulated as a function of the current density as:

$$T = T_r + k_{t1} j + k_{t2} j^2 \quad (13)$$

where k_{t1} and k_{t2} are properly fitted parameters. In the numerical implementation of such models, consistent analytical expressions can be used without introducing approximation if the electric current density j is chosen as the independent variable to compute all voltage terms. In order to deal with the non-invertible Butler-Volmer equation, a look-up table can be conveniently used.

3.2 Methanol fuel cells

DMFCs suffer from two fundamental problems: (i) the sluggish kinetics of the methanol electro-oxidation reaction and (ii) the high degree of permeation of the methanol through the membrane (crossover). Analytical and numerical models are necessary for better understanding the interactions between mass transfer and electrochemical phenomena, and for optimizing the power output and runtime to interface electronics [3].

3.2.1 DMFC analytical models

A schematic of a typical DMFC inside a cell stack is sketched in **Figure 2**. It consists basically of an anode flow channel (AFC), an anode diffusion layer (ADL), an anode catalyst layer (ACL), a proton exchange membrane (PEM), a cathode catalyst layer (CCL), a cathode diffusion layer (CDL), and a cathode flow channel (CFC). Analytical models of DMFCs account for the following phenomena: electrochemical reactions occurring at catalyst layers, protonic conduction, and methanol crossover across the PEM, diffusion of reactants in porous media layers, fluid flow inside channels, and coupling between heat and mass transfer. Small DMFCs for portable electronics usually include a fuel reservoir with no pumps for feeding the

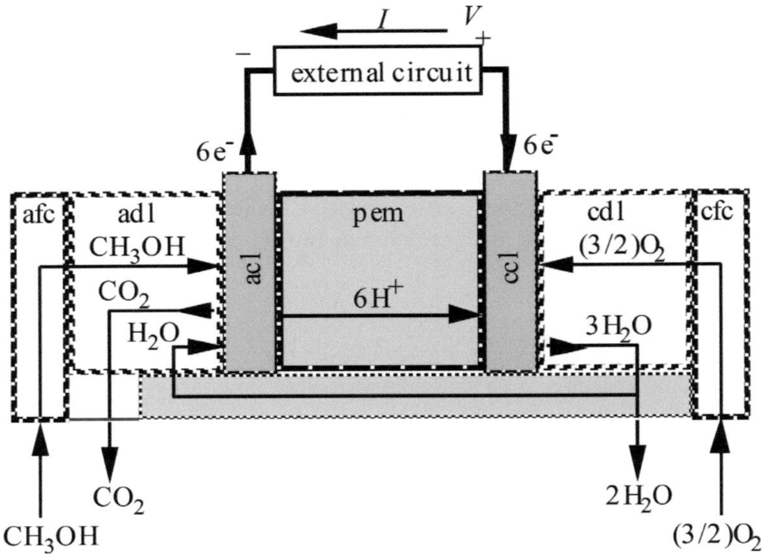

Figure 2.
DMFC schematic with reactant and species flows (a, anode; c, cathode; pem, proton exchange membrane; fc, flow channel; dl, diffusion layer; cl, catalyst layer).

cell and make use of atmospheric oxygen. Simplifying assumptions are typically made on both model geometry and physics to obtain the current-voltage characteristic in closed form. In 2-D models, methanol/oxygen concentrations and electric potentials depend also on the direction (y-axis) orthogonal to the species flow direction (x-axis). In 1-D models, this dependency is neglected by assuming constant concentrations along the y-axis.

The following assumptions are usually made in analytical DMFC models: (i) constant physical parameters; (ii) one-phase model (vapor phase is neglected); (iii) ideal and diluted solutions; (iv) homogeneous electrochemical reactions in the electrodes; and (v) negligible overpotential variation across catalyst layers and along the y-axis.

The basic equation for analyzing the cell voltage as a function of current density is:

$$V(J) = E_0 + \frac{\partial \mathrm{E}}{\partial \mathrm{T}}\Delta T - \eta_a(J) - \eta_c(J) - \frac{\delta_m}{\sigma_m}J - R_s J \tag{14}$$

where E_o is the (constant) standard cell potential, E is the fuel cell standard potential, T is the temperature, σ_m and δ_m are the electric conductivity and the thickness of the membrane, η is the activation overpotential, J is the current density, and R_s is the overall contact area specific resistance (ASR). This resistance per unit cross section, typically assumed constant, accounts for all resistances between gas diffusion layers and current collectors. The current density at the anode in Eq. (14) is related to the activation voltage overpotential at the anode η_a. In the simplifying assumption of the Tafel equation, it can be expressed as:

$$J_a = J_{a,\,ref}\frac{C_{ac}}{C_{ac,\,ref}}\,exp\left(\frac{\alpha_a F}{RT}\eta_a\right) \tag{15}$$

where $J_{a,\,ref}$ and $C_{ac,\,ref}$ are the reference current density and the reference concentration at the anode, respectively. The voltage overpotential η_a is required to overcome the activation energy of the electrochemical reaction, and causes an energy loss. A relationship similar to Eq. (15) holds for the cathode side, where the atmospheric oxygen is provided at the catalyst layer through the gas diffusion layer and the flow channel.

In the one-dimensional model proposed in [12], the cell performance is assessed by describing mass transport in the porous electrode structures, including methanol crossover, and the potential and concentration distributions in electrodes. Multiphase flow is accounted for in [13], where liquid-vapor mixtures at anode and cathode compartments are considered. A two-dimensional analytical model of a DMFC, which describes electrochemical reactions on the anode and cathode and main transport phenomena in the fuel cell including methanol crossover, diffusion of reactants in porous media layers, and fluid flow in the reactants distributor, is presented in [14].

Previous models, mathematically and physically based, cannot be used directly in CAD software for electronic circuits, which is used to design the DMFC-circuitry interface. By using the one-dimensional assumption, a lumped circuit model for simulating the DMFC runtime is derived in [15]. The methanol consumption at the reservoir is simulated by mass conservation equation and the charge/discharge electrode dynamics on the short time scale is simulated by a fictitious equivalent capacitance. A dynamic nonlinear circuit model for passive DMFC, including water mass flow and membrane hydration, is presented in [16]. By synthesizing physics equations in circuit form, this model is able to take into account mass transport, current generation, electronic and protonic conduction, methanol adsorption, and electrochemical kinetics. Adsorption and oxidation rates, which affect the cell dynamics, are modeled by a detailed two-step reaction mechanism. The fuel cell discharge and methanol consumption are computed by combining mass transport and conservation equations. As a result, the runtime of a DMFC can be predicted from current and initial methanol concentration.

3.2.2 DMFC semi-empirical models

Most DMFC performance models combine differential and algebraic equations with empirically determined parameters. Simpler empirical expressions can be derived from such models, allowing designers and engineers to predict the fuel cell performance as a function of different operating conditions (such as pressure, temperature, or fuel concentration). In [17], a semi-empirical approach to account for the limiting current behavior is proposed. The general expression for the polarization curve is based on analytical formulations of the catalyst layer reaction and includes the overpotential due to transport limitation in diffusion backing layer and the effect of methanol crossover. A simple semi-empirical model for evaluating the cell voltage of a DMFC is provided in [18] by using the semi-empirical Meyers-Newman rate equation for assessing the anode overpotential and a Tafel-type kinetic model for the cathode. Ohmic losses are accounted for by considering constant and uniform membrane conductivity. Such semi-empirical models are useful to describe the steady-state fuel cell behavior. Dynamic models can resort as well to semi-empirical relationships in order to limit the model complexity in time-domain simulations, notably when implementing models in real-time controllers. In [19], a completely different approach for extracting equivalent parameters is used. Instead of fitting current-voltage equations, a nonlinear equivalent circuit is derived from impedance fuel cell models. The equivalent circuit is composed of nonlinear electrical circuit elements that include resistors, capacitors, and inductors in order to simulate the DMFC performance in a wide operative range, including steady-state and transient conditions. A model of a micro-DMFC battery is developed in [20] to capture all pertinent dynamic and steady-state electrical performance parameters, including capacity and its dependence on current and temperature, open circuit voltage, methanol-crossover current, polarization curve and its dependence on concentration, internal resistance, and time-dependent response under various loading conditions. The main advantage of this model is that it is able to predict the DMFC runtime for a given usable energy capacity of the methanol reservoir. A 1-D analytical model for simulating both DMFC static and dynamic operations is interfaced to a PSO (particle swarm optimization) stochastic algorithm in order to maximize the battery duration while minimizing methanol crossover [21]. In [22], a semi-analytical model of a passive-feed DMFC with non-isothermal effects and charge conservation phenomenon is proposed. The 1-D model is suitable for predicting the current-voltage curve by resolving iteratively both (bi-phase) mass transfer, electrochemical, and heat transfer equations, starting from imposed cell current, ambient temperature, and methanol feed concentration. The phenomenon of cathodic mixed potential, which is related to methanol crossover is accounted for as well by the DMFC model. Adaptive operations for voltage stabilization are proposed in [23] by using a simplified semi-empirical model combined to an on-demand control system. A rapid response of DMFC (less than 5 s) to variations of operating parameters is experimentally observed, ensuring the applicability of the proposed adaptive control strategy.

3.3 Solid oxide fuel cells

A solid oxide fuel cell (SOFC) consists of an anode and a cathode with a ceramic electrolyte between them that transfers oxygen ions. Oxidant reduction occurs in the cathode catalyst layer, oxygen ions are transported through the electrolyte and oxidation of the fuel occurs at the anode catalyst layer. A SOFC typically operates between 700°C and 1000°C, that is, temperatures at which the ceramic electrolyte becomes sufficiently conductive of oxygen ions. In fact, the electrolyte ionic conductivity is a strongly increasing function of the operating temperature. Moreover, since the operating temperature accelerates electrochemical reactions, precious metal catalysts are not required to promote the reactions and cheaper materials such as nickel can be used as catalysts. In addition, the SOFC can be fed with conventional hydrocarbon fuels, reforming being performed inside the cell. SOFCs can be planar or tubular, with the former having gained increasing success because of easier manufacturing and higher performance.

Among the ceramic electrolytes used in SOFCs, the most important is yttria-stabilized zirconia (YSZ), which is the most conductive. The anode is generally made of nickel/yttria-stabilized zirconia cermet and the cathode is an LSM layer chemically expressed as La1−*x*Sr*x*MnO3. The open circuit potential of a SOFC is given by the Nernst equation, whereas the activation overpotentials in both electrodes are high, so that the electrochemical kinetics of the both electrodes can be approximated by the Tafel equation, with the concentration dependence of exchange current density given by [24]. Zirconia (zirconium oxide, $Zr\ O_2$) has a crystalline structure and is stabilized in the allotropic cubic form YSZ with the addition (doping) of Yttria molecules (yttrium oxide, Y_2O_3). Moreover, Yttria significantly increases oxygen holes, namely an Y^{3+} ion replaces a Zr^{4+} inducing an oxygen hole to maintain electrical neutrality and such oxygen holes facilitate the transport of O^+ ions in the lattice, by means of jumps. Doping with yttrium therefore dramatically increases the concentration of charge carriers (holes) and conductivity is proportionally improved. However, if the concentration of Yttria increases too much, the holes interact with each other thus reducing their mobility. Maximum conductivity is obtained with doping concentrations of about 8%: in such conditions, hole concentration is about 10^2 times greater than the zirconia intrinsic concentration. The probabilistic dependence of the charge carrier mobility on the energy yields an exponential dependence of the conductivity on temperature:

$$\sigma(T) = \sigma_o e^{-W_a/kT} \tag{16}$$

A typical conductivity value is $\sigma \cong 0.02$ S/cm at T = 1000 K, with activation energy W_a = 0.5–1.2 eV. This is a rather high conductivity value but not the highest among other types of FC electrolytes and, consequently, in SOFCs, ionic conduction losses are relatively higher.

4. Lumped parameter models

FC models must strike a balance between opposite requirements. On the one hand, they should be extremely rich in order to be able to represent the complete behavior of the cell by capturing the tridimensional distribution and time evolution of the physical quantities inside the cell. Such performance is achievable with a multiphysics, three-dimensional model, described by partial differential equations (PDEs) and characterized by a large number of physical parameters. On the other hand, they should be sufficiently simple to be included in other numerical procedure, such as a stochastic optimization loop, that is, the model should be numerically computable in a very short time, and they should be characterized by a relatively small number of parameters in order to avoid the *curse of dimensionality* issue. In such cases, a zero-dimensional stationary model is preferable since it avoids PDE numerical discretization and their inherent computational burden, allowing to run the algorithm on standard PCs.

4.1 Circuit models

The fuel-cell balance equations can be arranged so as to correspond to lumped equivalent circuits. Indeed, several of them can be identified, around the same basic concept, depending on the required level of accuracy and the behaviors to be highlighted.

In **Figure 3**, E is the open circuit voltage (the reversible voltage, i.e., the ideal electromotive force provided by the Nernst equation that depends on temperature and concentrations), R_{ion} is the electrolyte ionic resistance, R_a and R_c are the equivalent resistances of the electrode kinetics (at anode and cathode), equal to the

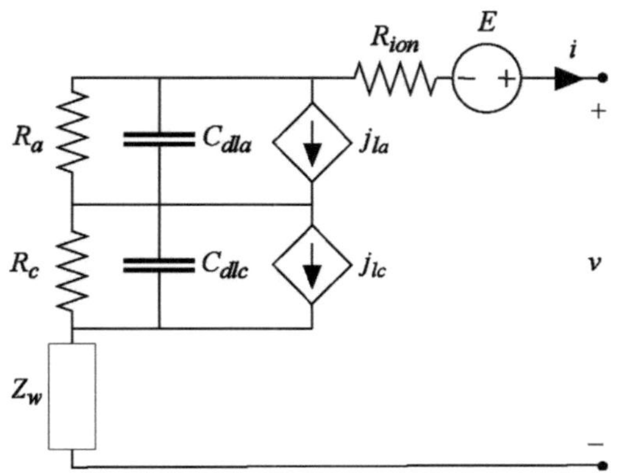

Figure 3.
Simple lumped equivalent circuit of a fuel cell.

activation overpotentials divided by the electrode-generated current. Depending on the analysis purpose, both the incremental and differential values can be used:

$$R = \frac{\eta_c}{i'} \text{ or } R = \frac{d\eta_c}{di'} \tag{17}$$

The latter resistance characterizes transient responses to small variation of the cell's working point. Expanding it in the neighborhood of a working point for which the Tafel approximation holds provides:

$$R = \frac{d}{di}\left(-\frac{RT}{\alpha nF}\ln j_0 + \frac{RT}{\alpha nF}\ln i\right) = \frac{RT}{\alpha nF\ i} = \frac{RT}{\alpha nF\ i_0 e^{\alpha nF\eta/RT}} \tag{18}$$

that is an adynamic circuit element, strongly dependent on the current or voltage. C_{dl} is the double layer capacity, related to TPB interfacial charge behavior, J_l are leakage current densities, due to various phenomena and particularly to the reactant crossover through the electrolyte, which make the output electric current smaller than the electrode-generated current (they are controlled source elements since such currents depend on the cell useful currents and on concentrations), Z_w is an element accounting for the losses due to concentration gradients. This is a dynamic element because it synthesizes the diffusion of reactant concentration at the electrodes, which are governed by the diffusion dynamics equations. In case of pulsating variations at an angular frequency ω around a steady-state working point, this is a Warburg nonlinear impedance that can be expressed as:

$$Z_w = \frac{RT}{A\left(nF\right)^2 \bar{c}\sqrt{D}\sqrt{\omega}} \tanh\left(\delta\sqrt{\frac{j\omega}{D}}\right) e^{j\frac{\pi}{4}} \tag{19}$$

where δ is the electrode thickness. By taking the limit $\omega \to 0$ it reduces to:

$$Z_w(0) = \frac{RT}{A\left(nF\right)^2 \bar{c}} \frac{\delta}{D} \tag{20}$$

The dynamic elements C_{dl} and Z_w limit the response dynamics of the FC. In other words, an FC is not able to face arbitrary power variations and when these are too rapid it is necessary to support the FC with a storage device capable of faster response. This solution is adopted in automotive FCs, which are coupled with a lithium-ion battery or a supercapacitor, sized to meet the fastest transients and capable of reversible operations, enabling regenerative braking, but unable to accumulate as much energy as the H_2 tank of the FC.

A more sophisticated equivalent circuit of a DMFC is presented in **Figure 4** in the case of a DMFC. Here, the dependences of the leakage currents on the reagent gradients and on the electrolyte currents have been separated and the concentration losses have been represented as voltage sources controlled by the methanol and oxygen gradients. Three separated equivalent circuits have been added, to represent the water, methanol, and oxygen behavior. Each of these includes capacitances, which account for the volumes of species accumulating at electrolyte, catalyst, and diffusion layers, whereas the controlled current sources describe drain and source at the catalyst layers and current-driven crossover.

In any case, equivalent circuits provide an approximated description of the complex reactions and transport events occurring inside the FC. They have the

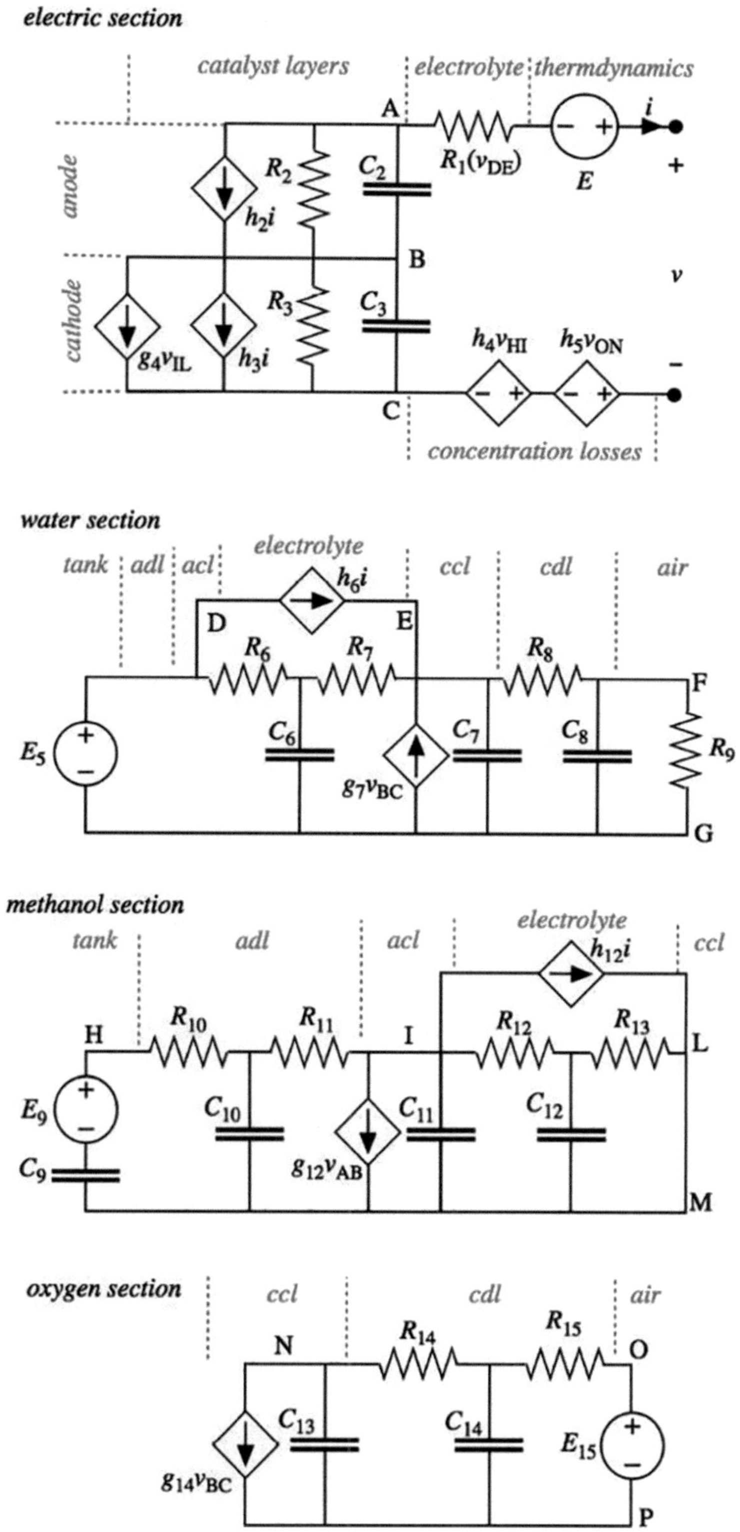

Figure 4.
Multiphysics equivalent circuits capable of simulating electrical, chemical, and mass transport interactions (courtesy of IEEE Trans. on Industrial Electronics).

merit to highlight separately single events (such as the ohmic losses in the electrolyte), thus allowing to easily study the effects of the variation of single quantities/parameters on the FC behavior. The main drawback of circuit models is that complex interactions and nonlinearity are not simulated in detail. More importantly, they can be implemented into circuit simulation software, to study the electrical interface of the FC with the power management electronics and system supervisor that is devoted to provide FC control, in order to study the overall system dynamics [16].

5. Distributed parameter models

The knowledge of how physical fields (electric field, current density, flow, velocity, temperature, species concentrations) are distributed within internal components constitutes a pivotal aspect in FC analysis and design, since gradients and irregularities hamper the achievement of optimal performance, but they can not be gripped by zero-dimensional, lumped models. On the other hand, distributed models have to cope with additional challenges deriving from the huge number of grid points needed for a complete tessellation of the multilayer 3-D domain, resulting in issues of "curse of dimensionality" which can hardly be faced without resorting to supercomputers. Parallel computing with domain decomposition can overcome this challenge if less powerful computers are used, by assigning one subdomain to each processor and implementing the few interactions between subdomains. The two electric potentials, for the electronic and electrolyte phases, are coupled by the surface overpotentials at the catalyst layers, where reaction kinetics is modeled by the Butler-Volmer equation. In the case of a one-dimensional formulation, Newton's method is an efficient algorithm to integrate the equations, with LU factorization used at each interaction. Nevertheless, in the case of 2-D and 3-D formulations, the sparse Jacobian matrix produced by Newton's method is too large to be efficiently handled. In this case, Gauss elimination with a generalized minimal residual subroutine (GMRES) preconditioned with a Gauss-Seidel block and a multigrid algorithm has proven to be more suitable to face the non-symmetric Jacobian matrix.

5.1 Proton exchange membrane fuel cells

Typical multiphysics coupled models include, among others, proton conduction, water and fuel transport, joule dissipation, and thermal diffusion. The models, typically discretized with the finite element method (FEM), pose significant numerical challenges. Some commercial simulation tools like COMSOL® Multiphysics allow the solution of general time-dependent systems of partial differential equations [25] and are therefore very useful tools for this class of problems. For the computation of the fluid-dynamic field, particularly in the case of turbulent motion at high Reynolds numbers, the finite volume method is also used. Ansys Fluent® is a commercial package based on this method particularly efficient in modeling fuel cells. PEM fuel cells, as the name implies, are based on proton-conducting polymeric membranes. The most commonly used material for their realization is persulfonated polytetrafluoroethylene, commercialized as Nafion® by Chemours. This material has a structure similar to the one of PTFE, but is functionalized with sulfonic acid groups providing charge sites for proton conduction [7]. If the membrane is properly hydrated, protons can form hydronium complexes which once freed from sulfonic acid groups can move through the membrane. In these conditions, that is, proper hydration, proton conduction strongly depends on the water content and the temperature of the membrane, and can reach values as high as 20 S m^{-1} at 100°C.

5.1.1 Electrical conductivity model

As briefly mentioned above, protonic conduction in Nafion® strongly depends on the temperature, since the mechanism is based on charged particles jumping from site to site, with a rate described by the diffusivity D. This statistical parameter depends on the activation barrier energy which exhibits an exponential dependence on the temperature T according to the law:

$$D = D_o \ e^{-W_{ai}/k\,T} \tag{21}$$

where D_o is a diffusivity reference value, W_{ai} is activation barrier energy, and k is Boltzmann's constant. The charged particle mobility μ is proportional to D according to the Einstein relation:

$$\mu = \frac{|z|FD}{RT} \tag{22}$$

with $|z|$ the ion charge number, so that the proton conductivity $\sigma = \rho_c\,\mu$ can be written as:

$$\sigma = \frac{(zF)^2 c D_o}{RT} \ e^{-W_{ai}/kT} = \sigma_o \ e^{-W_{ai}/kT} \tag{23}$$

being $\rho_c = |z|$, Fc the charge density, and c the molar concentration.

Apart from the temperature, the proton conductivity also depends on the water content in the membrane. A common modeling approach to represent such dependence is based on the hydration λ, that is, the ratio between the number of water molecules and the number of charge sites available for proton conduction. In the specific case of Nafion®, such ratio can be rewritten in a modified form using the water concentration c_w and the sulfonic acid concentration c_{as}, that is, $\lambda = c_w/c_{as}$. In absence of more sophisticated models, a linear dependence of conductivity on hydration can be assumed:

$$\sigma = \alpha \ \lambda = 0.5139 \ \lambda \quad S/m \tag{24}$$

where λ is derived from a correlation empirically derived for Nafion® [7]. Combining the above, the factorized expression of $\sigma(\lambda, T)$ is obtained:

$$\sigma(\lambda, T) = \alpha\lambda \ e^{\frac{W_{ai}}{k}\left(\frac{1}{303}-\frac{1}{T}\right)} \tag{25}$$

with W_{ai}/k = 1268 K for ions hopping. This is a more sophisticated model than the one expressed by Eq. (9). The conductivity σ influences the scalar potential φ according to the charge conservation equation in quasi-static conditions:

$$\nabla \cdot \sigma(\lambda, T)\nabla\varphi = 0 \tag{26}$$

which shows that the distribution of φ depends on λ and T.

5.1.2 Hydration model

A critical issue in modeling PEMFCs consists in providing an accurate description of the hydration effects, which rules proton conductivity [26]. The distribution of λ in the membrane can be computed resorting to specific equations at the surfaces and in the bulk. For the membrane bulk, two mechanisms are taken into account, namely electro-osmotic drag and back-diffusion, giving rise to the following expression of the water molar flow:

$$\boldsymbol{N}_w = \boldsymbol{N}_{we} + \boldsymbol{N}_{wd} = \xi\lambda \frac{\boldsymbol{J}}{|z|F} - D_w c_{as} \nabla\lambda \tag{27}$$

where $\boldsymbol{N}_i$ is the ionic molar flow vector and $\boldsymbol{J}$ the current density vector, and D_w is the water diffusivity in the membrane. This equation is nonlinear because D_w itself depends on λ and also on T. Such dependences can be expressed by factorizing the statistical mechanics exponential dependence on T with a polynomial regression obtained from experimental data:

$$D_w(\lambda, T) = \sum_{i=0}^{3} d_i \lambda^i e^{\frac{W_{aw}}{k}\left(\frac{1}{303}-\frac{1}{T}\right)} \tag{28}$$

where W_{aw}/k = 2416 K for water in Nafion® and $d_0 = 2.563671 \times 10^{-6}$, $d_1 = -0.33671 \times 10^{-6}$, $d_2 = 0.0264 \times 10^{-6}$, and $d_3 = 0.000671 \times 10^{-6}$ for D_w (in $cm^2\ s^{-1}$). The dynamics of $\boldsymbol{N}_w$ is ruled by Fick's second law, which can be written in terms of λ:

$$\nabla \cdot \boldsymbol{N}_w + c_{as}\, \partial_t \lambda = 0 \tag{29}$$

Letting Eq. (27) in Eq. (29) and assuming $|z|$ = 1 for protons provide the following diffusion equation:

$$\nabla \cdot D_w \nabla\lambda - \partial_t \lambda - \nabla \cdot \frac{\xi\lambda \boldsymbol{J}}{c_{as} F} = 0 \tag{30}$$

According to Maxwell's equations, the current density $\boldsymbol{J}$ can be expressed in terms of φ, so that *Eq.* (30) becomes:

$$\nabla \cdot D_w (\lambda, T)\nabla\lambda - \partial_t \lambda - \nabla \cdot \frac{\xi\lambda}{c_{as} F}\sigma(\lambda, T)\nabla\varphi = 0 \tag{31}$$

Imposing interfaces conditions at the electro-catalyst layers is troublesome since the precise distribution of λ at such surfaces cannot be determined with sufficient accuracy. However, hydration can be expressed in terms of a more easily representable quantity, that is, the water vapor activity (that is, the relative humidity). The relationship between these two physical quantities can be modeled by resorting to an empirical relationship [27], which in turn can be expressed as a function of the water vapor pressure by another empirical model.

5.1.3 Thermal model

Since most of quantities appearing in the conductivity and hydration models depend on temperature, it is also necessary to take into account the transient heat equation:

$$\rho c_p \partial_t T - \nabla \cdot k(\lambda) \nabla T - \sigma(\lambda, T) |\nabla \varphi|^2 = 0 \tag{32}$$

where ρ is the hydrated Nafion® density, c_p its specific heat, and $k(\lambda)$ the thermal conductivity. According to data given in [28], $k = 0.12 + 0.81\lambda$ (W m^{-1} K^{-1}) can be assumed. The last term at the left-hand side represents Joule's losses.

5.1.4 Coupled multiphysics model

The complete model to be solved is assembled from the above equations together with proper boundary (time-dependent Dirichlet and homogeneous Neumann) and initial conditions. The specific characteristics of the set of partial differential equations together with the high aspect ratio of the geometry (i.e., the very small thickness of the membrane compared to its extension in the plane) lead, after discretization with FEM, to a badly conditioned system of linear equations. The system tends to be quite large so that direct solvers may become inapplicable and it is also difficult to precondition so that iterative solvers tend to converge slowly or to fail altogether. Due to the strong nonlinearity of the complete coupled problem, a further numerical challenge concerns the nonlinear solver, typically the Newton-Raphson (NR) algorithm, which usually converges only, if at all, with strong under-relaxation.

The numerical solution of the final system may require substantial effort in correctly setting linear and nonlinear solver parameters to achieve convergence. Extreme care is needed in the choice of the drop tolerance if GMRES coupled to an ILU pre-conditioner with threshold is applied to solve the linear system arising in Newton's method. Numerical experiments have shown that the selection of the drop tolerance needed to obtain convergence is highly problem-dependent so that the only way to reliably obtain a solution was to apply an efficient direct solver (PARDISO®). This however limited the maximum admissible problem size on the available hardware. Furthermore, the fully coupled nonlinear system of equations can typically not be solved in its complete assembled form by the standard NR technique, even with strong under-relaxation. The problem was therefore solved with a further iterative loop, that is, a so-called "segregated solution," in which one or more blocks of equations are fed into the next one in a simple iteration until convergence. In our specific case, the electrical model formed the first block, and its solution was inserted into the thermal model; then, the solution of both was used in the diffusion model. An important technological problem in the construction of PEM fuel cells is the reproducibility of the production process in the case of large membranes. In particular, variations in the thickness can result in hot spots which can cause aging effects and may lead to membrane failure. In order to investigate such effects, a lens-shaped compression of 2 mm radius and 50 μm depth (1/4 of the total PEM thickness of 200 μm) was studied. Results show an increase in current density in the order of 100%, leading to strong localized overheating (**Figure 5**).

5.2 Direct methanol fuel cells

Spatially resolved analyses of DMFCs allow studying their limitations, like sluggish kinetics and methanol crossover. These models resort to a detailed description of the real device geometry and materials, and of local nonlinear physics phenomena such as heat, momentum, multicomponent mass transport, and electrochemical processes. Early distributed parameter models were multi-domain ones, in the sense that the problem variables were defined in separated domains by introducing appropriate internal boundary conditions. As an example, in [12], the equations

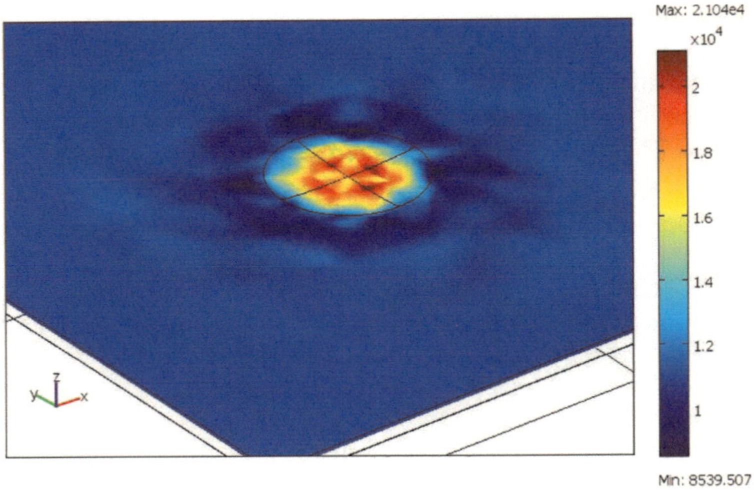

Figure 5.
Local increase of current density due to a manufacturing defect (courtesy of IEEE Trans. on Magnetics).

describing the concentration and potential distribution within the electrode were solved numerically using the finite difference method (FDM) and Newman's BAND algorithm for the resulting simultaneous nonlinear equations. After the introduction of computational fluid dynamics (CFD) in the simulation of fuel cells, mostly single-domain models have been developed. The main advantages over the multi-domain approach is that internal BCs and continuity conditions at each domain interface are not required, thus simplifying the model geometry construction and speeding up problem set-up into a commercial CFD code. Single-domain CFD models can be classified into two-dimensional (2-D) and three-dimensional (3-D) models, depending on the simplifying assumptions on geometry and on boundary conditions. 2-D models generally provide a strong reduction in terms of computational cost, but their solution is less accurate compared to 3-D models.

5.2.1 DMFC two-dimensional models

A two-phase, multicomponent-flow 2-D model of a DMFC that accounted for capillary effects in both anode and cathode backings was developed [29]. In addition to electrochemical reactions, this model takes into account diffusion and convection of both gas and liquid phases in backing layers and flow channels. The effect of mixed potential related to methanol oxidation at the cathode, as a result of methanol crossover caused by diffusion, convection, and electro-osmosis, is simulated as well. Multiphysics equations are solved after discretization by the finite volume method (FVM). Numerical results concerning polarization curves are validated by experimental measurements. The main contribution of this work is the two-phase flow modeling of the anode, that is, the gas phase at the anode saturated with water and methanol and the liquid phase saturated with CO_2. Numerical analysis shows that gas-phase transport is one of the major issues affecting the fuel cell performance. In [30], a similar two-phase, two-dimensional model is presented. A capillary pressure function is used in order to simulate the methanol adsorption of backing materials. In addition, detailed multistep reaction models for both ORR and methanol oxidation as well as the Stefan-Maxwell formulation for gas diffusion are proposed. The effect of methanol and water crossover trough the membrane is accounted for in [31]. The two-phase mass transport in the anode and cathode porous regions is formulated based on the classical multiphase flow in porous media. A micro-agglomerate model, that is able to reflect the effect of the microstructure of the catalyst layer on cell performance, is proposed. The resulting polarization curves and methanol crossover rates at different concentrations are in very good agreement with experimental data. In [32], a realistic passive liquid-feed DMFC in transient charge/discharge conditions is simulated. The main contribution of this work is that effects of feed methanol concentration in the reservoir and current density on both mass transport and performance are investigated. Analyses show that when the initial feed concentration in the reservoir decreases, methanol crossover is minimized, but the fuel cell runtime is shortened. Recent works, for example [33], provide a detailed description of the whole DMFC system, including reservoirs. By developing a transient multiphase model of a passive cell, the effects of operating current density, voltage, micro-porous layer, and methanol feeding condition are comprehensively investigated for the whole operating processes, that is, with the fuel tank evolving from full to empty. Results highlight that for all operating conditions, it is necessary to operate at moderate current density or voltage to limit the methanol crossover and ensure the energy conversion efficiency. A 2-D multiphase non-isothermal mass transfer model for the DMFC is presented in [34]. The model includes the reaction of methanol and oxygen at the anode and cathode and the diffusion of every component involved in the DMFC, such as water, oxygen, and methanol at the diffusion layer and methanol crossover. It is shown that a maximum output power can be achieved for optimal temperature and concentration values.

5.2.2 DMFC three-dimensional models

A few papers are reported on 3-D two-phase DMFC models, which can capture the species distributions and the transport limitations along any direction inside the DMFC. Ref. [35] proposes a 3-D, two-phase, multicomponent model. Catalyst layers are incorporated in the computational domain instead of being modeled as zero-thickness interfaces. This model includes the effects of the second phase on the reduction of active catalyst surface areas and the mixed potential effects due to methanol crossover. The amount of carbon dioxide obtained from 3-D models indicates that the porosity of the anode diffusion layer plays a very important role in the DMFC performance. With a low porosity, the produced carbon dioxide cannot be removed effectively from the catalyst layer, thus reducing the active catalyst surface area as well as blocking methanol from reaching the reaction area. Ref. [31] extends the 2-D two-phase mass transport model for liquid-feed DMFCs to a fully 3-D model. The two-phase mass transport in the anode and cathode porous regions is formulated based on the multiphase flow theory in porous media without defining the mixture pressure of gas and liquid and assuming a constant gas pressure in the porous regions. The interaction between the phases in this 3-D model is captured by taking into account the effect of non-equilibrium evaporation/condensation at the phase interface, as opposed to the assumption of other models of thermodynamic equilibrium condition between the phases. From 3-D analysis, it can be observed that methanol concentration in the diffusion layer is higher in the channel than under the ribs, demonstrating that the flow-field channels cause methanol to be distributed unevenly over the reaction area (**Figure 6**).

In [36], a commercial flow solver (i.e., Ansys Fluent®) is used to solve at the same time flow, species, and charge transport equations. 3-D simulations are carried

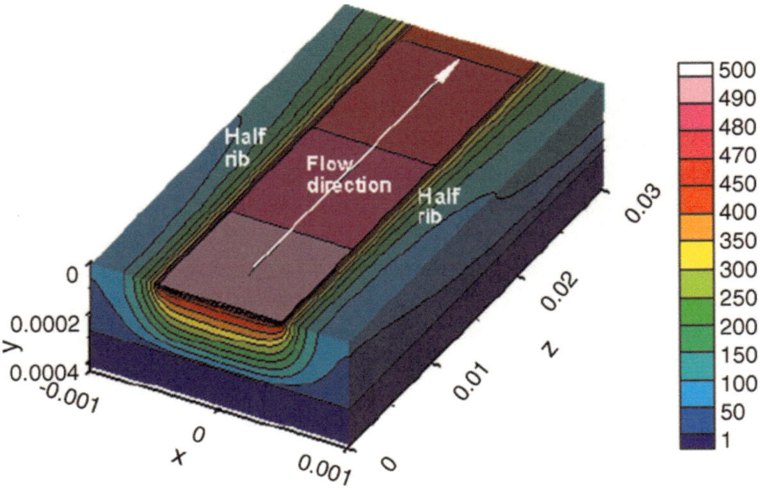

Figure 6.
Methanol concentration in the anode and membrane (unit: m^3/s) [30] (courtesy of Electrochimica Acta).

out in order to explore mass transport phenomena occurring in DMFCs for portable applications as well as to reveal an interplay between the local current density and methanol crossover rate. In [37], 3-D modeling is then extended to transient conditions. The authors note that cathode processes, for example oxygen and water transport coupled to electrochemical reaction, are inherently transient so that an unsteady-state model gives more accurate prediction than a steady-state model. Numerical simulations indicate that the cathode catalyst layer porosity has major effects on oxygen transfer and water removal. A three-dimensional multiphase model of DMFC is developed in [38], in which the Eulerian-Eulerian model is adopted to treat the gas and liquid two-phase flow in channel. By 3-D simulation, cell performance is found to be severely affected by accumulation of carbon dioxide mainly at the anode channel and by high-temperature operations. Ref. [39] shows that three-dimensional models are suitable for analyzing DMFC stacks with flowing electrolyte. A multiscale approach is therefore proposed in order the reduce the computational cost arising from 3-D modeling of the entire stack geometry. By this solution strategy, fully 3-D flow fields, backing layers, and membranes are numerically solved, whereas electrochemical reactions are analytically simulated.

It should be finally noted that multiphysics models coupling electrochemical reactions, methanol, water, and heat transport are still under investigation due to their high complexity.

5.3 Solid oxide fuel cells

Three different designs are used in planar SOFCs depending on their operating temperatures. High-temperature SOFCs (around 1000°C) usually present an electrolyte-supported structure, with thin electrodes (e.g., 50 μm) supported by a thick electrolyte (above 100 μm) [1]. The high temperature ensures so high a conductivity that the electrolyte resistance remains within acceptable values. In low-temperature SOFCs (though not less than 600°C), thinner electrolytes are used (e.g., 20 μm) and the cell is supported by either anode or cathode (300–1500 μm) with the other electrode being thinner (e.g., 50 μm). Understanding the multiphysics behavior is indispensable in identifying optimal design and operation of such SOFCs and a multiphysics numerical model is required at this purpose. Both

Fluent and STAR-CD have been applied with success to this task, providing precious information on the internal distribution of reactant and product, current density, temperature, and stress and, more generally, on the detailed operation of a SOFC. Models can include different fuels such as H_2 and CO and can take into account internal reforming by means of a catalytic chemical reaction [40]. Moreover, transport phenomena are not as complex as in a PEFC and DMFC.

Due to the high working temperatures, their gradients contribute to stress formation, which is a major technical issue of SOFC. Consequently, early modeling studies were aimed at predicting the current and temperature distributions, whereas flow and multicomponent transport were typically simplified. The subsequent use of CFD models has allowed more detailed three-dimensional multiphysics analyses. A Fluent-based CFD model has been developed by [41] to describe reactant flow, transport, and electrochemical reaction in a SOFC. STAR-CD was combined to an electrochemistry module by [42] to simulate a SOFC. Results showed that the co-flow reactant distribution at anode and cathode had the most uniform temperature distribution and the smallest thermal gradients. As a drawback, this approach treated the electrodes-electrolyte as a solid component, neglecting mass diffusion that is important in the case of thick electrodes. The effect of mass transport in a thick electrode has been analyzed by [43] in a two-dimensional study that couples mass and heat transport and included methane/steam reforming by means of a catalytic chemical reaction. Basically, in three-dimensional analysis, the governing equations for conservation of mass, momentum, species, thermal energy, electric charge, and electrochemical kinetics in anode and cathode of a SOFC are the same as those of a PEMFC, except that water transport through the electrolyte is not considered [44]. Orthogonal meshes are typically used in modeling planar SOFCs. A five-channel geometry can be modeled with $80 \times 80 \times 35$ mesh, resulting in 224,000 cells, on which the model can converge in some 300 iterations. Simulations of this kind reveal how current density increases in correspondence of the channels in the current collectors, where the electrodes receive more reactants. The effect of the flow channel design on the cell performance can be analyzed in detail with such numerical tools. Due to high operating temperature, experimental validation of the numerical model is particularly challenging and few works are reported in the literature on this topic.

6. Optimization and identification

6.1 Optimization

A number of nonlinear deterministic optimization methods (DOMs) have been applied to PEMFCs in the last decade, proving successful in dealing with specific tasks. Least squares methods have been applied to the estimation of single material parameters as well as parameters evolution under degradation events [2]. The gradient method has been exploited in the search for optimal designs and parameters evolution, such as cathode configuration optimization, geometric optimization, and flow field serpentine optimization [45]. A review of deterministic optimization methods used for identification problems in PEMFCs is given in [46].

Deterministic methods are known for their efficiency, that is, speed of convergence, but their applicability may be hindered, depending on the specific algorithm, by lack of flexibility in handling arbitrary constraints, sensitivity to noise in the objective function, possible need of function derivatives, and premature convergence to local minima. On the other hand, stochastic optimization methods (SOMs), in spite of their comparatively low efficiency, typically overcome the above-mentioned shortcomings of deterministic methods. Although the convergence to the global optimum for SOMs is only asymptotically guaranteed, there is abundant numerical evidence that very good solutions can be obtained for many problems, including FC design problems. It is worth noticing that a crucial feature of FC circuit models is that they avoid partial differential equations, thus resulting in numerical formulations with relatively low computational costs, which make them ideally suited for SOMs. It should also be noted that since optimization problems related to PEMFCs are characterized by highly nonlinear device models, the resulting objective functions subject to minimization have many local minima, which need to be all identified in the search for the global one. Therefore, for this type of problems, stochastic optimizers may end up being almost as efficient as deterministic ones. A further advantage, which however is also shared by some deterministic methods, is that stochastic optimizers can also deal with non-differentiable or fully discrete, optimization problems. In recent years, the application of stochastic methods for the solution of FC optimization problems has been constantly increasing, and interesting results have been reported, for example, with genetic algorithms (GA) [47], particle swarm optimization (PSO) [48], and differential evolution (DE) [49]. The above-mentioned methods are all population-based, that is, they explore several candidate solutions concurrently, which makes them ideally suited for parallelization. SOMs can also be combined among them or hybridized with DOMs in order to tailor their behavior to the specific optimization problem. Multiobjective stochastic approaches have also been recently investigated [50].

6.2 Identification

Since fuel cells present a stratified structure of thin layers made of different materials, analyzing their behavior requires the full characterization of these materials, that is, the determination of their chemical, physical, thermal, and electrical parameters. The identification of these parameters is crucial for guiding the research for advanced functionalized materials. These parameters are also needed in FC models, used in the fast exploration of different operating scenarios and in the research of optimized structural design and operating conditions [51]. The systems of equations involved (Nernst equation, Butler-Volmer equation, Darcy's equation, Fourier's law, Ohm's law, etc.) are strongly nonlinear, making the models extremely sensitive to parameter variations and uncertainties.

Unfortunately, they are hard to measure in real operating conditions and their identification still remains challenging when dealing with direct measurements. Careful ex situ measurements can be performed by means of a number of diagnostic techniques; however, the transferability of results to operative fuel cells raises a number of issues. Conversely, in situ measurements can provide meaningful operational values, but very few, often complicated and cumbersome, techniques are available to determine a limited number of parameters. A different approach consists in tackling the identification of multiple parameters by using a very large body of experimental data collected at different of physical conditions (e.g., temperature, pressure, and humidity). However, this approach suffers from the well-known "curse of dimensionality," that is, the problem becomes exponentially harder to solve as the number of parameters increases. This challenging problem can be approached with an optimization procedure (i.e., the search of the minimum of a function $f(x)$). When using optimization algorithms for model parameter identification, x is the n-dimensional vector of the unknown parameters to be identified and $f(x)$ consists in an error function that assesses the difference between the output of the parameter-based model and the measurements. The parameter identification problem is a constrained one, that is, the domain A where x values are defined is supplemented with a number of constraints and the problem is also typically burdened by model nonlinearity, as is the case of an FC model, which results in the non-convexity of f and consequent local minima. Moreover, large problems lead to high computational cost. Given the problems of "curse of dimensionality," presence of local minima, and computational costs, smart strategies are needed to find good solutions, if not the absolute best one, which actually may be impossible to identify.

In the last decade, stochastic methods have been applied to the study of FC parameter identification problems and their use has been strongly increasing in the last 3 years. Studies reported in literature typically aim at using stochastic methods in order to obtain a good fit of PEMFC polarization curves and usually resort to simplified empirical PEMFC models. Such models tend to be interpolatory in nature and contain enough parameters (e.g., 5–7) to allow for a good fit. Given their nature, fitting a set of empirical parameters to match a given polarization curve is a relatively easy task for most optimization procedures, but the usefulness of the obtained results is rather limited. The relatively small number of parameters also helps in avoiding so-called duplicity problems, that is, multiple distinct solutions achieving the same minimal values of f. However, duality is not crucial, because empirical parameters have no direct physical meaning. A more ambitious challenge consists in identifying several physical parameters of the materials of a PEMFC by means of an optimization approach. An algorithm of this kind, built over an early investigation on the capability of stochastic methods to deal with FCs [52], uses a detailed multiphysical performance model that employs such parameters and takes into account some physical control quantities [53]. A straightforward use of a stochastic optimizer with a large number of unknowns and a weakly constrained nonlinear problem can result in duplicity. In order to overcome it, a possible strategy consists in splitting the overall identification problem into a sequence of distinct identification sub-problems, each having a lower number of unknowns, and thus suffering less from the "curse of dimensionality." This approach basically relies on isolating a group of equations, dominated by some of the unknowns only, and has already been applied successfully to fuel-cell problems [54]. This behavior emerges, for example, since the parameters related to the activation losses and to the concentration losses, which are strongly nonlinear, prevail over the ohmic losses, which are linear at given hydration and temperature, and their identified values tend to vanish [55]. This behavior can be exploited by separately considering the typical three parts of the polarization curves. The experimental data obtained at low current density can be used to identify the parameters related with activation losses, while those obtained at high current density can been used to identify the parameters related with the concentration losses. Finally, experimental data at intermediate current density values can be used to identify the parameters related with the ohmic losses, which dominate in the central part of the polarization curve (**Figure** 7).

Some final considerations emerge: first of all, the accuracy and reproducibility the experimental data and the experimental conditions must increase with growing number of unknown parameters. This may be hard to obtain in the case of the polarization curves of PEMFCs, which depend on many factors related to both the samples under test and the experimental conditions, some of which are hard to control. Such difficulties can be mitigated by collecting more curves in the same nominal operating conditions and performing a statistical selection of the data. Moreover, enriched experimental plans may allow to identify also some parameters that have a minor effect on the polarization curve, for example, a set of experimental data obtained at different back pressures may allow the identification of the anodic exchange current density, which is otherwise masked by the larger effect of

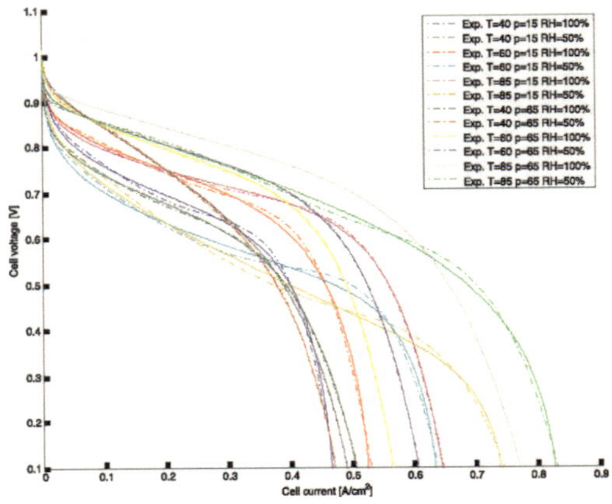

Figure 7.
Identification by means of a hybrid PSO-DE algorithm run simultaneously on all polarization curves. Dashed lines: experimental, continuous lines: optimized models (courtesy of Journal of Power Sources).

the cathodic one. On the other hand, a failure of the identification procedure, that is, a poor fit of the polarization curves with varying operating conditions, typically hints at weaknesses in the model.

7. Conclusion

The decarbonization of power grids and the development of electric vehicles and renewable resources are promoting researches on advanced fuel cells for both stationary and mobile applications. Several programs are running that aim at reducing costs and improve performance and duration. These researches can be strongly supported by numerical models which are the core of a computer-aided engineering approach capable of reducing tentative experimentation by selecting those solutions which result more competitive on the basis of analytical/numerical computations, both in terms of device design and its operation. Effective models must take into account all relevant chemical-physical-electrical quantities, their interdependence, and their evolutions. These models play an interactive game with diagnostic and measurement issues, because their reliability depends on their fitting to the cases under investigation and this occurrence involves the determination of the many parameters used in nonlinear equations. Design optimizations based on numerical procedures are strong tools not only in identifying device materials and geometries capable of competitive performance, but also in determining the whole FC system, including static converters and system supervisors.

Fuel cell computational modeling confirms to be an important topic of applied research, involving multiphysics, multiscale problems which are still challenging for the researchers working on this subject, both at the scientific and industrial level.

Acknowledgements

This work was supported by MAESTRA 2011—"From Materials for Membrane-Electrode Assemblies to Electric Energy Conversion and Storage Devices," the 2011 University of Padua Strategic Project cod. STPD11XNRY_002.

Acronyms

A	area (m^2)
a/c	anode/cathode (−)
C_{dl}	double-layer capacitance (F)
c_i	molar concentration (mol m^{-3})
c_p	specific heat capacity (J kg^{-1} K^{-1})
D	diffusion coefficient (m^2 s^{-1})
d_m	membrane thickness (m)
E_0	standard potential (V)
F	Faraday constant (A s mol^{-1})
j	electric current density (A m^{-3})
J	electric current density vector (A m^{-2})
k	Boltzmann constant (J K^{-1})
k_T	global thermal exchange coefficient (K W^{-1})
n	number of electrons in reaction (−)
N	molar flow density (mol m^{-2} s^{-1})
P_{diss}	dissipated power (W)
p_i	partial pressure (Pa)
R	universal gas constant (J mol^{-1} K^{-1})
T	temperature (K)
V	cell voltage (V)
W_a	activation energy (eV)
Z_w	Warburg impedance (Ω)
α	charge transfer coefficient (−)
δ	electrode thickness (m)
$\Delta\hat{s}$	molar entropic variation (J K^{-1})
η	overpotential (V)
κ	mass transfer coefficient (_)
λ	Nafion® hydration (−)
μ	charged particle mobility (m^2 V^{-1} s^{-1})
ξ	electro-osmotic drag coefficient (_)
ρ	mass density (kg m^{-3})
σ	electric conductivity (S m^{-2})
φ	electric potential (V)
ω	angular frequency (rad s^{-1})

Author details

Massimo Guarnieri*, Piergiorgio Alotto and Federico Moro
Dipartimento di Ingegneria Industriale, Università di Padova, Padova, Italy

*Address all correspondence to: massimo.guarnieri@unipd.it

References

[1] Wang CY. Fundamental models for fuel cell engineering. Chemical Reviews. 2004;**104**:4727-4766. DOI: 10.1021/cr020718s

[2] Carnes B, Djilali N. Systematic parameter estimation for PEM fuel cell models. Journal of Power Sources. 2005; **144**:83-93. DOI: 10.1016/j.jpowsour.2004.12.024

[3] O'Hayre R, Cha SW, Coltella W, Prinz FB. Fuel Cell Fundamentals. New York: John Wiley & Sons; 2009

[4] Bard AJ, Faulkner LR. Electrochemical methods. In: Fundamentals and Applications. New York: John Wiley & Sons; 2006

[5] Parthasarathy A, Srinivasan S, Appleby AJ, Martin CR. Temperature dependence of the electrode kinetics of oxygen reduction at the platinum/Nafion® interface – A microelectrode investigation. Journal of the Electrochemical Society. 1992;**139**(9): 2530-2537. DOI: 10.1149/1.2221258

[6] You L, Liu H. A two-phase flow and transport model for the cathode of PEM fuel cells. International Journal of Heat and Mass Transfer. 2002;**45**(11): 2277-2287. DOI: 10.1016/S0017-9310(01)00322-2

[7] Springer TE, Zawodzinski TA, Gottesfeld S. Polymer electrolyte fuel cell model. Journal of the Electrochemical Society. 1991;**138**: 2334-2341. DOI: 10.1149/1.2085971

[8] Sequeira C, Santos D. Polymer Electrolytes – Fundamentals and Applications. Abington, Cambridge, MA: Woodhead Publication; 2010

[9] Tsushima S, Hirai S. In situ diagnostics for water transport in proton exchange membrane fuel cell. Progress in Energy and Combustion Science. 2011;**37**:204-220. DOI: 10.1016/j.pecs.2010.06.001

[10] Vilekar SA, Datta R. The effect of hydrogen crossover on open-circuit voltage in polymer electrolyte membrane fuel cells. Journal of Power Sources. 2010;**195**(8):2241-2247. DOI: 10.1016/j.jpowsour.2009.10.023

[11] Mench MM. Fuel Cell Engines. Hoboken: Wiley; 2008

[12] Scott K, Taama W, Cruickshank J. Performance and modeling of a direct methanol solid polymer electrolyte fuel cell. Journal of Power Sources. 1997;**65**: 159-171. DOI: 10.1016/S0378-7753(97)02485-3

[13] Sundmacher K, Scott K. Direct methanol polymer electrolyte fuel cell: Analysis of charge and mass transfer in the vapor-liquid-solid system. Chemical Engineering Science. 1999;**54**:2927-2936. DOI: 10.1016/S0009-2509(98)00344-3

[14] Guo H, Ma C. 2-D analytical model of a direct methanol fuel cell. Electrochemistry Communications. 2004;**6**:306-312. DOI: 10.1016/j.elecom.2004.01.005

[15] Alotto P, Guarnieri M, Moro F. Optimal design of micro direct methanol fuel cells for low-power applications. IEEE Transactions on Magnetics. 2009a;**45**(3):1570-1573. DOI: 10.1109/TMAG.2009.2012745

[16] Guarnieri M, Di Noto V, Moro F. A dynamic circuit model of a small direct methanol fuel cell for portable electronic devices. IEEE Transactions on Industrial Electronics. 2010;**57**: 1865-1873. DOI: 10.1109/TIE.2009.2027916

[17] Kulikovsky AA. The voltage-current curve of a direct methanol fuel cell:

"Exact" and fitting equations. Electrochemistry Communications. 2002;**4**:939-946. DOI: 10.1016/S1388-2481(02)00494-0

[18] Dohle H, Wippermann K. Experimental evaluation and semi-empirical modeling of U/I characteristics and methanol permeation of a direct methanol fuel cell. Journal of Power Sources. 2004;**135**: 152-164. DOI: 10.1016/j.jpowsour.2004.04.014

[19] Wang Y, Au G, Plichta EJ, Zheng JP. A semi-empirical method for electrically modeling of fuel cell: Executed on a direct methanol fuel cell. Journal of Power Sources. 2008;**175**:851-860. DOI: 10.1016/j.jpowsour.2007.09.101

[20] Chen M, Rincón-Mora G. A compact electrical model for microscale fuel cells capable of predicting runtime and I-V polarization performance. IEEE Transactions on Energy Conversion. 2008;**23**:842-850. DOI: 10.1109/TEC.2008.926038

[21] Alotto P, Guarnieri M, Moro F. A coupled electro-chemical model of a direct methanol fuel cell for portable electronic devices. COMPEL. 2009b;**28**: 1005-1019. DOI: 10.1108/03321640910959062

[22] Mallick RK, Thombre SB, Kothekar KP. An integral mathematical model of liquid feed passive DMFCs with non-isothermal effects and charge conservation phenomenon. Journal of Cleaner Production. 2018;**182**:654-667. DOI: 10.1016/j.jclepro.2018.01.177

[23] Yang QW, Hu XQ, Lei XC, Zhu Y, Wang XY, Ji SC. Adaptive operation strategy for voltage stability enhancement in active DMFCs. Energy Conversion and Management. 2018;**168**: 11-20. DOI: 10.1016/j.enconman.2018.04.110

[24] Costamagna P, Honegger K. Modeling of solid oxide heat exchanger integrated stacks and simulation at high fuel utilization. Journal of the Electrochemical Society. 1998;**145**(11): 3995-4007. DOI: 10.1149/1.1838904

[25] COMSOL. Multiphysics. http://www.comsol.com. [Accessed: 15 March 2017]

[26] Zhang G, Jiao K. Multi-phase models for water and thermal management of proton exchange membrane fuel cell: A review. Journal of Power Sources. 2018;**391**:120-133. DOI: 10.1016/j.jpowsour.2018.04.071

[27] Haynes WM. Handbook of Chemistry and Physics. Boca Raton: CRC Press; 2016

[28] Khandelwal M, Mench MM. Direct measurement of through-plane thermal conductivity and contact resistance in fuel cell materials. Journal of Power Sources. 2006;**161**:1106-1115. DOI: 10.1016/j.jpowsour.2006.06.092

[29] Wang ZH, Wang CY. Mathematical modeling of liquid-feed direct methanol fuel cells. Journal of the Electrochemical Society. 2003;**150**(4):A508-A519

[30] Divisek J, Fuhrmann J, Gartner K, Jung R. Performance modeling of a direct methanol fuel cell. Journal of the Electrochemical Society. 2003;**150**: A811-A825. DOI: 10.1149/1.1572150

[31] Yang WW, Zhao TS, Xu C. Three-dimensional two-phase mass transport model for direct methanol fuel cells. Electrochimica Acta. 2007;**53**: 853-862. DOI: 10.1016/j.electacta.2007.07.070

[32] Xiao B, Faghri A. Transient modeling and analysis of a passive liquid-feed DMFC. International Journal of Heat and Mass Transfer. 2008;**51**: 3127-3143. DOI: 10.1016/j.ijheatmasstransfer.2007.08.022

[33] Guo T, Sun J, Deng H, Jiao K, Huang X. Transient analysis of passive direct methanol fuel cells with different operation and design parameters. International Journal of Hydrogen Energy. 2015;**40**:14978-14995. DOI: 10.1016/j.ijhydene.2015.09.040

[34] Ismail A, Kamarudin SK, Daud WRW, Masdar S, Hasran UA. Development of 2D multiphase non-isothermal mass transfer model for DMFC system. Energy. 2018;**152**: 263-276. DOI: 10.1016/j.energy.2018.03.097

[35] Ge J, Liu H. A three-dimensional two-phase flow model for a liquid-fed direct methanol fuel cell. Journal of Power Sources. 2007;**163**:907-915. DOI: 10.1016/j.jpowsour.2006.10.014

[36] Liu W, Wang CY. Three-dimensional simulations of liquid feed direct methanol fuel cells. Journal of the Electrochemical Society. 2007;**154**:B352-B361. DOI: 10.1149/1.2429041

[37] Guo H, Chen YP, Xue YQ, Ye F, Ma CF. Three-dimensional transient modeling and analysis of two-phase mass transfer in air-breathing cathode of a fuel cell. International Journal of Hydrogen Energy. 2013;**38**:11028-11037. DOI: 10.1016/j.ijhydene.2013.03.054

[38] Sun J, Zhang G, Guo T, Jiao K, Huang X. A three-dimensional multi-phase numerical model of DMFC utilizing Eulerian-Eulerian model. Applied Thermal Engineering. 2018;**132**: 140-153. DOI: 10.1016/j.applthermaleng.2017.12.037

[39] Colpan CO, Ouellette D. Three-dimensional modeling of a FE-DMFC short-stack. International Journal of Hydrogen Energy. 2018;**43**:5951-5960. DOI: 10.1016/j.ijhydene.2017.11.123

[40] van Biert L, Godjevac M, Visser K, Aravind PV. Dynamic modelling of a direct internal reforming solid oxide fuel cell stack based on single cell experiments. Applied Energy. 2019;**250**: 976-990. DOI: 10.1016/j.apenergy.2019.05.053

[41] Prinkey M, Gemmen RS, Rogers WA. Application of a new CFD analysis tool for SOFC technology. In: Proc. MECE 2001. New York, 369-4: ASME; 2001. pp. 291-300

[42] Recknagle KP, Williford RE, Chick LA, Rector DR, Khaleel MA. Three-dimensional thermo-fluid electrochemical modeling of planar SOFC stacks. Journal of Power Sources. 2003;**113**(1):109-114. DOI: 10.1016/S0378-7753(02)00487-1

[43] Ackmann T, de Haart LGJ, Lehnert W, Stolten D. Modeling of mass and heat transport in planar substrate type SOFCs. Journal of the Electrochemical Society. 2003;**150**(6): A783-A789. DOI: 10.1149/1.1574029#

[44] Wehrle L, Wang Y, Banerjee A, Brandon N, Deutschmann O. Dynamic modeling of reversible solid oxide cells. Chemie Ingenieur Technik. 2019;**91**: 833-842. DOI: 10.1002/cite.201800188

[45] Secanell M, Carnes B, Suleman A, Djilali N. Numerical optimization of proton exchange membrane fuel cell cathodes. Electrochimica Acta. 2007;**52**: 2668-2682. DOI: 10.1016/j.electacta.2006.09.049

[46] Petrone R, Zheng Z, Hissel D, Péra MC, Pianese C, Sorrentino M, et al. A review on model-based diagnosis methodologies for PEMFCs. International Journal of Hydrogen Energy. 2013;**38**:7077-7091. DOI: 10.1016/j.ijhydene.2013.03.106

[47] Ohenoja M, Leiviska K. Validation of genetic algorithm results in fuel cell model. International Journal of Energy Research. 2010;**35**(22):12618-12625. DOI: 10.1016/j.ijhydene.2010.07.129

[48] Li Q, Chen W, Wang Y, Liu S. Parameter identification for PEM fuel-cell mechanism model based on effective informed adaptive particle swarm optimization. IEEE Transactions on Industrial Electronics. 2011;**58**(6): 2410-2419. DOI: 10.1109/TIE.2010. 2060456

[49] Chakraborty UK, Abbott TE, Das SK. PEM fuel cell modeling using differential evolution. Energy. 2012; **40**(1):387-399. DOI: 10.1016/j. ijepes.2013.05.031

[50] Chen X, Li W, Gong G, Wan Z, Tu Z. Parametric analysis and optimization of PEMFC system for maximum power and efficiency using MOEA/D. Applied Thermal Engineering. 2017;**121**:400-409. DOI: 10.1016/j.applthermaleng.2017.03.144

[51] Alotto P, Guarnieri M, Moro F, Stella A. Multi-physic 3-D dynamic modeling of polymer membranes with a proper generalized decomposition model reduction approach. Electrochimica Acta. 2011;**57**:250-256. DOI: 10.1016/j.electacta.2011.07.019

[52] Alotto P, Guarnieri M. Stochastic methods for parameter estimation of multiphysics models of fuel cells. IEEE Transactions on Magnetics. 2014;**50**(2): 7017304. DOI: 10.1109/ TMAG.2013.2283889

[53] Guarnieri M, Alotto P, Moro F. Modeling the performance of H2-O2 unitized regenerative PEM fuel cells for energy storage. Journal of Power Sources. 2015;**297**(11):23-32. DOI: 10.1016/j.jpowsour.2015.07.067

[54] Úbeda D, Pinar FJ, Cañizares P, Rodrigo MA, Lobato J. An easy parameter estimation procedure for modeling a HT-PEMFC. International Journal of Hydrogen Energy. 2012;**37**: 11308-11320. DOI: 10.1016/j. ijhydene.2012.04.157

[55] Guarnieri M, Negro E, Di Noto V, Alotto P. A selective hybrid stochastic strategy for fuel-cell multi-parameter identification. Journal of Power Sources. 2016;**332**:249-264. DOI: 10.1016/j. jpowsour.2016.09.131

Chapter 3

F-diagram Research Method for Double Circuit Solar System with Thermosyphon Circulation

Yedilkhan Amirgaliyev, Murat Kunelbayev, Aliya Kalizhanova, Ainur Kozbakova, Salauat Daulbayev, Timur Merembayev and Didar Yedilkhan

Abstract

In the work herein we have conducted researches of F-diagram method for the solar system with thermosyphon circulation-diagram method is based on correlation of a lot of simulations, computed non-dimensional variables. Modeling conditions varied in corresponding ranges of thermosyphon circulation double circuit solar system practical constructions' parameters. By means of F-diagram method there have been computed the environmental monthly temperature values with correction index, which shows that the monthly average daily heating degree and direct solar radiation decrease according to weather conditions. It may be noted that monthly temperature load fraction increases along with the collector square growth. It demonstrates that the monthly temperature load fraction is higher in summer in Almaty city (Kazakhstan) (July—the highest value) and lower in winter months (January, February—the lowest value).

Keywords: flat solar collector, F-diagram method, environmental temperature, solar radiation, double circuit solar system, thermosyphon circulation

1. Introduction

Solar radiation and environmental temperature measuring represent a complicated problem, as it includes unpredictable weather conditions for designing the solar system heating, which involves solar system components' dimensions exact defining. There are three categories considered upon solar heating systems designing. The first category belongs to the system, in which the collector's working temperature is known or it might be assessed, for which radiation critical levels can be revealed.

The second category is based on the big amount of detailed simulations and presents the F-diagram method.

The third category includes the short circuit simulation, in which the modeling is fulfilled by means of meteorological data for representative days and it is called the sol cost method [1].

In this experiment [2], eight typical Taiwanese solar water heaters, which were connected in series, were considered. The degree of temperature stratification and consumption of thermosyphon in a horizontal tank is estimated. The system was tested without load, with interruptions and constant load conditions. The results showed that stratification of the state was observed in tanks under load.

The researchers compared the gap filler in the results on the scale of complexity in the simulation program from the programs F-diagrams [3].

In the study [4], referring to the method of F-diagrams calculated coefficient of thermal characteristics of the solar collector.

Usage of F-diagram method for active premises heating systems applying the working liquid is the thermal research basis [1]. Applying this method thereof we can assess the fraction of total heating load, which can be provided by the solar energy system. In the method the primary project variable is a collector's square, and the secondary variables are accumulator capacity, collector type, dimensions of heat exchanger load and collector and liquid flow speed. F-diagrams have been developed for three system standard configurations: liquid and air, used for premises heating (and hot water) and systems for only hot water [1]. In this study [5], solar systems are analyzed using the F-chart method in order to satisfy the hot water needs of hotels. The annual fraction and heat loads for different solar collector's area and number of people is estimated. Flat plates and vacuum tube collectors are compared and analyzed.

Kalogirou used for protection from freezing water ethylene glycol, necessary for solar heating system operation [6].

In order to assess the solar energy potential, falling onto the territory in any region, it is necessary to have data on the solar energy potential. Based on actual observations and theoretical computations generalization we can obtain the following data: annual and latitudinal movement of potential monthly and annual sums of direct solar radiation, falling onto the perpendicular surface under the clear sky conditions; the data on sun shine duration; sunshine daily move; radiation for the annual typical days; maps of distributing the average monthly radiation amounts for June and December on Kazakhstan territory [7].

In the article herein, using the F-diagram method, the authors have designed the solar heating system and definition of total thermal load fraction (load on household water and heating), which will be supplied with solar energy with a view to the family, consisting of six people in Almaty city. The research includes the influence of the collector's different squares and storage capacity per the collector square meter and collector tilter the load share, maintained by the solar energy. As well, there will be studied an annual behavior of solar load share and monthly load change within a year.

2. System description

System's standard configuration, used in the given work frame, to which the F-diagram method will be applied, has been shown on the **Figure 1**. The system herein makes use of the liquids (commonly, water or antifreeze solution) as a heat carrier and water as a storage medium [5]. Flat solar collectors with thermosyphon circulation are used for transforming the falling solar radiation into the thermal heat. The energy thereof is accumulated in the form of notable heat in the tank for the liquid storage and used, when the need arises, to provide premises and water heating.

Proposed system operation is executed as follows (**Figure 1**). The solar energy E with temperature t_0 is absorbed by solar collector (1) with temperature t_1, heating the flow, the solar energy moves through a translucent insulating glazing unit (2). The heat, received from the solar stream, heats the liquid in coils (3), which is

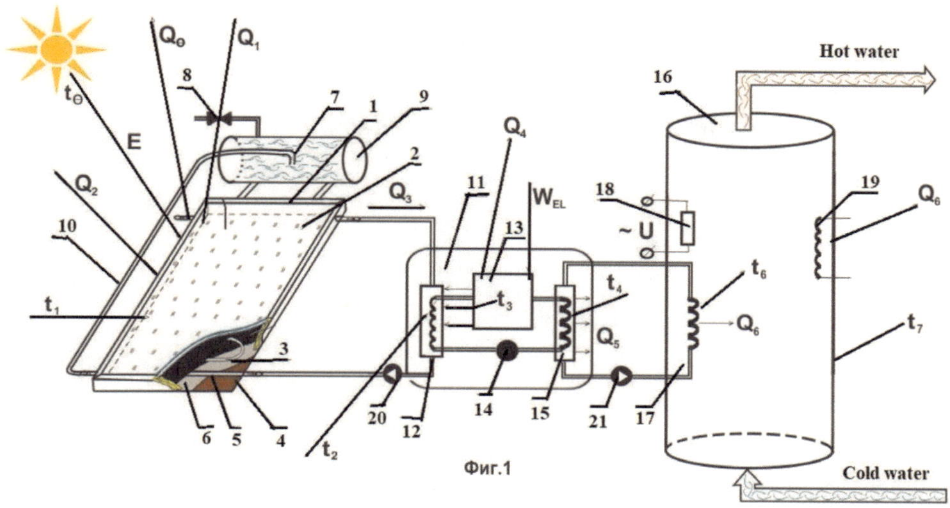

Figure 1.
Principal diagram of double circuit solar installation with thermosyphon circulation.

removed from a collector, and instead of it there incomes cold water from the water pipeline with a valve for cold water (8) and from the syphon of a dosimeter tank (7) there takes place constant thermosyphon circulation by means of circulation tube (10). Further the liquid enters a thermal pump (11), which consists of a condenser evaporator (12) with temperature t_2, with a heat exchanger in the form of a spiral, absorbing the heat carrier heat, lowers its temperature down the atmospheric temperature (Q_2) using the speed control valve (14), thereby serving to the heat additional absorption from the atmospheric air. The scheme also shows the solar irradiation, reflected from semi translucent cover (Q_0) and an absorbing panel surface (Q_1). In the thermal pump there is fulfilled a heat exchanger energy transfer, at respectively low temperature, to a condenser heat exchanger's heat carrier (15) in the form of a spiral with higher temperature t_2, which increases the square, as well, a heat exchanger intensity. To execute the cycle thereof we use a compressor (13) with temperature t_3, with an electric drive (17). Hereafter, by means of a condenser heat exchanger (15) with temperature t_4, the heat from a thermal pump (Q_5) is transferred to a heat exchanger's accumulator tank Q_6 with temperature $t_6$of the heating system (18). As the installation has two circuits, it is provided with automatic circulation pumps (19, 20) for liquid circulation between the solar collector and evaporator, condenser and accumulator tank. The water temperatures brought to the demanded technological level and supplied to a consumer for water provision and heating.

Figure 2 demonstrates the model of the flat solar collector. The main point and novelty is in the fact that in distinction from the known designing principle, the collector contains a translucent glazing unit (2) with double glass and reduced pressure, as well as a parametric frame (1). A wooden frame's bottom (7) has been made of plywood with 8 mm thickness and stuck a heat sealing film (5) with foil. In the gap between a glazing unit and frame's bottom there has been laid a flexible 4ϕ16 mm thin walled stainless corrugated pipe in the coil form. Pipe edges are attached to the inlet and outlet protruding tubes (6) (**Table 1**).

Figure 3 shows a flat solar collector's mockup. The solar collector is the main heat generating unit of the solar installation. To reach a preset aim we have developed a principally new flat solar collector, based on which there will be constructed various types solar installations, used for water heating and buildings and premises heating.

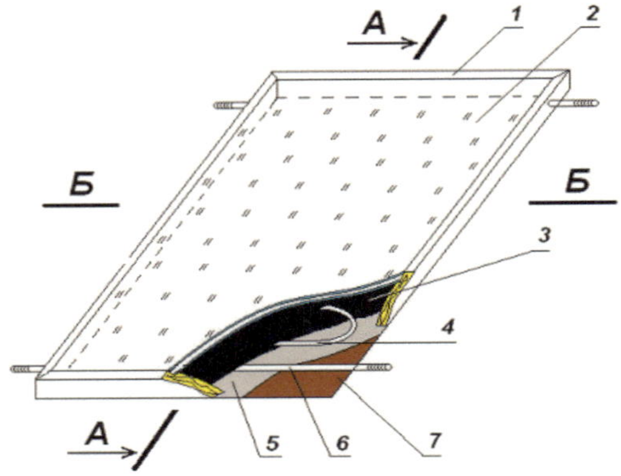

Figure 2.
Flat solar collector.

Parameters	Value
Absorbing plate material	Copper
Absorbing plate dimensions	2 m × 1 m
Plate thickness	0.4 mm
Glazing material	Hardened glass
Glazing dimensions	2 m × 1 m
Glazing thickness	4 mm
Insulation	Foam plex (foam polyurethane)
Collector tilt	45°
Absorber's heat transmission capacity	401 W/(m K)
Insulation heat transmission capacity	0.04 W/(m K)
Transmission-absorption factor	0.855
Sun visual temperature	4350 K
Atmospheric temperature	303 K
Radiation intensity	1000 W/m^2

Table 1.
Technical specifications of flat solar collector.

Table 2 represents the ranges of the basic project variables, used upon developing the correlations for liquid solar system heating [1].

The authors have elaborated a new computation methodology and selection of thermosyphon solar collector's geometrical parameters. As well, there has been shown the dependence of the tube's cross section on the flow time for different pressure values. Along with the syphon head increase there the liquid flow time grows as well. It is explained with the fact that the syphon hydraulic resistance increases along with pressure increase, which brings to the liquid speed reducing. For the first time there has been formulated the dependence, defining the fluid discharge time according to the solar collector's geometrical parameters. The methodology, having been elaborated, has allowed stating that the local hydraulic resistance and friction play a sufficient role in a heat carrier's expenditures [8]. Also we have considered the mathematical model of separate constructions and operation

Figure 3.
Mockup of flat solar collector.

Parameter	Range
$(\tau\alpha)_n$	0.6–0.9
$F'_R A_C$	5-120 m^2
U_L	2.1–8.3 W/m^{2*}K
β(collector slop)	30–90°
$(UA)_h$	83–667 W/K

Table 2.
Construction parameters ranges used upon developing F-charts for liquid systems [5].

mode of thermosyphon circulation double circuit solar collector. Proceeding from the analysis results we have managed to optimize individual structural elements, as well to prognoses the thermal regime and alternative solutions selection for designing the flat solar collectors and their operation regime selection [9].

3. F-chart method

Solar heat supply system's energy balance for the month period can be presented as [1]:

$$Q = Q_{h.ws} + E = \Delta U \tag{1}$$

where Q is the solar installation monthly heat production, $Q_{h.ws}$ is the hot water supply monthly load, E is the energy total amount, obtained within a month, and ΔU is the energy amount change in the accumulating unit.

At dimensions of accumulators, commonly used in the solar water supply systems, the difference ΔU is small comparing to Q, $Q_{h.ws}$ and E and can be adopted as equal to zero. Then Eq. (1) can be presented in the form of [1].

$$f = \frac{Q_{h.ws} - E}{Q_{h.ws}} = \frac{Q}{Q_{h.ws}} \tag{2}$$

where f is the fraction of the monthly thermal load, provided at the solar energy expense.

Straightforwardly, Eq. (2) cannot be used for computing f, as the value Q is the function of the falling radiation, environmental temperature and thermal loads. However, consideration of the parameters, the Q is dependent on, allows supposing, that the replacement rate of it can be empirically linked with two dimensionless complexes [1].

$$X = F_k K_k (T_a - T_b) \frac{\Delta t}{Q_{h.ws}} \tag{3}$$

$$Y = \frac{F_k \eta_0 E_k n_d}{Q_{h.ws}}, \tag{4}$$

where T_a is basic temperature, accepted as equal to 100°C, T_b is the average monthly temperature of outside air, °C, Δt is the time change, and E_k is the average monthly daily incoming of the total solar radiation falling onto the flat collector's inclined surface, J/(m^{2}*day).

F-diagram method is based on correlation of many simulations in terms of easily computed dimensionless variables. Modeling conditions varied in the corresponding ranges of the system's practical constructions parameters. Resulting correlations give to f the fraction of monthly heating load (in the case herein—the premises heating and hot water), provided with the solar energy, as the function of two dimensionless parameters. One of them is linked with the ratio of collector losses to the thermal loads (X), another—the ratio of the absorbed solar radiation to the thermal loads Y. Proceeding from the systems simulation, where was used the F-diagram, it has become possible to develop the correlation between dimensionless variables and f–monthly load fraction, transmitted by the solar energy. Dimensionless parameters X and Y are defined as follows [1]:

$$\mathrm{X} = \frac{\textit{Collector energy loss during a month}}{\textit{Total heating load during a month}} \tag{5}$$

$$\mathrm{Y} = \frac{\textit{Absorber solar radiation}}{\textit{Total heating load during a month}} \tag{6}$$

Parameters X and Y can be recorded as in Eqs. (3) and (4), respectively.

$$\mathrm{X} = \frac{A_C F^{/} U_{L(T_{ref}} - T_a)\Delta\tau}{L} \tag{7}$$

$$\mathrm{Y} = \frac{A_C F^{/} R(\breve{\tau}\alpha)\overline{\overline{H}}N}{L} \tag{8}$$

To simplify the computations the dimensionless parameters values X and Y in Eqs. (3) and (4) are usually placed as in the equations [1], respectively. The reason for the arrangement thereof consists in the fact that coefficients values (LR UF and n RF) are obtained from the results of the standard collector testing. The ratios F'R/FR correct various temperature gradients between collector and storage tank and they are computed with the techniques, generalized in [1]. The ratio (τα)/(τα) n is as well assessed with the techniques, given in [1].

$$X = F_R U_L \frac{F'_R}{F_R}\left(T_{ref} - T_a\right)\Delta\tau \frac{A_C}{L} \tag{9}$$

$$Y = F_R(\tau\alpha)n * \frac{F'_R}{F_R} * \frac{\overline{\overline{(\tau\alpha)}}}{(\tau\alpha)_n} * \overline{\overline{Ht}} * N * \frac{A_C}{L} \tag{10}$$

4. Performance of solar heating system on the liquid

In the section herein the specifications of the solar heating system, shown in **Figure 1** will have been analyzed applying F-diagram method, solar energy monthly fraction (monthly solar energy contribution), thermal load and annual solar energy contribution. Correlation X, Y and f in the equation form equals to [1].

$$f = 1029Y – 0.065X – 0.245Y^2 + 0.0018X^2 + 0.0215Y^3 \tag{11}$$

where $0 < Y < 3$ и $0 < X < 18$.

Due to the equation nature (11) it should not be used beyond the ranges, shown with curves in **Figure 4**. In case a reference point is out of the range, the chart might be used for extrapolation with satisfactory results [1]. For simplicity, the common method "degree-day" is used for calculating the monthly average load for premises heating necessary for the system in the framework of the research herein. The method of premise heating extent assessment in degrees-days is based on the principle that the need in energy to heat the premises, first and foremost, depends on the temperatures difference: in the premise and outside. It is assumed that monthly load for heating the buildings, premises, in which the temperature is maintained at 24°C is proportional to degree-days amount in a month DD [1].

$$L_S = (UA)h * DD \tag{12}$$

where Ls is the load for premises heating, and (UA)h is the multiplication of losses by the building square. For the research the building with (UA)h 467 W/m² °C has been taken from the building project. Days amount in degrees (DD) in one

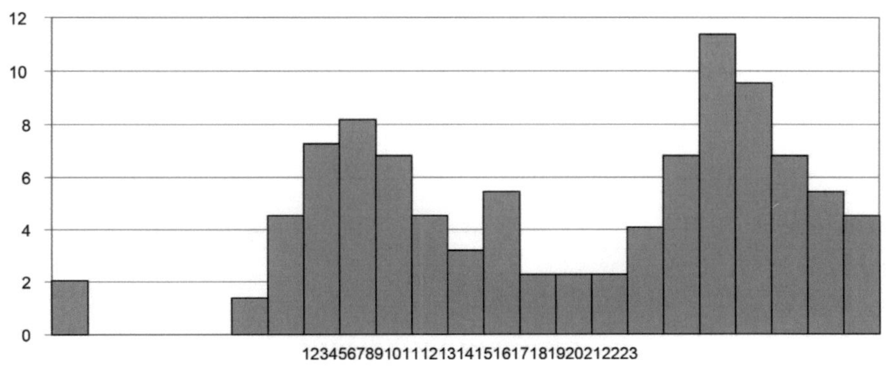

Figure 4.
Average monthly daily solar radiation for Almaty city.

day is the difference between 18.3°C and average daily atmospheric temperature (average of maximum and minimum atmospheric daily temperature). In case the average daily environmental temperature exceeds 18°C, the number of days in degrees is accepted being equal to zero [5]. For Almaty city the amount of days with a heating degree, monthly average daily solar radiation and environmental temperature are given in **Table 2**.

Another load, included into the research by F-diagram method is the load for water heating for household consumption (amount of energy, necessary for domestic water heating). It much depends on the building inhabitants life style. Average assumed water need and its consumption in Almaty constitutes 300 l per a person per day [1]. Monthly load for water heating, Lw.

$$L_w = N * N_p * V * (T_w - T_m) * \rho * C_p \quad (13)$$

where N is the days number in a month; Np is the people number in the family; T_w is the minimal hot water permissible temperature: it is ~60°C [1], V is the daily water consumption per a person in m^3, T_m is the temperature of the main feed water (°C), ρ is the water density in kg/m^3, and C_p is the water specific heat capacity (4190 J/kg*°C). Monthly total load (L) represents the total load for the building heating (L_S) and loads for household water heating (L_w), as in the work [1].

$$L = L_S + L_w \quad (14)$$

Monthly total load fraction, incoming from the solar heating system and water heating, shown in **Figure 1**, is given as a function of dimensionless parameters X and Y, defined in Eqs. (1) and (2) and in **Figure 3**. In order to define f, the heating load share, provided by the solar energy within a month, values X and Y are computed for collector and thermal load (**Table 3**). F-value is defined at X and Y junction point in **Figure 3**. It is done for every month of the year. Contribution of solar energy for a month is multiplication by total heating load L for a current month. Share of annual thermal load, provided by solar energy, represents the sum

Month	$\overline{\overline{T_a}}$ (°C)	$\overline{\overline{H}}$ (MJ/m²*day)
January	−17	−16.0
February	−20	−18.63
March	7	6.2
April	12	10.5
May	18	18.4
June	25	19.0
July	30	20.0
August	28	19.12
September	22	17.00
October	15	13.0
November	7	6.2
December	−10	−9.2

Table 3.
Heating degree and monthly average daily temperature and global radiation in Almaty city.

of solar energy monthly contribution, divided by the annual load, as in the following equation:

$$F = \frac{\sum f_i L_i}{\sum L_i} \tag{15}$$

Let us consider the solar system's heat supply computation method for the conditions, when the hot water supply load is prevailing or the only. Both the pipe water temperature *T*x.в, and minimal permissible hot water temperature *T*г. influence at the system's characteristics. As the average working temperature in the system, and consequently, the heat losses from the collector depend on *T*x.в and *T*г.в, it is worth to suppose that expression of *X complex*, characterizing the heat loss from the collector, might be updated in the way, to account the impact of *T*x. and *T*г.в. If to multiply monthly X values by correction factor, defined by the expression given below, then the F-method of solar heating and hot water supply liquid systems computation can be used for defining the monthly F-values, achieved in the solar hot water supply systems. Correction factor for the hot water supply systems [1].

$$\frac{X_c}{X} = \left(\frac{M}{75}\right)^{-0.25} \tag{16}$$

at

$$37.5 < M < 300.$$

Capacity of the systems with other values can be assessed from F-chart, changing Y applying the correction factor of the loading heat exchanger Yc/Y, as denoted in Eq. (17) or in **Figure 4** [1].

$$\frac{Y_C}{Y} = 0.39 + 0.65e^{\left(-0.139\,\frac{(UA)_h}{\varepsilon C_{min}}\right)} \tag{17}$$

at

$$0.5 < \left(\frac{\varepsilon L C_{min}}{(UA)h}\right)$$

5. Result and discussion

In this section we discuss the demanded thermal load, fraction of the load, supplied by the solar energy system for various collector zones and parametric researches.

Figure 5 shows that the chart of monthly values from correction factor by means of F-diagram method shows that the heating extent, monthly average daily temperature and direct solar radiation lower dependent on the weather conditions.

It is clearly seen from **Figure 6** that changed correction factor Yc/Y from environmental temperature has an exponential function, which according to a correction factor, increases the annual fraction of the load, provided by the solar energy.

It can be seen from **Figure 7** that the monthly load fraction increases along with the increase of a collector square. It also shows that the monthly fraction is higher in

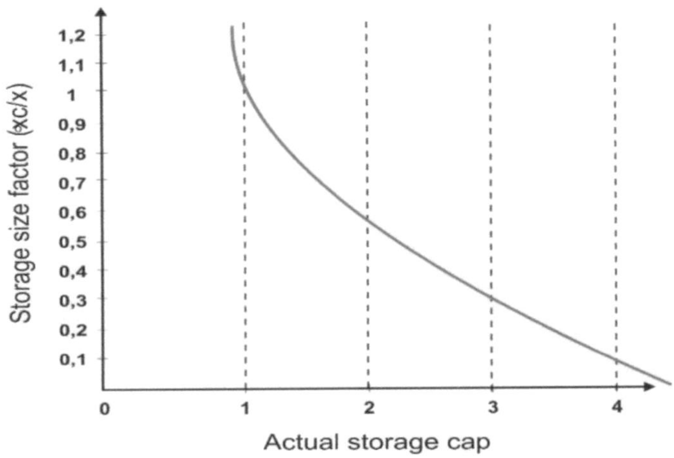

Figure 5.
Dependence of monthly values on correction factor for hot water supply systems.

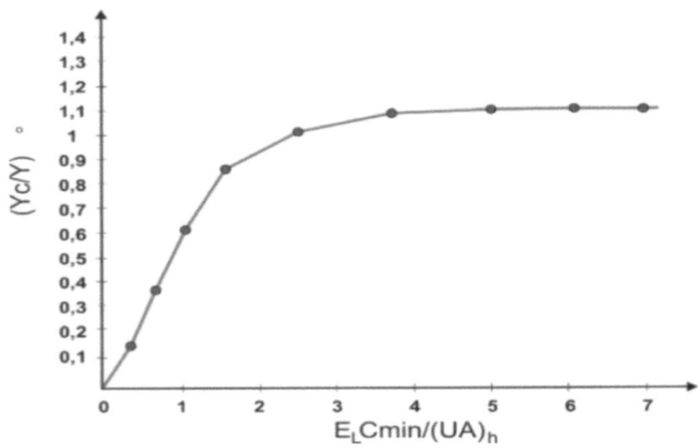

Figure 6.
Dependence of monthly values on the altered correction coefficient Yc/Y for hot water supply systems.

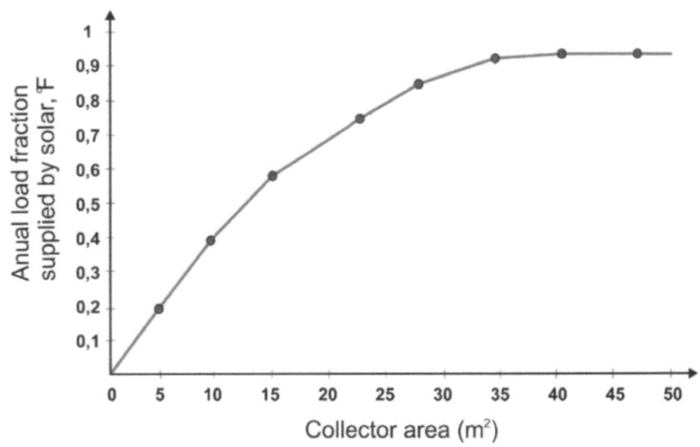

Figure 7.
Dependence of monthly load fraction change on collector's different squares within a year for hot water supply systems.

summer months in Almaty city (Kazakhstan) (July—the highest value) and lower in winter months (January, February—the lowest value).

In **Figure 8** it is observed that the lowest heating load is in the summer months of the year. It is the time, within which the need in heating load is minimal. This figure also denotes that the load for buildings heating is accessible only within 3 months, namely, in December, January and February. In the remaining time the load for heating also equals to zero. It is a very interesting result due to the fact that the load corresponds to the winter peak demand, during which the load for buildings heating is necessary, and the result thereof also provides important advantage from economic point of view at the expense of fuel-electricity total cost reduction that, otherwise, could be spent for the energy, necessary for heat supply, required in winter period.

Figure 9 shows that the annual load fraction increases for a little and the biggest fraction is in the collector's larger square. In particular, if the storage capacity exceeds approximately 50 l of water per m^2 of collector's square, there is only the insufficient increase (upgrading) in the annual load fraction, provided with the solar energy.

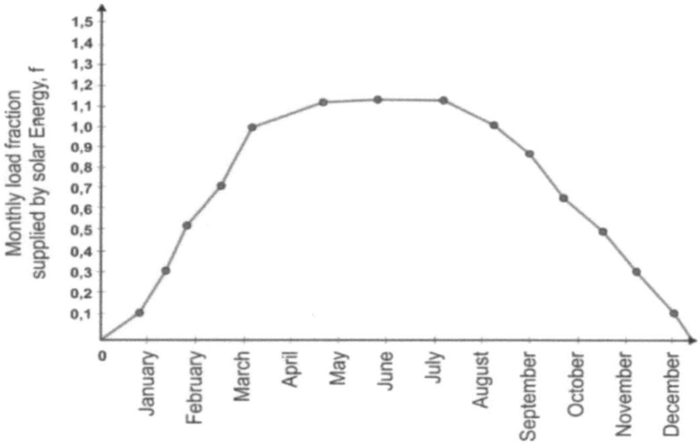

Figure 8.
Change of premises and hot water heating load depending on the month of the year.

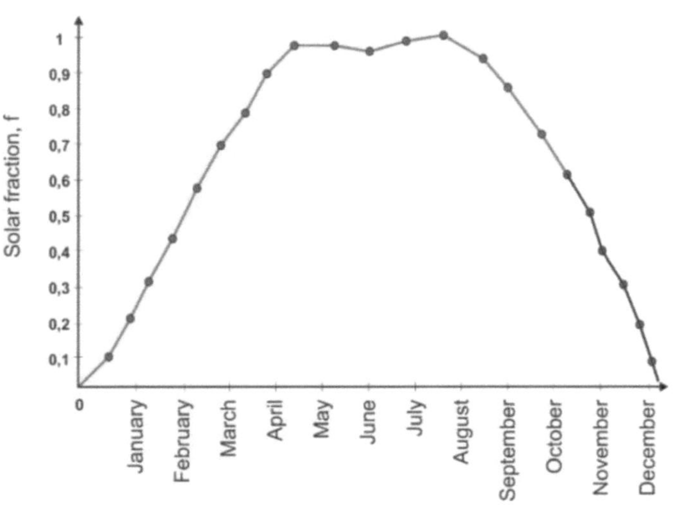

Figure 9.
Influence of actual storage capacity in liter to m^2 of collector's square per the annual load fraction, provided with the solar energy.

6. Conclusion

In the work herein by means of F-diagram method we have conducted the development of liquid solar heating system and assessment of the total thermal load fraction (load on domestic water and heating), provided by the solar energy for a family of six people in Almaty city (Kazakhstan). Proceeding from the executed research, it appears that, the more a collector's square, the more annual solar energy load fraction. From experimental data it is seen that total solar radiation in summer months is higher, in winter months-lower, as well, by means of F-diagram method it has been computed that the thermal load's peak value is in January, while the minimum load value is in summer months of the year. It proves that the thermal load change is in the phase with the need in thermal load. Due to collector's tilt angles the load increases from minimum value in March and reaches the maximum value in July. It has been defines that for Almaty city (Kazakhstan) the annual optimal angle tilt of collector's configuration, which provides the maximum solar load share comprises about 45°.

Acknowledgements

The work was supported by the project of the Ministry of Education and Science of the Republic of Kazakhstan No. 055236693; Contract No. 318 of March 30, 2018

Author details

Yedilkhan Amirgaliyev, Murat Kunelbayev*, Aliya Kalizhanova, Ainur Kozbakova, Salauat Daulbayev, Timur Merembayev and Didar Yedilkhan
Institute Information and Computational Technologies CS MES RK, Almaty, Kazakhstan

*Address all correspondence to: murat7508@yandex.kz

References

[1] Duffie JA, Beckman WA. Solar Engineering of Thermal Processes. 4th ed. Hoboken, New Jersey: John Wiley & Sons, Inc; 2013

[2] Liu Y-M, Chung K-M, Chang K-C, Lee T-S. Performance of thermosyphon solar water heaters in series. Energies. 2012;**5**:3266-3278. DOI: 10.3390/en5093266

[3] Dickinson EW. Solar Energy Technology Handbook. CRC Press/Taylor and Francis; 2018. pp. 12-42

[4] Bloss WH, Pfisterer F. Advanced in solar energy technology. In: Proceedings of the Biennial Congress of the International Solar Energy Society; Hamburg, Federal Republic of Germany; September 13–18, 1987. Elsevier; 2013

[5] Koce A, Atmaca I, Ertekin C. A comparison of flat plate and evacuated tube solar collectors with f-chart method. Journal of Thermal Science and Technology. 2015;**35**(1):77-86

[6] Kalogirou S. Solar Energy Engineering: Processes and Systems. Elsevier; 2009

[7] Amirgaliyev YN, Kunelbayev M, Wójcik W, et al. Solar-driven resources of the Republic of Kazakhstan. News of the National Academy of Sciences of the Republic of Kazakhstan, Series of Geology and Technical Sciences. 2018;**3**(430):18-27

[8] Amirgaliyev Y, Kunelbayev M, Wójcik W, et al. Calculation and selection of flat-plate solar collector geometric parameters with thermosiphon circulation. Journal of Ecological Engineering. 2018;**19**(6):176-181

[9] Amirgaliyev Y, Kunelbayev M, Kalizhanova A, Auelbekov O, Katayev N, Kozbakova A. Theoretical and mathematical analysis of double circuit solar station with thermo siphon circulation. Journal of Polytechnic-Politeknik Dergisi. 2019;**22**(2):485-493

Chapter 4

Einstein's Equation in Nuclear and Solar Energy

Ancuta M. Magurean, Octavian G. Pop, Adrian G. Pocola, Alexandru Serban and Mugur C. Balan

Abstract

Starting from the equation of Einstein ($E = m \cdot c^2$), the chapter proposes a simple and fundamental presentation of the fission and fusion principles, together with some of their applications: nuclear reactors and nuclear propulsion vessels and submarines. Fission and fusion are chosen between the multiple forms of energy, as being the most important forms of nuclear energy, directly related with the equation of Einstein. Some characteristics of solar energy, produced from the fusion process inside the Sun, are deducted from the same equation of Einstein: thermal power of solar radiation; specific power of solar radiation; surface temperature of the Sun; solar constant on different planets, etc. The yearly variation of the solar radiation on each planet of the solar system is also presented.

Keywords: Einstein equation, fission, fusion, solar energy, solar constant, solar radiation

1. Introduction

The energy is a form of a manifestation of matter in motion, the widely used definition of which is that the energy of a system is its capacity to perform mechanical work, when it passes from an existing state to a reference state [1, 2].

The energy, as defined in the literature, as well as on numerous websites, in various languages of international circulation, is the ability of a physical system to produce mechanical work [3].

Occasionally it is mentioned in the definition of energy, the capacity of a physical system to produce heat, as well.

However, these definitions refer only to the production or conversion of mechanical work or heat, but these represent only two of the many forms of existing energy [4].

The notion of energy is much more complex being obviously associated with other systems besides the physical ones, namely, biological and chemical systems, etc. [5].

Sometimes the literature considers that the energy is involved in all processes that require any kind of change or conversion, being responsible for the production of those changes [6].

It can be even considered that the matter itself is a "condensed" form of energy and that energy is stored in the atoms and molecules of which matter is composed [7–9].

The goal of the study is to investigate the connections between the fission, fusion, and solar energy through Einstein's equation.

2. Einstein's equation

The connection between energy and matter is represented by Albert Einstein's famous equation introduced as [10]:

$$E = m \cdot c^2 \tag{1}$$

where E is the energy; m is the mass; c is the speed of light.

Albert Einstein was a theoretical physicist of Jewish ethnicity, which was born in Germany, stateless since 1896 and later Swiss citizen, in 1899. He immigrated in 1933 to the USA, being naturalized as an American citizen in 1940. He was a university professor at Berlin and Princeton, and he is the author of the theory of relativity and one of the brightest scientists of mankind. In 1921 he was awarded with the Nobel Prize in Physics.

The above relationship could be associated with the beginning of the universe and, at least on an empirical level, can support the theory of the appearance of matter in the universe, when after the original explosion (called the "Big Bang"), an enormous amount of energy was transformed into matter. This process can also be correlated with the so-called "information preservation" principle, which in a simplified way shows that evolutionary processes cannot create the information required to generate biological evolution. The law of information preservation was enunciated by biologist Peter Medawar in his work *The Limits of Science*, in 1984 [11].

Through various processes, the high amount of energy contained in the atoms (especially in the nuclei) can be released and used for various purposes, and as a result of these processes, the matter which is used as an "energy source" is significantly transformed.

Two of the most representative examples of these types of transformations are the production of energy by nuclear fission (the break of element nuclei), respectively, and by nuclear fusion (the recombination and joint of element nuclei).

Both processes are accompanied by matter conversions into high amounts of energy.

Fusion is even the energy source of the stars, of which category the Sun belongs.

Both the fission and the fusion are characterized by the breakage of some types of bonds, existing initially at the level of the nuclei and the construction of other types of bonds, in the new nuclei formed as a result of these processes.

It is mentioned that iron (Fe) and nickel (Ni) are the chemical elements which have the highest breaking energies of the nucleus. In other words, for these two elements, the greatest amount of energy must be consumed for breaking the nucleus. The nuclei of all other elements "break" more easily [12].

3. Einstein's equation and the nuclear fission

Nuclear fission is a nuclear reaction or a process of radioactive decay, after which the atomic nucleus splits into lighter nuclei.

Usually, through the fission (breaking/disintegration) of nuclei heavier than iron, more energy than is necessary to maintain the cohesion of the newly formed (lighter) nuclei is released. Consequently, through the fission of "heavy" nuclei, energy can be obtained.

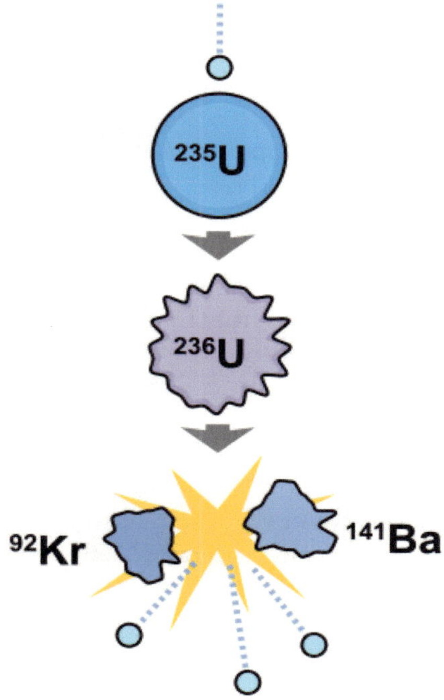

Figure 1.
Scheme of the fission reaction of uranium in krypton and barium (https://en.wikipedia.org/wiki/File:Nuclear_fission.svg).

Note: For the fission of nuclei lighter than iron, in order to maintain the cohesion of the newly formed nuclei, more energy is required than can be released through "the break" of the initially existing nucleus. Therefore, for the fission of light nuclei, energy input from the outside is required.

Usually, the fission results in nuclei with close mass, the ratio of the masses of the nuclei formed by fission being of maximum 2 or 3 [13].

Nuclear fission of heavy elements was discovered in 1938 by Lise Meitner, Otto Hahn, Fritz Strassmann, and Otto Robert Frisch [14, 15].

Modern fission, artificially produced, is usually initiated using a neutron that is "embedded" in a nucleus and disrupts its balance.

The energy released in fission reaction was calculated for the first time in [16].

Figure 1 shows a scheme of the induced fission reaction to uranium, a reaction used in nuclear power plants.

- **Comments**:
  - The kinetic energy of the neutron absorbed by the U235 nucleus causes the formation of the U236 nucleus, which is unstable and fragment (fission) in Kr92 and Ba141.
  - Following the reaction, in addition to Kr92 and Ba141, three neutrons, Γ (gamma) radiation (not shown in the figure), and a very large amount of energy are obtained.
  - Instead of uranium, various plutonium isotopes can be used as fuel.
  - The presence of gamma radiation requires the protection of the nuclear reactor against the emission of radiation of this type, since they are harmful to life.

The energy released in the fission reaction can be calculated with Einstein's Eq. $E = \Delta m \cdot c^2$, where Δm is the mass difference in the mass difference after the fission [17].

The most important applications of fission are:

- Electricity generation (in nuclear power plants).

- Propulsion of ships and submarines.

Figure 2 shows the scheme of a nuclear power plant.

The Romanian nuclear power plant from Cernavodă was designed with five reactors, of which only two are currently operating, each with a net power of ≈ 655 MW, respectively, and a total capacity of 706 MW. Currently, this plant provides approx. 18% of Romania's electricity demand.

The nuclear power plant from Cernavodă is CANDU type, a name derived from "Canada Deuterium Uranium." The reactor uses natural uranium (0.7% U^{235}) as fuel, respectively, and heavy water (D_2O) as neutron moderator and primary cooling agent. The notion of "neutron moderator" refers to the lagging of neutrons resulting from fission (thermal neutrons), in order to increase their efficiency in producing new fission reactions. CANDU reactor technology has been used in all nuclear power plants in Canada and in countries such as India, Pakistan, Argentina, South Korea, China, and Romania.

The scheme of the CANDU reactor is shown in **Figure 3**.

Figure 4 shows the first nuclear-powered aircraft carrier, undertaken by fission. This is the USS Enterprise aircraft carrier, built in 1964, currently decommissioned (since December 1, 2012), which has remained the longest ship in the world (342 m) to date, followed by the 10 US aircraft carriers in the "Nimitz" class, manufactured between 1975 and 2009 (333 m).

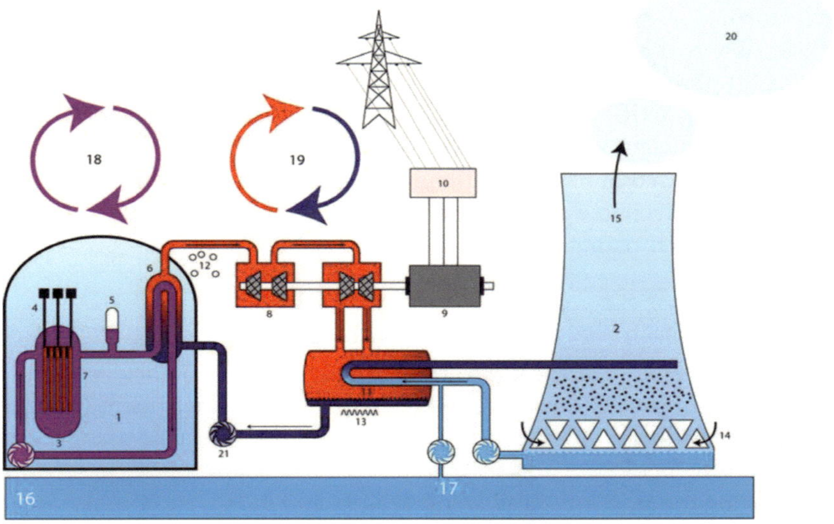

Figure 2.
Schematic diagram of a nuclear power plant (http://commons.wikimedia.org/wiki/File:Nuclear_power_plant-pressurized_water_reactor-PWR.png). 1, reactor block; 2, cooling tower; 3, reactor; 4, control rods (fission inhibitor); 5, pressurized tank for the primary coolant; 6, steam generator; 7, fuel bars; 8, turbines; 9, electric generator; 10, high voltage transformer; 11, condenser; 12, steam; 13, condensate; 14, cooling air; 15, hot air with high humidity; 16, water source; 17, cooling water outlet; 18, primary circuit (heavy water D_2O in the case of natural/enriched uranium used as fuel); 19, secondary circuit (water H2O); 20, water vapor evacuated into the air; 21, recirculation pump.

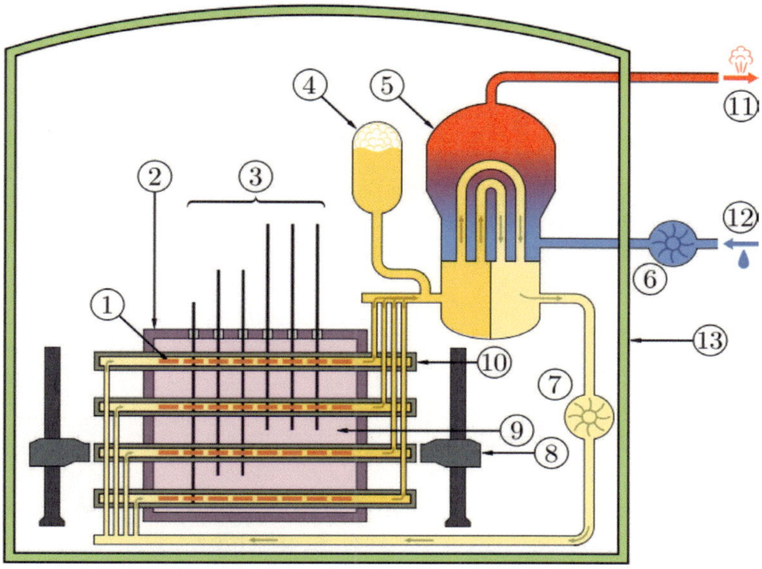

Figure 3.
Scheme of the CANDU reactor (http://ro.wikipedia.org/wiki/Fişier:CANDU_Reactor_Schematic.svg). 1, fuel rods; 2, reactor shell; 3, control rods (fission inhibitors); 4, pressurized tank for primary coolant (D_2O); 5, steam generator; 6, secondary circuit pump; 7, primary circuit pump (D2O); 8, nuclear fuel loading (replacement) system; 9, neutron moderator (D_2O); 10, pressurized tubes; 11, steam; 12, condensate; 13, reactor block.

Figure 4.
USS Enterprise aircraft carrier, the first nuclear-powered ship (http://lanterloon.com/wp-content/uploads/Aircraft_carriers_USS_Enterprise.jpg).

Currently, it started the replacement of the carriers belonging to "Nimitz" class, with the carriers of the "Gerald R. Ford" class, or the "Ford" class, which have a length of 337 m. The construction of the first aircraft carrier belonging to the Ford class began on November 8, 2005.

- **Comments**:

    - The name of the "Nimitz" class for the 10 operational aircraft carriers of the US Navy is used in honor of Admiral Chester W. Nimitz, commander of the Pacific fleet, for the US Navy during World War II. Admiral C.W. Nimitz was the last five-star admiral (general) of the US Army.

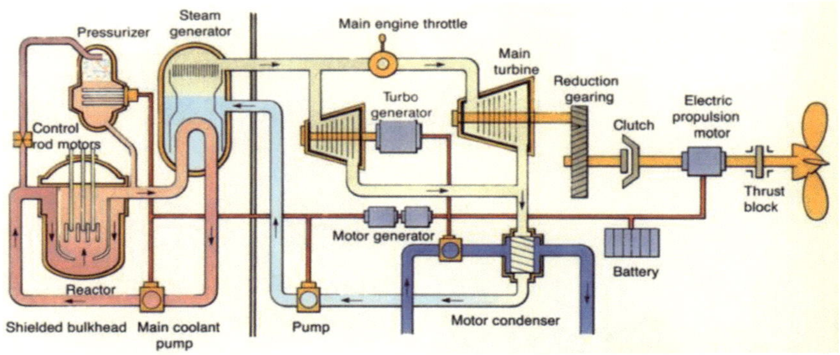

Figure 5.
Scheme of the nuclear propulsion system of ships and submarines (http://www.subadventures.net/Sub_04_719_files/image018.jpg).

- A list of the longest ships in the world is available on the Internet: http://en.wikipedia.org/wiki/List_of_longest_naval_ships.
- The inscription on the main deck was made by the aircraft carrier's crew to mark 40 years of naval nuclear propulsion.
- The aircraft carrier USS Enterprise was decommissioned due to the long period of operation of the nuclear propulsion system and due to the equipment on the main deck, which allowed the radar position to be detected on the aircraft carrier. The aircraft carriers of the "Nimitz" class are of "stealth" type (hidden or not detectable on the radar).

Figure 5 presents the scheme of operation for the nuclear propulsion system of ships and submarines.

4. Einstein's equation and the nuclear fusion

Nuclear fusion is a nuclear reaction that causes two or more nuclei to collide at very high speeds and merge to form a new type of atomic nucleus. Sometimes, the energy needed to initiate this process is provided by a very high "working" pressure (e.g., inside the stars this pressure is provided by the gravity determined by their mass).

By fusion (the union of nuclei) easier than iron, more energy than is necessary is produced, in order to form bonds for the newly formed nucleus. As a result, by the fusion of "light" nuclei, energy can be obtained.

Note: For fusion of nuclei heavier than iron, in order to achieve the necessary connections to maintain the cohesion of new nuclei, external energy contribution (consumption) is required.

Figure 6 shows a scheme of the fusion reaction between a deuterium atom (H^2) and a tritium atom (H^3), after which a helium (He^4) atom is formed. From the reaction also results a neutron.

The amount of energy produced is 17.59 MeV = $2.8{\cdot}10^{-12}$ J, which is in agreement with Einstein's equation, considering the equivalent loss of mass as a result of the fusion reaction [18–20].

The only application of the artificially produced fusion is the hydrogen bomb.

Figure 7 shows the explosion of the first hydrogen bomb, whose code name was "Ivy Mike" (November 1, 1952).

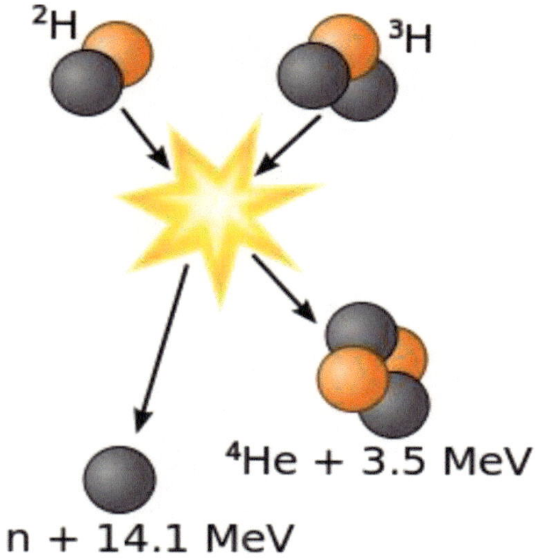

Figure 6.
Scheme of the fusion reaction between deuterium and tritium (https://en.wikipedia.org/wiki/File:Deuterium-tritium_fusion.svg).

Figure 7.
Explosion of the first hydrogen bomb (https://en.wikipedia.org/wiki/File:IvyMike2.jpg).

One of the most important scientific research projects, which aim to obtain energy through fusion, for peaceful use, is the International Thermonuclear Experimental Reactor (ITER) [21].

The project is carried out in collaboration with many countries: European Union countries, the USA, Japan, Russia, China, South Korea, and India.

Figure 8 presents the small-scale model of the ITER fusion reactor.

The ITER fusion reactor was designed to produce 500 MW of final energy and will consume approx. 50 MW to ensure its own energy consumption.

The construction of the ITER complex began in 2013, with the cost of construction now reaching $ 16 billion USD, almost three times more than originally expected.

The infrastructure is expected to be completed in 2025, the commissioning of the reactor to be completed in the same year, as well. Plasma experiments should start in 2025, and deuterium-tritium fusion experiments should begin in 2027.

Figure 8.
Small-scale model of ITER (https://upload.wikimedia.org/wikipedia/commons/thumb/7/75/ITER_Exhibit_%2801810402%29_%2812219071813%29_%28cropped%29.jpg/250px-ITER_Exhibit_%2801810402%29_%2812219071813%29_%28cropped%29.jpg).

The project contributes to the implementation of the results obtained in decades of research, in an experimental installation, which will allow the transition to a commercial installation.

5. Einstein's equation and the solar energy

The most representative example of fusion is represented by the reactions inside the Sun.

The Sun represents the energy source of the Earth [22], contributing to maintain the planet's temperature well above the value of almost 0 K, encountered in the interplanetary space and is the only source of energy capable to sustain life on Earth [23].

The Sun can be considered as a sphere with a diameter of approximatively 1.4 million km, more precisely 1.39×10^9 m [24], at a distance of approx. 150 million km from the Earth, 1.5×10^{11} m [24]. This distance is so high that two straight lines that start from one point on the Earth's surface to two diametrically opposite points on the solar disk form an angle of approximately half a degree. Under these conditions, although solar radiation is emitted in all directions, it can be considered that the solar rays that reach the Earth's surface are parallel [25].

In the core of the Sun, continuous nuclear fusion reactions occur, by which hydrogen is converted into helium. Currently, the mass composition of the Sun is approx. 71% hydrogen, 27.1% helium, 0.97% oxygen, and other elements in lower concentrations [26].

The rate of conversion of hydrogen into helium is approx. 4.26 million tons per second [27, 28]. This flow of substance is continuously transformed into energy. It is estimated that at this rate, in the next 10 million years, approximatively 1% of the current amount of hydrogen will be consumed, so there is no imminent danger of depletion of the Sun's energy source. The lifetime of the Sun is estimated at approximatively 4–5 billion years.

Considering the mass flow of solar substance that is consumed continuously turning into energy m = 4.26 million t/s = $4.26 \cdot 10^9$ kg/s, the thermal power of the solar radiation emitted as a result of this process (P) can be calculated starting from the famous equation of Einstein for energy calculation (E):

$$E = m \cdot c^2\ [J]; P = \dot{m} \cdot c^2\ [W]. \tag{2}$$

where c = 300,000 km/s = $3{\cdot}10^8$ m/s is the speed of light.

Substituting the relation of the thermal power of the radiation emitted by the Sun, we obtain:

$$P = 4.26 \cdot 10^9 \cdot 3^2 \cdot 10^{8\cdot 2} = 38.34 \cdot 10^{25}\,W. \tag{3}$$

The specific power of the radiation emitted by the Sun (P_S), representing the power of the radiation emitted by the surface unit, can be calculated with the relation:

$$P_S = P/S_S\ [W/m^2]. \tag{4}$$

where $S_S = 6.08{\cdot}10^{12}\,km^2 = 6.08{\cdot}10^{18}\,m^2$ is the total surface area of the Sun.

Replacing it, we obtain:

$$P_S = 38.34 \cdot 10^{25}/6.08 \cdot 10^{18} = 63.059 \cdot 10^6\,W/m^2 = 63.059\,MW/m^2. \tag{5}$$

For comparison, it is mentioned that the maximum power developed by the Renault engine K7M (1.6 MPI), which equips some models of the Renault Group cars, is 64 kW, at a maximum speed of 5500 rpm. Thus, the specific power of the radiation emitted by the Sun (P_S) is approximately equivalent to that of 1000 engines that equip these cars, which operate at maximum speed. Considering that the length of a car is 4.25 m, those 1000 cars placed one after the other, in a straight line, "bar to bar" would stretch 4.25 km. Also for comparison, the net power of a nuclear reactor from Cernavodă (655 MW) represents about 10 times more than the specific power of the radiation emitted by the Sun (63 MW/m^2). In other words, every square meter of the Sun's surface emits energy characterized by a thermal power approximately equivalent to one tenth of the power of a reactor from Cernavodă.

Since the Sun emits radiation over all wavelengths, it can be considered an absolute black body [29, 30], and the power emitted in the unit of time, on the surface unit, by an absolute black body (P_S) depends only on its temperature and can be calculated according to Boltzmann's law, with the relation:

$$P_S = \sigma \cdot T^4\ [W/m^2]. \tag{6}$$

where σ is Boltzmann's constant: $\sigma = 5.67{\cdot}10^{-8}\ [W/m^2K^4]$; T is the absolute temperature of the black body (of the Sun) [K].

Using the above relationship, the value of the Sun's surface temperature can be determined as:

$$T = \sqrt[4]{\frac{P_S}{\sigma}}\ [K] \tag{7}$$

Replacing it, we obtain T = 5774 K ≈ 5500°C.

This value corresponds to that indicated by most bibliographic sources, which also confirms that all undertaken calculations are correct.

The core temperature of the Sun is estimated to vary between (8 and 40)·10^6 K [14].

It can be considered that the solar radiation is emitted uniformly in all directions and can be found throughout the solar system. The intensity of the available solar radiation due to this mechanism obviously depends on the distance to the Sun, and the thermal power of the solar radiation is evenly distributed on spherical surfaces, with the Sun in the center.

On these considerations, the thermal power of the radiation emitted by the Sun ($P = 38.34 \cdot 10^{25}$ W) can be calculated with the relation:

$$P = I_S \cdot S_S \text{ [W]}. \tag{8}$$

where I_S [W/m^2] is the intensity of radiation available on the surface unit of a sphere with the Sun in the center; S_S [m^2] is the surface of the sphere on which the intensity of the solar radiation is calculated.

By using the relation presented above, the intensity of the solar radiation related to the surface unit of a sphere having the Sun in the center (I_S) can be calculated with the relation:

$$I_S = P/S_S \text{ [W/m}^2\text{]}. \tag{9}$$

where $S_S = 4 \cdot \pi \cdot D^2$ [m^2].

Replacing it in the previous relationship, we obtain:

$$I_S = P/(4 \cdot \pi \cdot D^2). \tag{10}$$

Thus, the intensity of the available solar radiation at the upper limit of the Earth's atmosphere can be calculated using the previous relation, considering that D is the distance between the Earth and Sun and D = 149,597,871 km = $1.496 \cdot 10^8$ km = $1.496 \cdot 10^{11}$ m:

$$I_S = 38.34 \cdot 10^{25}/(4 \cdot \pi \cdot 1.1496^2 \cdot 10^{11 \cdot 2}) = 1.364 \cdot 10^3 \text{ W/m}^2. \tag{11}$$

The intensity of the available solar radiation at the upper limit of the Earth's atmosphere is referred to as the solar constant [31].

The value of the solar constant calculated previously corresponds to the value adopted by the World Radiation Center, of 1367 W/m^2. This value is also reported by numerous bibliographic sources. The value of the solar constant, which is determined by measurements undertaken by satellites, underwent several corrections over time, as can be seen in **Table 1**.

The value of the available solar radiation at the upper limit of the terrestrial atmosphere suffers throughout the year small variations of approx. ± 3%, mainly due to fluctuations in the distance between the Earth and the Sun [24].

Value [W/m^2]	Year	Author	Ref.
1323	1940	Moon	[32]
1355	1952	Aldrich and Hoover	[33]
1396	1954	Johnson	[31]
1353 ± 1.5%	1971	NASA	[34]
1373 ± 2%	1977	Frohlich	[35]
1368	1981	Willson	[36]
1367–1374	1982	Duncan et al.	[37]
1367 ± 1%	—	World Radiation Center	[24]

Table 1.
Accepted values over time for the solar constant.

6. Conclusions

Even if nuclear and solar energies seem to be different domains, the study proved that fission, fusion, and solar energy can be connected and have in common the famous equation of Einstein ($E = m \cdot c^2$).

Both in fission and fusion, the mass varies during the reactions, and it was highlighted that the mass variation and the released energy are related by the equation of Einstein.

The same equation was also applied to the mass flow of solar substance that is continuously consumed in the solar fusion reactions, and starting from this point, it was possible to calculate important parameters such as the energy and the power emitted by the sun.

Following this new approach, it was possible to determine the temperature of the sun's surface and the solar constant, both being in agreement with the values provided in literature.

It can be concluded that fission, fusion, and solar energy are linked together by the equation of Einstein.

Author details

Ancuta M. Magurean, Octavian G. Pop, Adrian G. Pocola, Alexandru Serban and Mugur C. Balan*
Technical University of Cluj-Napoca, Cluj-Napoca, Romania

*Address all correspondence to: mugur.balan@termo.utcluj.ro

References

[1] Moradpour H, Nunes RC, Abreu EMC, Neto JA. A note on the relations between thermodynamics, energy definitions and Friedmann equations. Modern Physics Letters. 2017;**32**(13):1750078. DOI: 10.1142/S021773231750078X

[2] Grad GA, Paston SA, Sheykin AA. Definitions of energy for the description of gravity as the splitting theory. Journal of Physics: Conference Series. 2018;**1038**:012007. DOI: 10.1088/1742-6596/1038/1/012007

[3] Swenson R. Order, evolution, and natural law: Fundamental relations in complex system theory. In: Negoita CV, editor. Cybernetics and Applied Systems. Boca Raton, USA: CRC Press; 2018. DOI: 10.1201/9781482277180

[4] Koohi-Fayegh S, Rosen MA. A review of energy storage types, applications and recent developments. Journal of Energy Storage. 2020;**27**:101047. DOI: 10.1016/j.est.2019.101047

[5] Madarasan T, Balan MC. Termodinamica tehnica (in Romanian). Cluj-Napoca, Romania: Sincron; 1999

[6] Gilliland M. Energy Analysis: A New Public Policy Tool. New York, USA: Routledge; 2019

[7] Essler FHL. Applications of integrable models in condensed matter and cold atom physics. Les Houches. Lecture Notes. 2019:106. DOI: 10.1093/oso/9780198828150.003.0007

[8] Bagdoo R. Scenario for the origin of matter (according to the theory of relation). Journal of Modern Physics. 2019;**10**:163-175. DOI: 10.4236/jmp.2019.102013

[9] Roy S. Time evolution of the matter content of the expanding universe in the framework of Brans-Dicke gravity. Research in Astronomy and Astrophysics. 2019;**19**(4):61(14pp). DOI: 10.1088/1674-4527/19/4/61

[10] Einstein A. Ist die trägheit eines körpers von seinem energieinhalt abhängig? Annals of Physics. 1905;**323**(13):639-641

[11] Dembski WA. Conservation of information made simple. In: Evolution News & Science Today. 2012

[12] Fewell MP. The atomic nuclide with the highest mean binding energy. American Journal of Physics. 1995;**63**:653-658. DOI: 10.1119/1.17828

[13] Saha GB. Fundamentals of Nuclear Pharmacy. Springer International Publishing; 2010. XVII, 428 p. DOI: 10.1007/978-3-319-57580-3

[14] Meitner L, Frisch OR. Disintegration of uranium by neutrons: A new type of nuclear reaction. Nature. 1939;**143**:239-240. DOI: 10.1038/143239a0

[15] Hahn O, Strassmann F. Über den Nachweis und das Verhalten der bei der Bestrahlung des Urans mittels Neutronen entstehenden Erdalkalimetalle (on the detection and characteristics of the alkaline earth metals formed by irradiation of uranium with neutrons). Die Naturwissenschaften. 1939;**27**:11-15. DOI: 10.1007/BF01488241

[16] Fames MF. Energy released in fission. Journal of Nuclear Energy. 1969;**23**:517-536

[17] Lo CY. The invalid speculation of $m = E/c^2$, the Reissner-Nordstrom metric, and Einstein's unification. Physics Essays. 2012;**25**:49-56. DOI: 10.4006/0836-1398-25.1.49

[18] Eddington AS. On the radiative equilibrium of the stars. Monthly Notices

of the Royal Astronomical Society. 1916;**77**:16-35. DOI: 10.1093/mnras/77.1.16

[19] Eddington AS. The internal constitution of the stars. The Scientific Monthly. 1920;**11**:297-303. DOI: 10.1126/science.52.1341.233

[20] Shultis JK, Faw RE. Fundamentals of Nuclear Science and Engineering. CRC Press Inc; 2002. 524 p. ISBN: 978-0-8247-0834-4

[21] Campbell DJ. The physics of the international thermonuclear experimental reactor FEAT. Physics of Plasmas. 2001;**8**:2041-2049. DOI: 10.1063/1.1348334

[22] Kren AK, Pilewskie P, Coddington O. Where does Earth's atmosphere get its energy? Journal of Space Weather and Space Climate. 2017;**7**:16 p. DOI: 10.1051/swsc/2017007

[23] Glaser PE. Power from the sun: It's future. Science. 1968;**162**:857-861. DOI: 10.1126/science.162.3856.857

[24] Duffie J, Beckman WA. Solar Engineering of Thermal Processes. 2nd ed. New York: Wiley; 1991. xxiii, 919 p. ISBN: 0471510564

[25] Reifsnyder WE. Radiation geometry in the measurement and interpretation of radiation balance. Agricultural Meteorology. 1967;**4**:255-265. DOI: 10.1016/0002-1571(67)90026

[26] Chaisson E, McMillan S. Astronomy Today. 6th ed. Benjamin-Cummings Publishing Company; 2007. 848 p. ISBN-10: 0132400855

[27] Machacek J, Prochazka Z, Drapela J. The temperature dependent efficiency of photovoltaic modules—A long term evaluation of experimental measurements. In: Hammons TJ, editor. Renewable Energy. InTech; 2009. pp. 415-446. DOI: 10.5772/7359

[28] Miramonti L. Solar neutrinos: From their production to their detection. In: Proceedings of Science 4th School on Cosmic Rays and Astrophysics; 25-04 September; Sao Paulo. 2010

[29] Martyn DF. Solar radiation in the radio spectrum I. radiation from the quiet sun. Proceedings of the Royal Society of London. 1948;**193**:44-59. DOI: 10.1098/rspa.1948.0032

[30] Badescu V. Spectrally and angularly selective photothermal and photovoltaic converters under one-sun illumination. Journal of Physics D: Applied Physics. 2005;**38**:13 p. DOI: 10.1088/0022-3727/38/13/014

[31] Johnson FS. The solar constant. Journal of Meteorology. 1954;**11**:431-439

[32] Moon P. Proposed standard solar-radiation curves for engineering use. Journal of the Franklin Institute. 1940;**230**:583-617. DOI: 10.1016/S0016-0032(40)90364-7

[33] Aldrich LB, Hoover WH. The solar constant. Science. 1952;**116**:3. DOI: 10.1126/science.116.3024.3

[34] NASA SP-8005—Solar Electromagnetic Radiation; 1971

[35] Frohlich C. In: White OR, editor. The Solar Output and Its Variation. Boulder: Colorado Associated University Press; 1977. pp. 93-109

[36] Willson RC. Solar total irradiance observations by active cavity radiometers. Solar Physics. 1981;**74**:217-229. DOI: 10.1007/BF00151292

[37] Duncan CH, Willson RC, Kendall JM, Harrison RG. Latest rocket measurements of the solar constant. Solar Energy. 1982;**28**:385-387. DOI: 10.1016/0038-092X(82)90256-0

Chapter 5

How to Build Simple Models of PEM Fuel Cells for Fast Computation

Jonathan Deseure

Abstract

Hydrogen is one of the leading candidates in the search for an alternative to fossil hydrocarbon fuels. The spread of these technologies requires a real-time control of generator performances. Artificial intelligence (AI) and mathematic tools can make smarter the smart grid. The electrochemical modeling can be coupled successfully with artificial intelligent approach, if these models can be quickly computed with a large numerical stability. This chapter shows a methodology to build this kind of modeling work. Thanks to a simplified but physically reasonable model of PEM fuel cell, we will show that the reactant access (oxygen) or water management (a product of the reaction) and the reaction rate can be easily described with low computing time consuming. In addition, the artificial neural network could be trained with a reduced amount of data generated by these cell models.

Keywords: electrochemical modeling, PEMFC, AI

1. Introduction

In the current context of the spread of renewable energies, these are by nature variable, therefore subject to both daily and seasonal intermittencies. Electrochemical devices have been successful in proving their applicability in terms of energy storage (power to gas) [1]. Controlling in real time, predicting the performance is the advantage that electrochemical generators can offer.

In addition, electricity consumption and production must, at every moment, be in perfect adequacy with the demand of the users. However, this demand is variable and cannot be completely regulated. The production must be able to adapt instantly to the demand, to preserve the stability of the network. Thus, exchanging information between the production and storage sites becomes a major issue. This will result in a strategy of predictive actions in a power grid strongly constrained by intermittent sources of energy. The interpretation and exchange between blocks of storage and energy supplies will be the key to a decentralized energy production and smart grid [2, 3]. According to Ramchurn et al. [4] the grand challenge for artificial intelligence is to put the smarts into the Smart Grid.

Electrochemical modeling can provide "smart" tools for smart grid. The establishment of a mathematical model of an electrochemical generator is handled by the scientific culture of the researcher who establishes it. Therefore, a great deal of

IntechOpen

subjectivity appears in any modeling work, the approach of a mechanic/energy specialist, an electrochemist, or a physicochemist will differ mainly in the basic assumptions of modeling (model simplifications). However, whatever the cultural origin of the modeler, the numerical resolution of a multiphysical problem makes it possible to assure three major functions in the phase of development [5]:

- Assistance with the understanding of experimental results
- Study and optimization of design
- The prediction of performances

The modeling of an electrochemical system involves the mathematical expressions of the physical phenomena that take place there (a priori). Obviously, all model representations only offer a fragmentary assessment of the real systems. These various representations are distinguishable by their scales of time and space. However, the notion of adapted or appropriate modeling remains subjective. Indeed, as described in the literature about fuel cell models [6, 7], each of the approaches has limitations of description or prediction, and their main interest is to highlight one specific process. Despite this subjectivity, the model must prove its validity.

The validity of the model could be named external, i.e., related to theories, concepts, assumptions, and experimental data. Thus, the model is theoretically valid if it accepts theories or models already validated. In addition, if the model well matches to its potential of scientific explanation (the state of the art), one will qualify its heuristic validity. However, building a model cannot be done without solving it in all its intended range. Consequently, it is also necessary to define criteria of internal validity, which are criteria of evaluation of the model independent of the theories, results, and hypotheses. The algorithm (solver) must be appropriate, and the evaluation errors must remain within "valid" limits.

External validations that may be acceptable include empirical validity (the model corresponds to the available data) or pragmatic validity (the model satisfies the intended use).

A fuel cell is a nonlinear and strongly coupled dynamic system. It is a multi-input multi-output system based on multiphase flow, electrochemical reactions, and heat transfer. For example, the control strategies of PEMFC can be built on a prediction of the future output of the system to compute the current control action [8]. In practice, the current control action is obtained by solving online an optimization problem. The aim of the optimization problem is to find the optimum of a cost function that minimizes the mean squared difference between predicted outputs and target values.

To spare computation time due to computing of multiphysic fuel cell models, artificial intelligence (AI) techniques are useful as alternate approaches to conventional multiphysic modeling: e.g., artificial neural network (ANN) simulator could be employed to predict the fuel cell behavior [9–12]. The ANN could be trained with a reduced amount of data generated by a validated cell model [13]. Once this network is trained, it can predict different operational parameters of the fuel cell reducing the computation time [14]. This strategy has many possibilities [15]: spectroscopic analysis, prediction of reactions, chemical process control, and the analysis of electrostatic potentials. The ANN is trained to learn the internal relationships from data. These data may be taken from the real process even if there are noisy. The database should have a significant size and contain the maximum combination of inputs-outputs covering the studied domain. Therefore it is possible to generate database from model [14].

As shown below, electrochemical modeling can be coupled successfully with artificial intelligent approach; thus in the following subsections, we provide basic mathematical models of fuel cells. These models can be quickly computed with a large numerical stability.

2. Main electrochemical phenomenon

An electrochemical cell is characterized by the I-V (current–voltage) behavior: the current that passes across the cell to the applied cell voltage. The I-V relation depends on various physical phenomena and is fundamental to achieve efficiently electrochemical conversion. When current is drawn using computational tools, the current density may not be uniformly distributed on the electrode surfaces. The performance and lifetime of electrochemical cells, such, is often improved by a uniform current density distribution. Therefore, it is necessary to optimize the current distribution. The electric current is a flow of electric charges: through the electrolyte between the anode and the cathode in the ions form and within the wires and current collectors/electrode materials in the electrons form. When the overall current of the cell is equal to zero, for example, on the disconnection of an electrode from the power supply or the load, the cell voltage is equal to UOCV the open circuit voltage:

$$I = 0; U_{cell} = U_{OCV} = E_{a,0} - E_{c,0} \tag{1}$$

where $E_{a,0}$ *and* $E_{a,0}$ are the potential of each electrodes at OCV. It is shown that the U_{OCV} is related to the difference of free enthalpy differences of each reaction, involving the number n of electrons exchanged and the Faraday constant:

$$U_{OCV} = \frac{\Delta G_i}{nF} \tag{2}$$

where ΔG_i is the Gibbs energy of the species i with specified cell temperature (T) and cell pressure (P). At a non-zero current, the cell voltage of an electrochemical reactor (electrolyzer) is greater than U_{OCV}, and electrochemical generator (fuel cell or battery) is smaller than U_{OCV} due to the various irreversibilities standing in the electrochemical conversion. The electrode potentials differ from the equilibrium values $E_{a,0}$ and $E_{a,0}$, and this difference is called overvoltage:

$$\eta_a = E_a - E_{a,0} \tag{3}$$

$$\eta_c = E_c - E_{c,0} \tag{4}$$

According to this description, an anode overvoltage is positive, while the cathode overvoltage is negative in all cases. The overvoltages depend on the current density at the electrode, depending on the involved electrochemical reactions, the electrode materials, and several operating conditions: concentration species, flow rate, etc. The current density is an extensive quantity that can be defined in any point of the electrochemical device, i.e., at the surface of the electrodes and through the electrolyte. In addition Ohm's law expresses the current density according to the local potential gradient, using the conductivity as follows:

$$\vec{I} = -\sigma \overrightarrow{\nabla E} \tag{5}$$

where σ is the ionic conductivity of an electrolyte. In the case of a one-dimensional system where the electrolyte is confined in a finite space between two surface electrodes S and separated from each other by the distance e, this potential difference (called ohmic drop η_{ohm}) can be expressed as follows:

$$\eta_{ohm} = -\frac{e}{\sigma S} I \tag{6}$$

where J is the total current through electrochemical device: J is the average current density I multiplied by the electrode surface S. For other geometries, it will be necessary to calculate η_{ohm} from the relation (Eq. 5) by integration. The ohmic drop in the electrolytic solution of an electrochemical cell is one of the parameters to be evaluated to optimize the cell efficiency. This limitation is also called primary current distribution. Thus, the cell potential is calculated from following expression:

$$U_{cell} = U_{OCV} - \eta_a + \eta_c - \eta_{ohm} \tag{7}$$

Inside the nonaqueous electrochemical device (**Figure 1**), the primary current distribution is well controlled, because a solid electrolyte is confined in a finite space between two surface electrodes and only the misalignment of both electrodes could affect the current distribution through the electrolyte. In nonaqueous case, the optimization endeavor is devoted to secondary (electrochemical activation) and tertiary (mass transport limitation) current distributions. In this context, fuel cells are the best example to scrutinize energetic balance of electrochemical devices.

The main advantage associated with fuel cells is that they are not limited by Carnot efficiency. Besides, no moving parts are required to convert thermal energy into mechanical energy. The energy release from interatomic bonds of the reactants is converted efficiently into electrical energy. In this document, we consider proton-exchange membrane fuel cells (PEMFCs): a PEMFC consists of a polymer electrolyte sandwiched between two electrodes to form a membrane electrode assembly

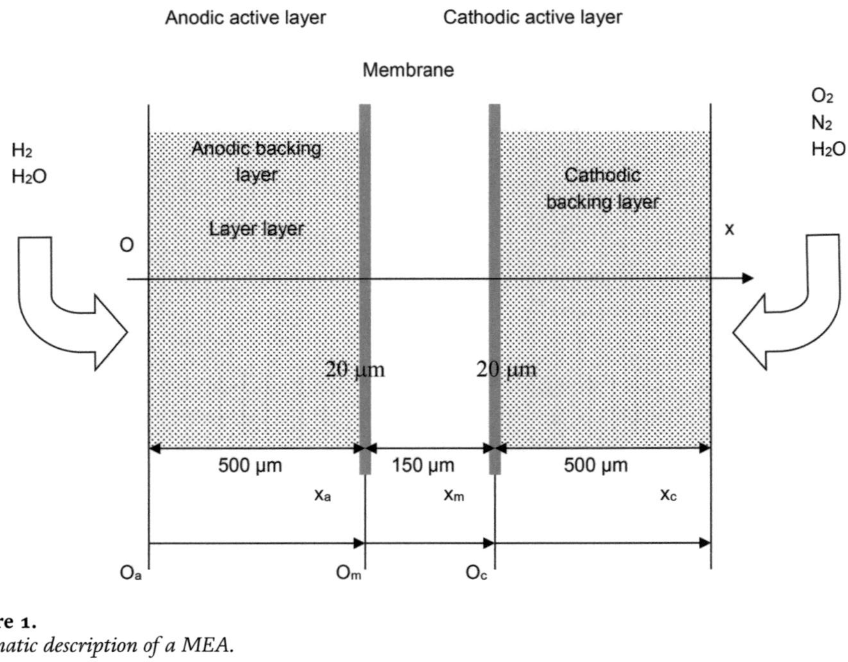

Figure 1.
Schematic description of a MEA.

(MEA), placed between two graphite bipolar plates, which feed the device with gases and cool it down. At the anode, fuel H_2 is oxidized, liberating electrons and producing protons. The electrons flow to the cathode via an external circuit, where they combine with the proton and the dissolved oxidant O_2 to produce water and heat:

$$H_2 \leftrightarrows 2H^+ + 2e^- \quad (8)$$

$$2H^+ + 2e^- + \tfrac{1}{2}O_2 \leftrightarrows H_2O \quad (9)$$

The efficiency of fuel cell is easily computed according to:

$$\psi \frac{U_{OCV}}{U_{cell}} \quad (10)$$

Proton transfer from the anode to the cathode via the membrane closes the electrical circuit. A careful water management is required to ensure that the membrane remains fully hydrated in order to improve the ionic conductivity and to avoid the electrode flooding (which occurs when an excess of liquid water restrains the active species access to the active layer). The gas diffusion electrodes (GDEs) are made up of two distinct areas: an inactive area (backing layer) and an active area (active layer) which is a place of the electrochemical and chemical reactions (**Figure 1**). Therefore, in the first approach it is possible to observe separately secondary current distribution (only in the active layer) and tertiary current distribution (only in the backing layer).

PEMFC and all solid electrolyte cells are the model devices in order to sketch the primary (in the polymeric membrane), the secondary (in active layer), and tertiary (in backing layer) current distributions. In addition, operating cell potential and each overvoltage could be compared to thermodynamic potential (Gibbs energy) to access enthalpic balance.

3. One-dimensional modeling of electrochemical membrane cell

A single cell can be described schematically as an assembly of several layers constituting four distinct areas. A fuel cell is a composite structure of anode, cathode, and electrolyte. Good electrochemical performance of the cell requires effective electrocatalysts. On both sides of the cell, the interconnect plates cumulate three functions: current collector, gas feeding via the gas channels, and thermal control thanks to cooling water channels. Flow field is used to supply and distribute the fuel and the oxidant to the anode and cathode electrocatalysts, respectively. The distribution of flow over the electrodes should ideally be uniform to try to ensure a uniform performance of each electrode across its surface. Thus, it is possible to develop a single-cell 1D model. Although most of the models used are one-dimensional, they correctly predict the electrochemical behavior of membrane electrode assembly.

In order to build a 1D model, a particular attention is required on water flux. The water managements is the key issue in single-cell modeling. In order to predict cell performance, the single-cell model must take into account gas diffusion in the porous electrodes, water diffusion, and electroosmotic transport through the polymeric membrane. Ramousse et al. [16] have developed one-dimensional coupled charge and mass transfer model in the electrodes. The three main types of model can be employed to evaluate the overvoltages at both electrodes:

- The Tafel law results from experimental observations: this simple and widely used description is based on the phenomenological Eq. (12).
- The classical catalyst layer model exhibited in the previous section (Appendix 1).
- The agglomerate model takes into account the existence of gas pore and the ohmic drop in the agglomerate. This model results to considering a cylindrical model of pores inside the electrode [17, 18].

In these three models of active layer, the authors have computed the same values of kinetic coefficients b_i and $i_{o(a,c)}$ (**Table 1**).

The simulations presented below describe, in steady state, one-dimensional mass transfer in the whole cell and charge and mass transfers in the electrodes. Then, only a numerical solution can be accessed. Thanks to numerical computation the cell overpotential $\eta_{a/c}$ and the ohmic drop through the membrane η_{ohm} can be simulated as a function of current density.

The total current is then calculated by integrating the local current density all over the MEA surface:

Parameter	Values
γ: roughness factor (−)	100
δ: gas backing layer thickness (m)	230×10^{-6}
L_{CA}: catalyst Layer thickness (m)	10×10^{-6}
δ_2: size of the agglomerate (m)	3×10^{-6}
L_m: *membrane* thickness (m)	125×10^{-6}
ε^{GDL}: gas backing layer porosity	0.8
ε^{CL}: active layer porosity	0.5
j_{Oc}: exchange current density (A/cm^2)	4×10^{-7}
j_{Oa}: exchange current density (A/cm^2)	1×10^{-2}
b_c: Tafel slope (mV/dec)	120
b_a: Tafel slope (mV/dec)	30
ν_e: total number of exchanged electrons	4
$D^{eff}_{H_2/H_2O}$: effective diffusion coefficients (m^2/s)	1.63×10^{-4}
$D^{eff}_{O_2/H_2O}$: effective diffusion coefficient (m^2/s)	3.20×10^{-5}
$D^{eff}_{O_2/N_2}$: effective diffusion coefficients (m^2/s)	2.41×10^{-5}
$D^{eff}_{H_2O/N_2}$: effective diffusion coefficients (m^2/s)	3.35×10^{-5}
D^{CL}: diffusion coefficient in the CA (m^2/s)	$10^{-9} \times (\varepsilon^{CA})^{1.5}$
EW: equivalent weight (kg/mole)	1.1
ρdry: dry *Nafion* density, (kg/m^3)	2020
$\tau 0$: osmotic coefficient	2.5/22
D_m: effective diffusion coefficients of water in the membrane (m^2/s)	3×10^{-10}

Table 1.
Common parameter values of PEMFC cells.

$$I = \iint_S i\, dS \tag{11}$$

Figure 2 exhibits the evolutions of the cathodic and anodic overpotentials, calculated in the same operating conditions, as functions of the current density. There are great differences between the overpotentials predicted by the three models, which emphasize the important influence of the active layer on fuel cell performances. The porous catalyst layer model (agglomerate model) and nonporous catalyst layer models require only intrinsic parameters: active layer thickness, platinum load (roughness factor), and kinetic coefficients (exchange current density). The polarization curves obtained with the porous and nonporous models exhibit Tafel behavior. Due to the slow diffusion of active species in the homogeneous solid phase, the nonporous model tends to overestimate the overpotentials. Particular attention should be paid to the use of the Tafel law. The effective platinum area being higher than the MEA section and the exchange current density i0 should be multiplied by a geometric parameter such as the roughness factor γ (Eq. 12) to provide corrected Tafel mode l as follows:

$$\eta_{a/c} = b_{a/c} \log \frac{i}{\gamma j_{0.a/c}} \tag{12}$$

In **Figure 2**, the anodic prediction is incoherent because the anodic overvoltage is negative. Therefore, the Tafel law is only adapted to high current density, far from thermodynamic equilibrium.

However, Singh et al. [19] suggested that a multidimensional model could improve the description, since gas composition and temperature vary along the feeding channels. In addition, these phenomena become critical for large cell areas used to obtain great current intensities. The flow field allows gas to flow along the length of the electrode while permitting mass transport to the electrocatalyst normal to its surface. One of the simplest flow field designs consists of a series of narrow parallel rectangular channels where fuel or oxidant is fed at one end and removed from the opposite side. **Figure 3** shows a schematic illustration of the PEMFC that consists of a membrane sandwiched between two gas diffusion electrodes, this assembly being pressed between two current collectors. Within a PEMFC, reactant depletion and cathodic water production occur along the length of

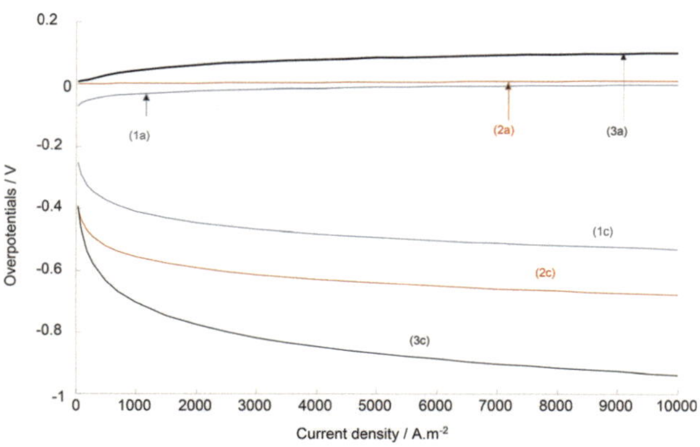

Figure 2.
Influence of electrode description on overvoltages à T = 60°C. (a) Anodic overvoltage; c, cathodic overvoltage; 1, Tafel law corrected with the roughness factor; 2, porous catalyst layer model (agglomerate model); 3, nonporous catalyst layer model.

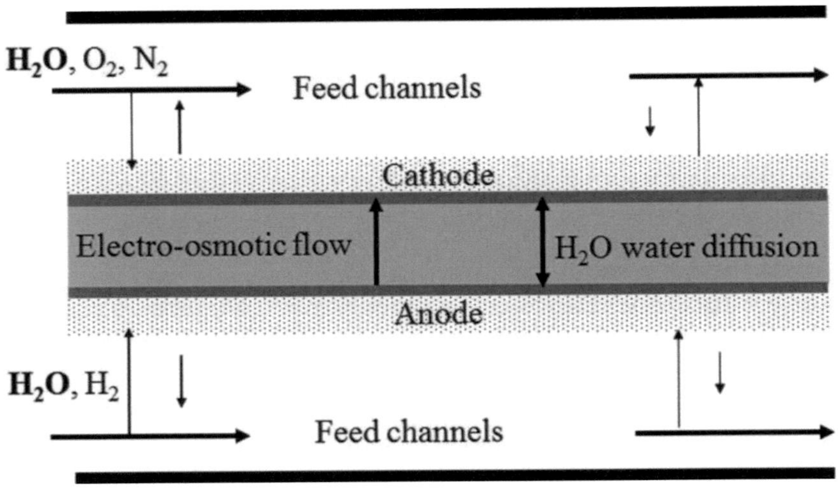

Figure 3.
Scheme of a proton exchange membrane fuel cell and the equivalent electrical circuit for the whole PEMFC for DC and AC solutions.

the fuel cell due to the electrochemical processes. Gas supply to the catalytic active site takes place in the porous gas diffusion electrode that contains a GDL consisting of hydrophobic gaseous pores and a reaction region (CL). Namely, slow diffusion in the GDL and CL can induce oxygen depletion in the case of air cathode. The membrane, commonly Nafion®, acts as both a separator and an electrolyte. Water transport across the membrane results from the electroosmotic water dragging with proton migration from anode to cathode and water "back-diffusion" that reduces the concentration gradient. During the operation, these effects can be responsible of either membrane dehydration or flooding on the porous electrode.

A finite volume method using a computational grid [20] can be used to solve mass, charge, energy, momentum balances including transport through porous media, and chemical and electrochemical reactions within the porous electrodes in a gas diffusion electrode model. Freshly, Zhang et al. [21] have developed a two-phase multidimensional model to properly handle mass transport. The results of these computations reveal that liquid water transport inside and across the polymeric membrane plays an important role in PEM fuel cell water distributions. In addition, it was shown that increasing the contact angle at GDL/channel interface is found to be able to improve the water removal process in channels. However, the resolution of fuel cell models requires important calculation times to obtain the detailed variations of flow field, species concentrations, temperature, liquid saturation, and electronic and ionic phase potentials.

4. Simple multidimensional modeling

4.1 Pseudo 2D modeling

To cut down these difficulties, the pseudo 2D as 1D + 1D approach implies that (i) meander-like channel is replaced with the straight one and (ii) water and oxygen concentrations in the channel provide "boundary conditions" for local transport of these species across the cell. PEMFC cell model, which couples 1D transport across the cell with 1D description of the flow in the feed channel, has been proposed [22, 23]. An approach is to model the flow as an ideal gas in a straight channel of cross-section and ignore the influence of viscosity. Dohle et al. [24] have proposed a pseudo two-dimensional model of the cathode compartment of a PEMFC. The model is based on the continuity equations for the gases and assumes constant pressure and plug flow conditions. The influence of hydrodynamics was not considered. Regardless of the latter point, the model shows some interesting effects for a simple, parallel flow field channel configuration. The mole fraction change along the gas channel length results from the normal molar flux N_i^{GDL} in the y-direction through the gas diffusion layer (GDL) allowing to the electrochemical processes in the active layer of the GDL and the corresponding current density changes. Note that this normal flux also depends on z. The mass balance for a single-phase species i (= O_2, H_2, N_2) may be written as:

$$\frac{dF_i}{dz} = N_i^{GDL} \tag{13}$$

where F_i is respectively the molar flow rate for the component Xi. Both oxygen and hydrogen fluxes are proportional to current density, while the nitrogen flux is zero because nitrogen is neither consumed nor produced in the fuel cell:

$$N_i^{GDL} = \frac{\nu_i i}{\nu_e F} \tag{14}$$

For the specific case of PEMFC, water produced at the cathode can flow through the gas diffusion layer to reach the gas channel and transport through the membrane:

$$N_{H_2O}^{GDL} + N_{H_2O}^{m} = \frac{i}{2F} \tag{15}$$

Even if water transport processes into the membrane are not well understood, the phenomenological model of water transport in the membrane takes into account water diffusion and water electroosmotic fluxes [25].

4.2 Pseudo 2D modeling: closed-form expression

Dohle et al. [24] have proposed an analytical solution of the oxygen profile along the channel. This model assumed that oxygen reduction is the determining rate step and neglects water transport. The oxygen concentration profile is given by:

$$[X_{O_2}] = [X_{O_2}]_0 \left(1 - (1 - \zeta)\frac{z}{L_\zeta}\right) \tag{16}$$

where L_ζ is the characteristic length of oxygen consumption.

$$L_\zeta = 4F\frac{h\, v_0 [X_{O_2}]_*}{\gamma\, j_{0,c}} \left(\frac{[X_{O_2}]_0}{[X_{O_2}]_*}\right)^{1-\zeta} exp\left(-\frac{2.3\,\eta_c}{b_c}\right) \tag{17}$$

where h is the channel height and v_0 is the inlet flow velocity, assuming a kinetic Tafel law for oxygen reduction (Eq. (12)). Note that for $\gamma = 1$, the oxygen profile distribution becomes:

$$[X_{O_2}] = [X_{O_2}]_0\, exp\left(-\frac{z}{L_1}\right) \tag{18}$$

The mean current density in the cell is defined as an average over the channel length L:

$$i = i_{lim}\left[1 - \left(1 - (1-\zeta)\frac{z}{L_\zeta}\right)^{\frac{1}{1-\zeta}}\right] \tag{19}$$

The limiting current density (i_{lim}) represents the limiting current density, which is attained when all the available oxygen is consumed:

$$i_{lim} = 4F\frac{h\, v_0[X_{O_2}]_0}{L} \tag{20}$$

The cathode overpotential is then given by:

$$\eta_c = \frac{b_C}{2.3}\, ln\left[\frac{K_\zeta}{1-\zeta}\left(1 - \left(1 - \frac{\bar{i}}{i_{lim}}\right)\right)^{1-\zeta}\right] \tag{21}$$

where:

$$K_\zeta = 4F\frac{h\, v_0[X_{O_2}]_*}{\gamma\, j_{0,c}\, L}\left(\frac{[X_{O_2}]_0}{[X_{O_2}]_*}\right)^{1-\zeta} \tag{22}$$

Based on the previous physical models, the experimental polarization curves can be approximated. The fitting parameters are generally the exchange current density, the Tafel slope, and the limiting current density. **Figure 4** displays the effect of the channel length. The decrease in channel length L increases the limiting current density (I_{lim}). The decrease in L not only increases I_{lim} but improves cell performance in the whole range of current densities. Physically, for lower L, oxygen concentration in the channel is more uniform, and this improves the overall performance per unit cross-section area.

This single-cell model exhibits that the reactive gas diffusion becomes a limiting step. However, the water management and electrochemical kinetics play a critical role. For a given cell voltage U_{cell} assumed to be constant along the gas channel, the

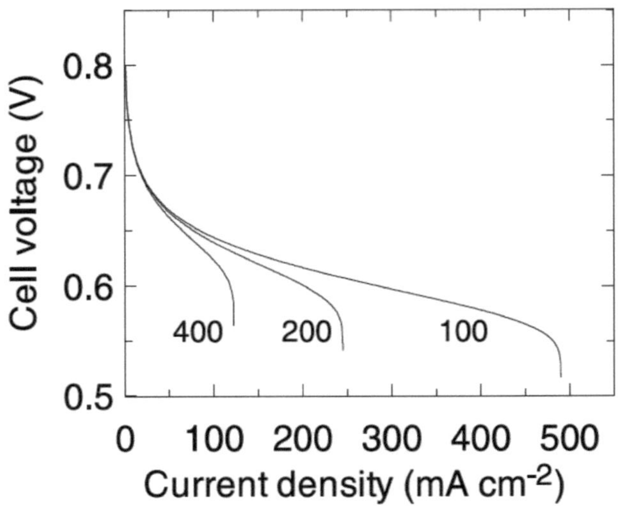

Figure 4.
Voltage–current curves for the three indicated values of channel length L (cm) [24].

current density distribution Iz must take in to account the membrane conductivity $\sigma_{m,z}$, which depends on membrane water content. Therefore, accurate numerical approaches should be proposed in order to calculate the electrical characteristics (current density, ohmic resistance, overpotential) as well as gas concentration distributions along the gas channel.

4.3 Continuous stirred tank reactor approach to fuel cell

A first rigorous 3D modeling of PEM single cell using CFD programs based on commercial finite volume technic solver has been developed by He et al. [26], to solve the fully coupled governing equations. The model assumes that liquid film is formed on the electrode surface during liquid water condensation and computes the diffusion flux, electroosmotic drag force, and water back-diffusion in order to assess to water management. According to *Zhang and Jiao* [27], despite the multiplicity of 3D CFD models available in the literature, a satisfactory 3D multiphase CFD model which is able to simulate the detailed gas and liquid two-phase flow in channels and reflect its effect on PEM fuel cell performance precisely still was not real, because it is difficult to solve coupling physics and computation amount is real barrier. Nandjou et al. [28] have proposed a pseudo 3D model to reduce the computational cost (i.e., the transport equations are formulated using a pseudo 3D approximation and coupled to an analytical electrochemical model at the catalyst layers/membrane interface). Nevertheless, the experimental measurement cannot easily emphasize or "validate" the computing results. Krewer et al. [29] have showed that 3D CFD results could be compared to experimental results via assuming that the experimental RTD of the inlet and outlet pipes can be used as input signal for the CFD simulations. Diep et al. [30], thanks to RTD experiments, have shown that a reduction in the unit cell model dimensionality to 1 + 1 or 1 + 0 D based on scaling arguments and contrasts with higher dimensional 3D models is accurate. **Figure 5** exhibits a comparison between the model and the outlet RTD. Good agreement between laminar flow 1+ 0 D-based model and experimental data was found.

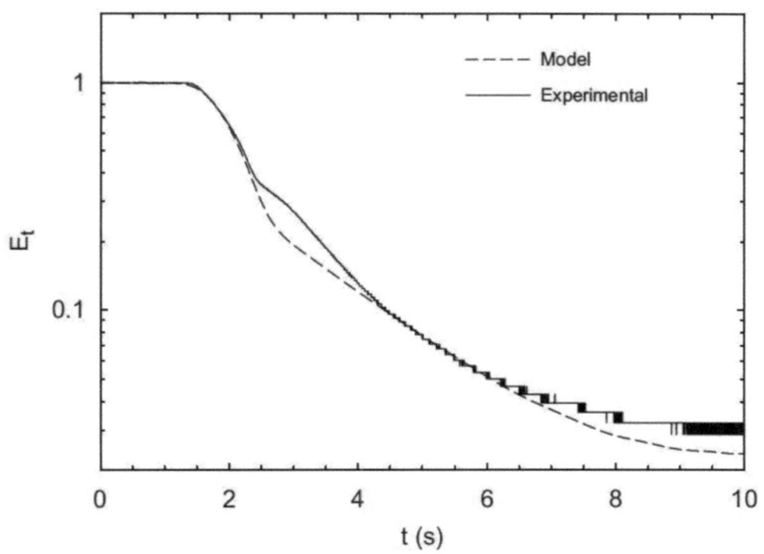

Figure 5.
Comparison between tracer step outlet RTD model fuel cell measurements and one-dimensional numerical model computations (cathode) [30].

On the other hand, to simplify the hydrodynamic description (modeling) or experimental observations, it is possible to employ the ideal reactor models. The continuous stirred tank reactor (CSTR) model does not require a long computing time. Benziger et al. [31] have employed this method to explore the auto-humidification of the PEMFC. In **Figure 6**, the anode and cathode chambers in the PEM fuel cell are modeled as two stirred tank reactors (STR). The previous models can only be considered as approximate, and more refined approach shall be preferred, in particularly for larger fuel cells.

Boillot et al. [32] have demonstrated that the bipolar plate is similar to a series of continuous stirred tank reactors. These authors have studied various models with or without exchange zones. They have observed that best fitting was obtained by the simple model of CSTR in series and the optimal number of CSTR (J) varied from four to six in the range investigated (**Figure 7**).

Deseure [33] proposed series of CSTR to solve the mass balance equation inside the gas channel (**Figure 8**). The electrode is divided into *k* elementary units. With such a hypothesis, gas transport in the gas channel is assumed to be infinitely fast, leading to homogeneous gas composition for each CSTR. In particular, the partial pressure of a component in the elementary unit (*k*) is equal to that of the outlet and

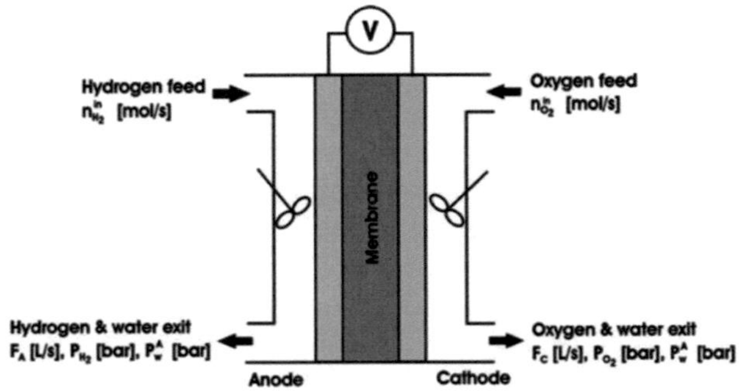

Figure 6.
STR PEM fuel cell model [31].

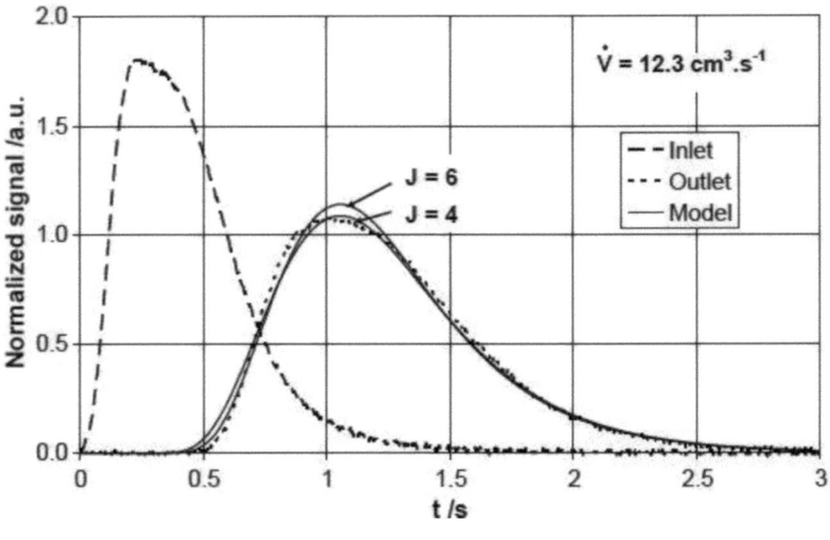

Figure 7.
RTD measurements and modeling of the column flow pattern (signals in arbitrary units) [32].

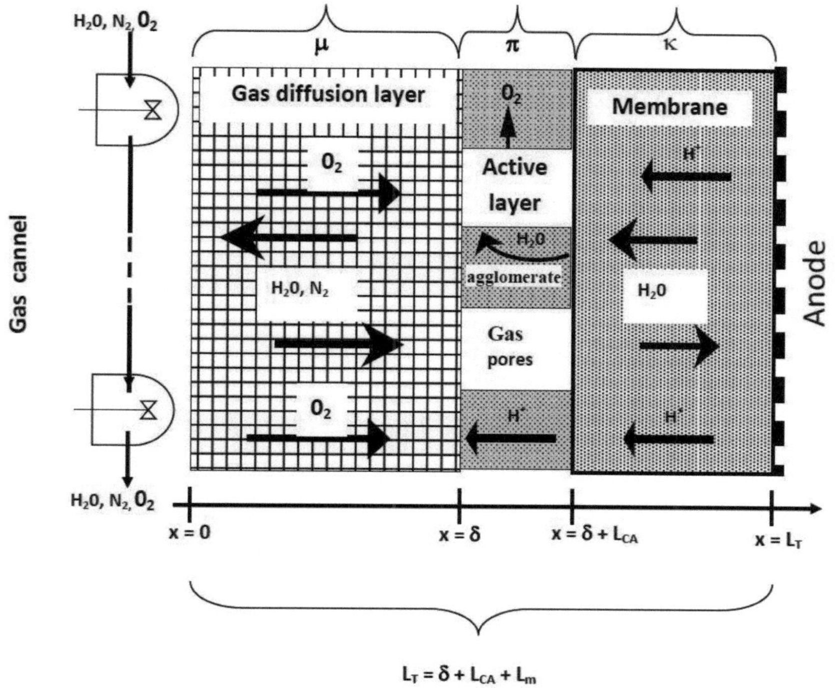

Figure 8.
Schematic description of the PEMFC half cell (cathode) [33].

to that of the elementary unit (k + 1) inlet. The resolutions of mass balances are numerically obtained thanks to the continuity of the flux between the GDL, the CL, and the membrane.

The electrode is divided into k elementary units, where the gas transport in the gas channel is assumed to be infinitely fast leading to homogeneous gas composition for each CSTR. Partial pressure of a component in the elementary unit (k) is equal to the outlet one and equal to the elementary unit (k + 1) inlet one. Like the 1D + 1D approach, all elementary units have the same voltage U due to the high electrical conductivity of bipolar plate and backing layer. Therefore, partial pressures of each component in each elementary unit vary and make necessary to solve mass balances and current density in each elementary unit at the same time. The current–voltage expression used is the one proposed previously in Section 2.1. The molar balance equation of each CSTR with electrode geometric area S^k for each component in gas phase in steady state is written as:

$$F_{H_2O}^{k,in} - F_{H_2O}^{k,out} = \xi^k S^k N_{H_2O}^{GDL,k} \tag{23}$$

$$F_{O_2}^{k,in} - F_{O_2}^{k,out} = S^k N_{O_2}^{GDL,k} \tag{24}$$

$$F_{N_2}^{k,in} - F_{N_2}^{k,out} = 0 \tag{25}$$

$$S^T = \sum_k S^k \tag{26}$$

The term ξ^k represents the ratio of gas molar flux divided by liquid molar flux that is estimated using a numerical optimization ($1 \leq \xi^k \geq 0$). The global mass balance for each CSTR is given by:

$$F_T^{k,in} - F_T^{k,out} = S^k\left(N_{O_2}^{GDL,k} + N_{H_2O}^{GDL,k}\right) \tag{27}$$

and the gas flux of consumption and production are expressed as:

$$N_{H_2O}^{GDL,k} = \frac{i^k}{2F} - N_{H_2O}^{m,k} \tag{28}$$

$$N_{O_2}^{GDL,k} = -\frac{i^k}{4F} \tag{29}$$

To satisfy saturated vapor pressure of water, the following expression is required:

$$P_{sat} = e^{\left(13.66 - \frac{5096.23}{(T+273.15)}\right)} \tag{30}$$

The resolution of mass balance is numerically obtained thanks to the continuity of flux between the backing layer, the active layer, and the membrane. The molar flux density of oxygen and nitrogen are nil at membrane/electrode layer interface, and the molar flux density of water are given by the solution of the equation (Eq. (15)). Thus, the sum of diffusion and electroosmotic fluxes yields in steady state a first-order differential equation:

$$N_{H_2O}^{m,k} = \frac{\tau_0 i^k}{F}\left[\lambda_a^k + \frac{\lambda_c^k - \lambda_a^k}{1 - e^{k_m^k L_m}}\right] \tag{31}$$

$$k_m^k = \frac{EW\tau_0 i^k}{\rho_{dry} D_m F} \tag{32}$$

λ_c^k and λ_a^k stand for polymer water content at the membrane/active layer interface respectively at cathode and anode. The thermodynamic equilibrium between water vapor in the backing layers and liquid water in the polymer has been assumed using Eq. (30).

The ionic conductivity of polymeric membrane could be accessed thanks to the Neubrand [34] equation as follows:

$$\sigma_{H^+} = e^{\left(-E_A\left(\frac{1}{T} - \frac{1}{353}\right)\right)}\left(0.0013\lambda^3 + 0.0298\lambda^2 + 0.2658\lambda\right) \tag{33}$$

$$E_A = 2640e^{(-0.6\lambda)} + 1183 \tag{34}$$

λ_c^k and λ_a^k are determined using sorption equilibrium of [35]:

$$\lambda = 0.3 + 10.8\left(\frac{Py_{H_2O}}{Psat}\right) - 16\left(\frac{Py_{H_2O}}{Psat}\right)^2 + 14.1\left(\frac{Py_{H_2O}}{Psat}\right)^3 \tag{35}$$

The set of these equations could be solved by iterative methods and have needed only these inlet conditions $[X_{O_2}]_*$, $[X_{N_2}]_*$, $[X_{N_2}]_*$ and $F_{O_2}^{in,0}$, $F_{N_2}^{in,0}$, $F_{H_2O}^{in,0}$, where $F_{O_2}^{in,0} + F_{N_2}^{in,0} + F_{H_2O}^{in,0} = F_T^{in,0}$. The conventional way to investigate the influences of the mass flux on electrochemical behavior uses the inlet molar flux, which are proportional to the current density and the stoichiometry at the cathode. This flux is currently expressed using the stoichiometry oxygen coefficient r_{O_2}, and this expression is given by:

$$F_{O_2}^{in,0} = r_{O_2} S^T \frac{I}{4F} \tag{36}$$

with I as the average current density of i^k given by Eq. (17). In order to initialize the computing, the initial inlet molar flux density is obtained considering only one STR according to Benziger et al. [31] investigations. This model does not take into account the liquid water apparition; according to this assumption the values of r_{O_2} are calculated in order to obtain a gaseous condition. A simple mass balance could estimate the minimal air flux or r_{O_2} value for non-apparition of liquid water on the cathode:

$$r_{O_2}(T) = 0.21 \left[\frac{2 + \frac{P}{P - P_{sat}(T)}}{\frac{P}{P - P_{sat}(T)} - \frac{HR_C P}{P - HR_C P_{sat}(T)}} \right] \tag{37}$$

This mass balance equation considers that the produced water by oxygen reduction is fully exhausted by the gas flux. However, this relation does not take into account the water flux provided from anode side: for this reason, the stoichiometry oxygen coefficient is proposed at high value. Indeed, **Figure 4** shows the limit of liquid water apparition as a function of water activity (HR_C) of inlet gas. In any simulated cases, the selected values of r_{O_2} allow to avoid the flooding. To insure no saturation by water in all studied cases, the feeding flux in simulations could be proposed at high level such as presented in **Figure 9**.

4.4 Results of continuous stirred tank reactor approach to fuel cell

The aim of these simulations is to emphasize the dependence of the locale electrochemical behavior on overall electrical performance. Three kinds of humidification mode are investigated: the auto-humidification, the humidification of inlet gas, and the humidification via the membrane. The steady-state equations of each elementary unit are solved thanks to the "simplex" numerical optimization method. First, the effect of water and ion transport within the membrane has been

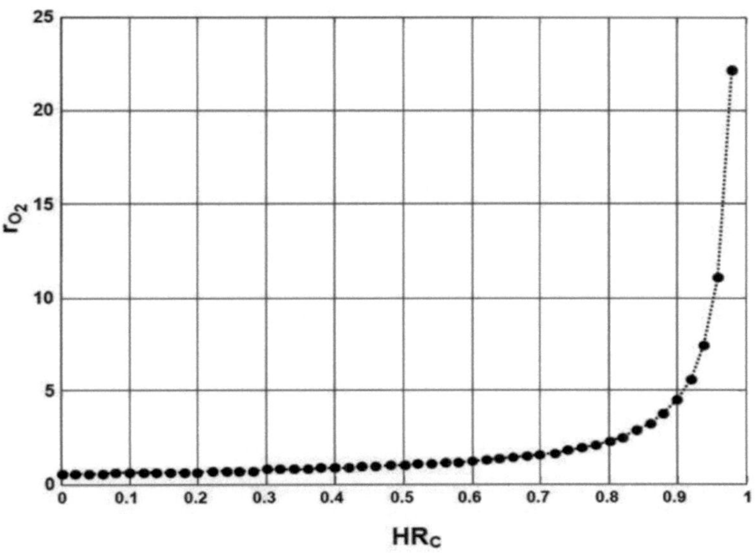

Figure 9.
The minimal stoichiometry oxygen coefficient (r_{O_2}) to obtain a single-phase condition as a function of INLET HR_C with t = 60°C without membrane water exchange.

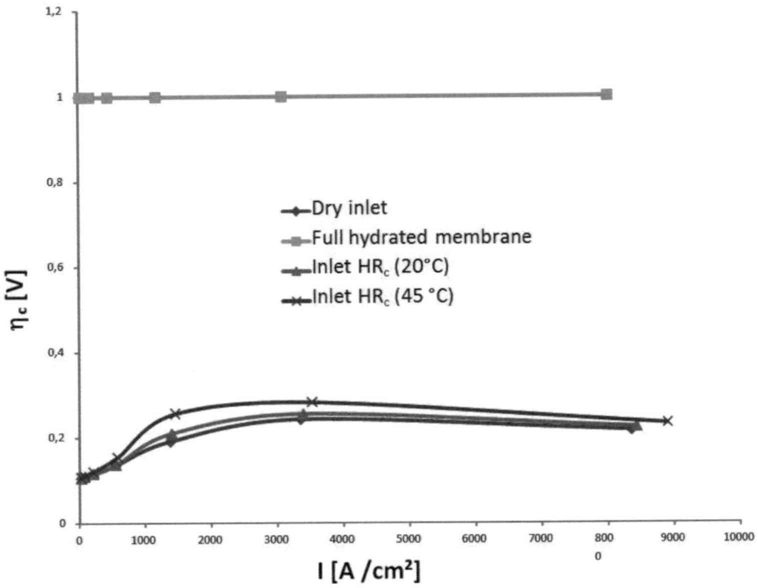

Figure 10.
Membrane ratio conductivity as a function of hydration process and current density considering a series of two CSTR, r_{O2} = 2.2 and T = 60°C.

investigated; **Figure 9** shows the minimal stoichiometry oxygen coefficient r_{O_2} to obtain a single-phase condition without membrane water exchange. Therefore it is possible to scrutinize the average ohmic drop within the membrane versus humidification process. The ohmic behavior appears as linear contribution of overall polarization curves. **Figure 10** shows thee ratio of ideal conductivity of fully hydrated Nafion® membrane (Eq. (35)) divided per average ionic conductivity trough the membrane according to the set parameter of gathered in Table. Considering a series of two CSTR with r_{O2} = 2.2, it was observed that the auto-humidification or the humidification of inlet gas involves similar ion conductivity performance. Of course, a humidification of inlet gas at 45°C improves the effective membrane conductivity, but the impact of water production inside catalyst layer is more significant.

In the cases of dry inlet condition which corresponds to a PEM fuel cell operating with pure hydrogen and oxygen, the stoichiometric effects have been simulated by Han and Chung [36]. The results of these computations have shown decreasing membrane ratio conductivity with increasing stoichiometry oxygen coefficient r_{O_2} and the main effect of water production. Therefore as highlighted by Nguyen and White [37], the membrane ratio conductivity is expected to depend on gas distribution. **Figure 11** shows a decreasing effective conductivity with increasing number of CSTR series: it is worth mentioning that piston flow distribution (20 CSTR) increases the ohmic drop and the static mixing (1 or 2 CSTR) improves membrane conductivity.

In **Figure 12** the opposite effect is observed for r_{O2} = 2.2; gas access is limited when static mixing is developed (two CSTR). Then, cathodic overvoltage is higher considering only 2 CSTR than using a series of 20 CSTR. Nevertheless, ionic access to the catalyst areas is modeled here, and for a large amount of gas flow (r_{O2} = .11), piston flow distribution increases the cathodic overvoltage.

At this stage, it is important to study the distribution of current density. In **Figure 13**, the simulations show the effects of oxygen gas access. The gas distribution is the main control on current densities along the gas channel. At the cathode inlet, dry gas feeds the fuel cell; the water content in the cathode stream increases

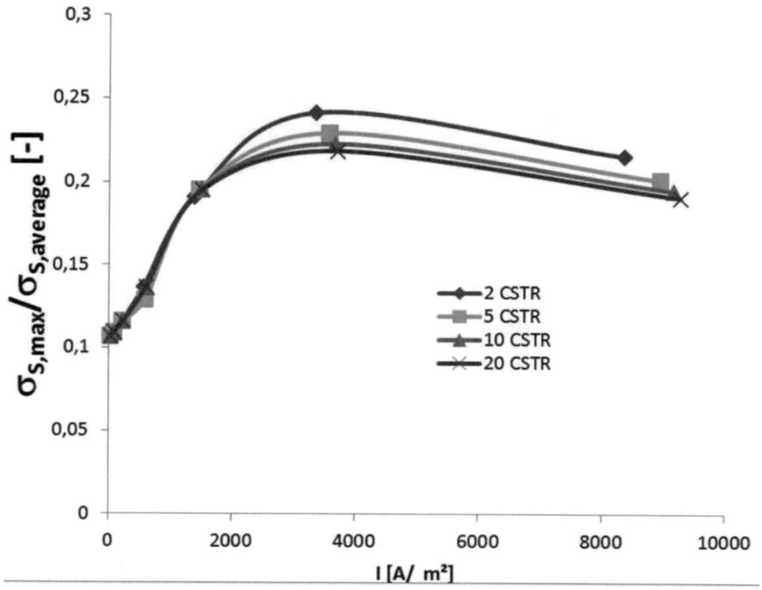

Figure 11.
Membrane ratio conductivity as function of CSTR series approaches and current density, considering r_{O2} = 2.2, T = 60°C with dry inlet conditions.

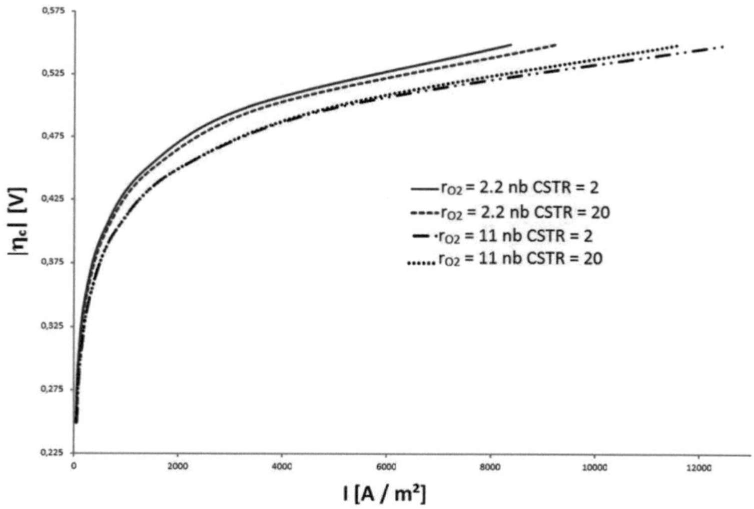

Figure 12.
Simulation of cathodic overpotential with varying CSTR approach and stoichiometry oxygen coefficient, considering r_{O2} = 2.2, T = 60°C with dry inlet conditions.

with decreasing number of CSTR because the amount of water transported back to the anode by back-diffusion is lower than convective water flux. For higher water content in the cathode side, diffusion limitations are increased. Consequently the local current density decreases, and the net water flux across the membrane also decreases, resulting in a lower depletion rate of water from the anode gas stream, a lower production rate of water in the cathode, and a lower depletion rate of hydrogen and oxygen.

However, near the inlet of the cell, as the current density is high and cathodic water concentration is low, the electroosmotic drag coefficient is higher, and ionic access limits the reaction rate. This observation is possible in case of piston flow (20

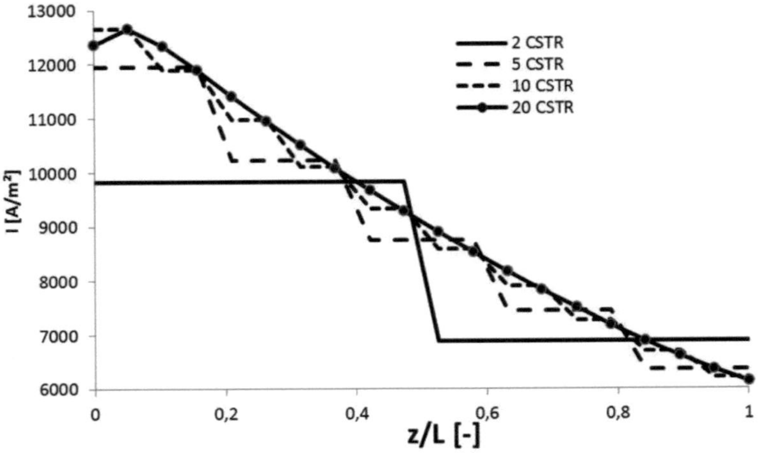

Figure 13.
Simulation of local current density distribution along the channel (r_{O2} = 2.2 for dry conditions at T = 60°C with $|\eta_c|$ = 0.55 V).

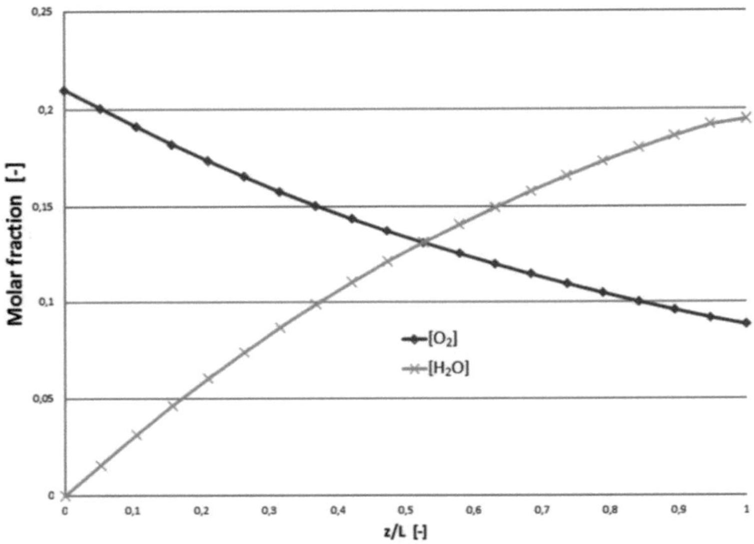

Figure 14.
Simulation of local oxygen (■) and water (×) molar fraction distribution along the channel (r_{O2} = 2.2 for dry conditions at T = 60°C with $|\eta_c|$ = 0.55 V).

CSTR). In **Figure 14**, at low stoichiometric value, a depletion of reactants can be observed close to the outlet, thus generating a cathodic overpotential increase and a dramatically current density decrease in this part. It is interesting to note that water molar fraction increases at the cathode side in connection to the electroosmotic drag of water, from anode to cathode, and the insufficient "back-diffusion" from the cathode to anode. Water transport and production at the cathode result in increasing water content at the cathode side along the gas channel while it decreases at the anode side.

This study has drawn attention to the relative advantages of using a CSTR description. Moreover liquid water phase only exists at the cathode side, thus leading to the polymer drying out at the anode and consequently increasing ohmic resistance. In the future, it will be important to include the flooding or partial flooding effects that were not taken into account in this study.

DOI: http://dx.doi.org/10.5772/intechopen.89958

5. Conclusion: single-cell modeling results and limits

We have shown, through a simplified but physically reasonable model of PEM fuel cell steady-state multiplicity, caused by reactant access (oxygen) or water management (a product of the reaction) and the reaction rate. Of course, catalyst loading is critical but 0-D models do not have enough accuracy to perform realistic predictions. Closed-form model (pseudo 2D) can provide relevant simulations but does not take into account water management. The succinct analogy with CSTR can provide fruitful analysis of water management through the fuel cell. This approach was settled with a view to determining the impact of gas distribution in the gas channel using a fluid dynamics observation (i.e., RTD) coupled with the usual electrochemical model. Water production and removal are analogous to heat production and removal. Therefore this analogy (energy balance) can be added to the present electrochemical pseudo 2D approach of mass balance (CSTR-PEM).

Operating conditions that can threaten the life of the PEM cell are not easy to detect; using simple models for fast computation with A.I. shall avoid the problematic lack of large experimental databases. Predictive models and A.I. can take up the challenge of "Smart Grid".

List of symbols

a_i	the specific surface area (m^{-1})
b	Tafel slope ($V\ dec^{-1}$)
C_{dl}	double-layer capacitance ($F\ m^{-2}$)
c_T	total gas concentration ($mol\ m^{-3}$)
d	pore diameter (m)
D_i	diffusion coefficient ($m^2\ s^{-1}$)
E_0	equilibrium potential (V)
EW	dry membrane weight per mole of sulfonate group ($kg\ mol^{-1}$)
F	Faraday's constant ($C\ mol^{-1}$)
F_i	molar flow rate for the component X_i ($mol\ s^{-1}$)
h	enthalpy ($J\ mol^{-1}$)
I	current (A)
$\mathbf{i}$	current density ($A\ m^{-2}$)
j	faradaic current density ($A\ m^{-2}$)
L_{CL}	CL thickness (m)
M_i	molecular weight of i ($kg\ mol^{-1}$)
$\mathbf{N}_i$	flux density of each dissolved species ($mol\ m^{-2}\ s^{-1}$)
R_i	homogeneous reaction rate ($mol\ m^{-3}\ s^{-1}$)
r_i	heterogeneous reaction rate ($mol\ m^{-2}\ s^{-1}$)
T	temperature (K)
t	time (s)
V	potential (V)
$\mathbf{v}$	bulk velocity ($m\ s^{-1}$)
$[X_i]$	total gas concentration of the species ($mol\ m^{-3}$)
x_i	mole fraction of the species X_i
Z	impedance (Ω)
z_i	charge (−)
ΔS	entropy ($J\ mol^{-1}\ K^{-1}$)
δ	GDL thickness (m)
ε	porosity (−)

Φ electrostatic potential
Γ_i surface concentration of a solute species i (mol m^{-2})
γ roughness factor
η overpotential (V)
λ water content of the membrane (−)
ν_e exchange number of electron (−)
ν_i stoichiometric coefficient of species i (−)
ρ density (kg m^{-3})
σ conductivity of the solution (S m^{-1})
τ tortuosity (−)
τ_0 electroosmotic drag coefficient (−)

A. Appendix

A.1 The classical catalyst layer model

Under steady-state conditions, the dimensionless equations relating the concentrations of species (Eq. (17)) are obtained by setting $\partial/\partial t = 0$:

$$\frac{\partial^2 X_i}{\partial Y^2} + a_e \frac{\nu_i j_{0,a/c} L^2}{|\nu_e| F D_i^{eff} [X_i]_*} exp\left(\pm 2.3 \frac{\eta_{a/c}}{b}\right) X_i = 0 \tag{A1}$$

By introducing the dimensionless parameter $X_i = [X_i]/[X_i]_*$ and $Y = y/LCL$, the Thiele' modulus (mL) appears as follows:

$$(mL)^2 = a_e \frac{|\nu_i| j_{0,a/c} L^2}{|\nu_e| F D_i^{eff} [X_i]_*} exp\left(\pm 2.3 \frac{\eta_{a/c}}{b}\right) = 0 \tag{A2}$$

Note that this dimensionless parameter mL enables us to compare diffusion resistances with regard to kinetic resistance on the active layer performances. The stationary problem has an analytical solution in the regime in which the effect of charge transport can be neglected. The steady-state formulation of the mass balance equation within the CL (Eq. (33)) leads to an expression of the geometrical current density i in the CL versus the overpotential:

$$i = \gamma j_{0,a/c} exp\left(\pm 2.3 \frac{\eta_{a/c}}{b}\right) \frac{[X_i]_0}{[X_i]_*} \frac{tanh\,(mL)}{mL} = 0 \tag{A3}$$

where γ (=$a_e \times L_{CL}$) is the roughness factor. This model predicts two distinct regions arising from control by kinetics and the other due to the control of both kinetic and diffusion. Under kinetics control, for low current density corresponding to $mL << 1$, concentration depletion in the active layer does not occur, and the current density remains equal to the kinetic one:

$$i = \gamma j_{0,a/c} exp\left(\pm 2.3 \frac{\eta_{a/c}}{b}\right) \frac{[X_i]_0}{[X_i]_*} = 0 \tag{A4}$$

Conversely, under kinetics and diffusion control corresponding to $mL >> 1$, since $tanh(mL)$ converges to unity, the current density can be approximated by [38]:

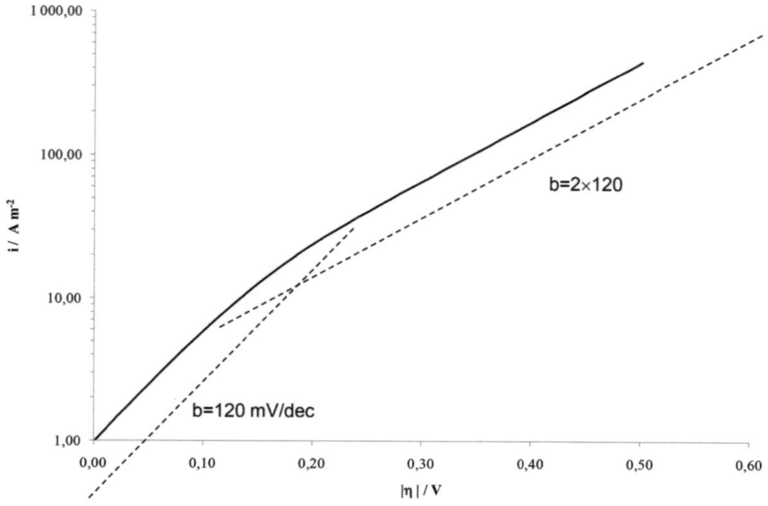

Figure A1.
Simulated i - |η| curve for a catalyst layer.

$$i = \sqrt{\frac{|\nu_e| F D_i^{eff} c_{i,0}\, \gamma\, j_{0,a/c}}{|\nu_i| L}}\, exp\left(\pm 2.3 \frac{\eta_{a/c}}{2\, b}\right) \frac{[X_i]_0}{[X_i]_*} = 0 \tag{A5}$$

A catalyst layer controlled by both kinetics and diffusion is analogous to the classical chemical engineering problem of porous catalyst pellet with a first-order reaction controlled by mass transport limitation of reactant characterized by Thiele's modulus. This explanation also holds for fuel cell model where the active layer may be considered to be a single pellet or agglomerate. When both kinetic and diffusion in the active layer are controlling the electrochemical process, it is well-known that limiting current behavior does not occur, but rather a doubling of the effective Tafel slope on a steady-state polarization curves and a first-order dependence on the concentration [38]. The shapes of the concentration profiles depend strongly on parameter mL [39]. The model allows us to understand easily how the diffusion limitation within the catalyst layer acts on the DC responses. Under kinetic control ($mL < <1$), the diffusion is fast enough to supply the reacting species up to the catalyst surface in the GDE, and the concentration profile remains close to a constant value in the catalyst layer. Conversely, with increasing diffusion limitation (i.e., increasing parameter mL), an increasing concentration gradient is predicted. **Figure A1** shows the influence of the diffusion limitation within the active layer on the polarization curve. Under kinetic and diffusion control in the active layer, a doubling of the apparent Tafel slope is observed in the DC response. Simulation was performed by considering parameters summarized in **Table 1**.

$$\mathrm{b} = 2 \times 120\ \mathrm{mV/dec}.$$

$$\mathrm{b} = 120\ \mathrm{mV/dec}.$$

Author details

Jonathan Deseure
LEPMI, CNRS, Grenoble Institute of Engineering (INP), University of Grenoble Alpes, University Savoie Mont Blanc, Grenoble, France

*Address all correspondence to: jonathan.deseure@univ-grenoble-alpes.fr

References

[1] Niakolas DK, Daletou M, Neophytides SG, Vayenas CG. Fuel cells are a commercially viable alternative for the production of "clean" energy. Ambio. 2016;**45**(S1):32-37

[2] Yoldaş Y, Önen A, Muyeen SM, Vasilakos AV, Alan İ. Enhancing smart grid with microgrids: Challenges and opportunities. Renewable and Sustainable Energy Reviews. 2017;**72**: 205-214

[3] Nair AS et al. Multi-agent systems for resource allocation and scheduling in a smart grid. Technology and Economics of Smart Grids and Sustainable Energy. 2018;**33**:15. https://doi.org/10.1007/s40866-018-0052-y

[4] Ramchurn SD, Vytelingum P, Rogers A, Jennings NR. Putting the "smarts" into the smart grid: A grand challenge for artificial intelligence. Communications of the ACM. 2012; **55**(4):86

[5] Bove R, Ubertini S. Modeling solid oxide fuel cell operation: Approaches, techniques and results. Journal of Power Sources. 2006;**159**(1):543-559

[6] Biyikoglu A. Review of proton exchange membrane fuel cell models. International Journal of Hydrogen Energy. 2005;**30**(11):1181-1212

[7] Fadzillah DM, Rosli MI, Talib MZM, Kamarudin SK, Daud WRW. Review on microstructure modelling of a gas diffusion layer for proton exchange membrane fuel cells. Renewable and Sustainable Energy Reviews. 2017;**77**: 1001-1009

[8] Damour C, Benne M, Kadjo J-JA, Deseure J, Grondin-Perez B. On-line PEMFC control using parameterized nonlinear model-based predictive control. Fuel Cells. 2014; **14**(6):886-893

[9] Pourkiaei SM, Ahmadi MH, Hasheminejad e SM. Modeling and experimental verification of a 25 W fabricated PEM fuel cell by parametric and GMDH-type neural network. Mechanics & Industry. 2016;**17**(1):105

[10] Ou S, Achenie LEK. Artificial neural network Modeling of PEM fuel cells. Journal of Fuel Cell Science and Technology. 2005;**2**(4):226

[11] Damour C, Benne M, Lebreton C, Deseure J, Grondin-Perez B. Real-time implementation of a neural model-based self-tuning PID strategy for oxygen stoichiometry control in PEM fuel cell. International Journal of Hydrogen Energy. 2014;**39**(24):12819-12825

[12] Bicer Y, Dincer I, Aydin M. Maximizing performance of fuel cell using artificial neural network approach for smart grid applications. Energy. 2016;**116**:1205-1217

[13] Arriagada J. Artificial neural network simulator for SOFC performance prediction. Journal of Power Sources. 2002;**112**(1):54-60

[14] Grondin D, Deseure J, Ozil P, Chabriat J-P, Grondin-Perez B, Brisse A. Solid oxide electrolysis cell 3D simulation using artificial neural network for cathodic process description. Chemical Engineering Research and Design. 2013;**91**(1): 134-140

[15] Seyhan M, Akansu YE, Murat M, Korkmaz Y, Akansu SO. Performance prediction of PEM fuel cell with wavy serpentine flow channel by using artificial neural network. International Journal of Hydrogen Energy. 2017; **42**(40):25619-25629

[16] Ramousse J, Deseure J, Lottin O, Didierjean S, Maillet D. Modelling of

heat, mass and charge transfer in a PEMFC single cell. Journal of Power Sources. 2005;**145**(2):416-427

[17] Siegel NP, Ellis MW, Nelson DJ, von Spakovsky MR. Single domain PEMFC model based on agglomerate catalyst geometry. Journal of Power Sources. 2003;**115**(1):81-89

[18] Gloaguen F, Convert P, Gamburzev S, Velev OA, Srinivasan S. An evaluation of the macro-homogeneous and agglomerate model for oxygen reduction in PEMFCs. Electrochimica Acta. 1998;**43**(24): 3767-3772

[19] Singh D, Lu DM, Djilali N. A two-dimensional analysis of mass transport in proton exchange membrane fuel cells. International Journal of Engineering Science. 1999;**37**(4):431-452

[20] Macedo-Valencia J, Sierra JM, Figueroa-Ramírez SJ, Díaz SE, Meza M. 3D CFD modeling of a PEM fuel cell stack. International Journal of Hydrogen Energy. 2016;**41**(48):23425-23433

[21] Zhang G, Fan L, Sun J, Jiao K. A 3D model of PEMFC considering detailed multiphase flow and anisotropic transport properties. International Journal of Heat and Mass Transfer. 2017; **115**:714-724

[22] Lazar AL, Konradt SC, Rottengruber H. Open-source dynamic Matlab/Simulink 1D proton exchange membrane fuel cell model. Energies. 2019;**12**(18):3478

[23] Oliveira V, Falcao D, Rangel C, Pinto A. Heat and mass transfer effects in a direct methanol fuel cell: A 1D model. International Journal of Hydrogen Energy. 2008;**33**(14): 3818-3828

[24] Dohle H, Kornyshev AA, Kulikovsky AA, Mergel J, Stolten D. The current voltage plot of PEM fuel cell with long feed channels. Electrochemical Communications. 2001; **3**(2):73-80

[25] Springer TE. Characterization of polymer electrolyte fuel cells using AC impedance spectroscopy. Journal of the Electrochemical Society. 1996;**143**(2):587

[26] He M, Huang Z, Sun P, Wang e C. Modeling and numerical studies for a 3D two-phase mixed-domain model of PEM fuel cell. Journal of the Electrochemical Society. 2013;**160**(4):F324-F336

[27] Zhang G, Jiao K. Multi-phase models for water and thermal management of proton exchange membrane fuel cell: A review. Journal of Power Sources. 2018; **391**:120-133

[28] Nandjou F, Poirot-Crouvezier J-P, Chandesris M, Bultel Y. A pseudo-3D model to investigate heat and water transport in large area PEM fuel cells—Part 1: Model development and validation. International Journal of Hydrogen Energy. 2016;**41**(34): 15545-15561

[29] Krewer U et al. Direct methanol fuel cell (DMFC): Analysis of residence time behaviour of anodic flow bed. Chemical Engineering Science. 2004;**59**(1): 119-130

[30] Diep J, Kiel D, St-Pierre J, Wong A. Development of a residence time distribution method for proton exchange membrane fuel cell evaluation. Chemical Engineering Science. 2007;**62**(3):846-857

[31] Benziger J, Chia E, Moxley JF, Kevrekidis IG. The dynamic response of PEM fuel cells to changes in load. Chemical Engineering Science. 2005; **60**(6):1743-1759

[32] Boillot M, Didierjean S, Lapicque F. Residence time distributions of gases in lab-scale polymer electrolyte membrane fuel cells (PEMFC). Chemical

Engineering Science. 2005;**60**(4): 1187-1192

[33] Deseure J. Coupling RTD and EIS modelling to characterize operating non-uniformities on PEM cathodes. Journal of Power Sources. 2008;**178**(1): 323-333

[34] Aubras F et al. Two-dimensional model of low-pressure PEM electrolyser: Two-phase flow regime electrochemical modelling and experimental validation. International Journal of Hydrogen Energy. 2017;**42**(42):26203-26216

[35] Hinatsu JT. Water uptake of perfluorosulfonic acid membranes from liquid water and water vapor. Journal of the Electrochemical Society. 1994; **141**(6):1493

[36] Han I-S, Chung C-B. Performance prediction and analysis of a PEM fuel cell operating on pure oxygen using data-driven models: A comparison of artificial neural network and support vector machine. International Journal of Hydrogen Energy. 2016;**41**(24): 10202-10211

[37] Nguyen TV. A water and heat management model for proton-exchange-membrane fuel cells. Journal of the Electrochemical Society. 1993; **140**(8):2178

[38] Perry ML. Mass transport in gas-diffusion electrodes: A diagnostic tool for fuel-cell cathodes. Journal of the Electrochemical Society. 1998;**145**(1):5

[39] Bultel Y, Genies L, Antoine O, Ozil P, Durand R. Modeling impedance diagrams of active layers in gas diffusion electrodes: Diffusion, ohmic drop effects and multistep reactions. Journal of Electroanalytical Chemistry. 2002; **527**(1–2):143-155

Section 2

Applications

Chapter 6

Improving Heat-Engine Performance via High-Temperature Recharge

Jack Denur

Abstract

Perfect (reversible) cyclic heat engines operate at Carnot efficiency. Perfect reversible) nonheat engines and noncyclic heat engines operate at unit (100%) efficiency. But a usually necessary, although not always sufficient, requirement to achieve reversibility is that an engine must operate infinitely slowly, i.e., quasi-statically. And infinitely slow operation, which implies infinitesimally small power output, is obviously impractical. Most real heat engines operate, if not at maximum power output, then at least closer to maximum power output than to maximum efficiency. Endoreversible heat engines delivering maximum power output operate at Curzon-Ahlborn efficiency. Irrespective of efficiency, engines' work outputs are in almost all cases totally frictionally dissipated as heat immediately (e.g., an automobile operating at constant speed) or on short time scales. But if a heat engine's work output must be frictionally dissipated, it is best to dissipate it not into the cold reservoir but at the highest practicable temperature. We dub this as high-temperature recharge (HTR). This is not always practicable. But if it is practicable, it can yield improved heat-engine performance. We discuss improvements of the Carnot and Curzon-Ahlborn efficiencies achievable via HTR, and show consistency with the First and Second Laws of Thermodynamics. We reply to criticisms of HTR.

Keywords: Carnot efficiency, Curzon-Ahlborn efficiency, entropy, First Law of Thermodynamics, Second Law of Thermodynamics, heat engines, high-temperature recharge (HTR)

1. Introduction, overview, and general considerations

Perfect (reversible) cyclic heat engines operate at Carnot efficiency [1–7]. Perfect (reversible) nonheat engines and noncyclic (necessarily one-time, single-use) heat engines operate at unit (100%) efficiency. A simple example of a noncyclic heat engine is the one-time expansion of a gas pushing a piston. (If the expansion is isothermal, the heat is supplied from the internal energy of a reservoir; if is adiabatic, the heat is supplied from the internal energy of the gas itself. A polytropic expansion is intermediate between these two extremes.) Other examples include rockets: the piston (payload) is launched into space by a one-time power stroke (but typically most of the work output accelerates the exhaust gases, not the payload) and firearms: the piston (bullet) is accelerated by a one-time power stroke and then discarded (but some, typically less than with rockets, of the work output accelerates the gases resulting from combustion of the propellant). (Some rocket engines, e.g., those employed in the Space Shuttle and by SpaceX, can be refurbished and reused, but both the first use and each subsequent refurbishment and reuse constitute a necessarily one-time, single-use of a *non*cyclic rocket heat engine.)

But a usually necessary [1–7], although not always sufficient, requirement to achieve reversibility is that a heat engine, whether cyclic or noncyclic—indeed any engine, heat engine or otherwise—must operate infinitely slowly, i.e., quasi-statically [1–7]. And infinitely slow operation, which implies infinitesimally small power output, is obviously impractical [1–7]. Indeed, some types of friction, such as sliding and rolling friction, do *not* vanish as the speed of operation becomes infinitely slow [8–11] [see also Ref. [1], Sections 5-2, 6-1, 6-2, 8-1, 11-1, and 11-2; Ref. [2], Section 4.2 (especially the 3rd and 4th paragraphs) and Figure 4.3; Ref. [3], Problem 4.2-1; and Ref. [4], Section 3.6]. In such cases, reversibility does *not* obtain *even with* infinitely slow, i.e., quasi-static, operation [8–11]. By contrast, for example, in cosmology, reversibility may in some cases obtain even with processes occurring at finite rates [12], but of course this is not relevant with respect to practical heat engines.

Most real heat engines operate, if not at maximum power output, then at least closer to maximum power output than to maximum efficiency. Assuming endoreversibility (irreversible heat flows directly proportional to finite temperature differences but otherwise reversible operation), at maximum power output cyclic heat engines operate at Curzon-Ahlborn efficiency [13–15] (see also Ref. [3], Section 4-9). The work outputs of heat engines—indeed of *all* engines, heat engines or otherwise—are in almost cases totally frictionally dissipated as heat immediately or on short time scales [16, 17]. For example, an automobile's cyclic heat engine's work output in initially accelerating the automobile is typically frictionally dissipated only a short time later the next time the automobile decelerates; its work output while the automobile travels at constant speed is immediately and continually frictionally dissipated. [Rare exceptions include, for example, a noncyclic rocket heat engine's work output being sequestered essentially permanently as kinetic and gravitational potential energy in the launching of a spacecraft (but typically most of the kinetic energy accelerates the exhaust gases, not the payload) and a cyclic heat engine's work output being sequestered for a long time interval as gravitational potential energy in the construction of a building.]

We note that the *work* output of *any* engine (heat engine or otherwise) can be dissipated *only via friction*. Additional losses can, and almost always if not always, also occur, for example, irreversible heat losses engendered by finite temperature differences (no insulation is perfect). [The Curzon-Ahlborn efficiency [13–15] (see also Ref. [3], Section 4-9) takes into account losses due to irreversible heat flows directly proportional to finite temperature differences but assumes otherwise reversible operation.] But such heat losses are *not* work. An engine's *work* output per se can be dissipated *only via friction*. This is true because work is a *force* exerted through a distance: thus work can be dissipated *only* by an opposing *force* that is *non*conservative. And *non*conservative *force* is *friction*. [It might be contended that, ultimately, friction is the electromagnetic force, which is conservative. But for all typical macroscopic motions, for which the kinetic energy in any given degree of freedom greatly exceeds $k_B T$ (k_B is Boltzmann's constant, T is the temperature), friction is *effectively non*conservative.]

But if a heat engine's work output must be frictionally dissipated, it is best to dissipate it not at the temperature of its cold reservoir but instead at the highest practicable temperature. This is consistent with the Second Law of Thermodynamics, which allows frictional dissipation of work into heat at *any temperature* [1–7] (in Ref. [6], see pp. 11–12, 60–65, and 263–265, especially pp. 263–265). The entropy increase resulting from frictional dissipation of work W at temperature T, namely $\Delta S = W/T$, decreases monotonically with increasing T but is positive for any finite T—and the Second Law requires only that $\Delta S \geq 0$ [1–7].

We should emphasize that the entropy increase $\Delta S = W/T$ associated with work W being frictionally dissipated at temperature T is *always* $\Delta S = W/T$ and hence *always* decreases monotonically with increasing T. Whether the coefficient of friction is small or large *makes no difference* in $\Delta S = W/T$. Of course, all other things being equal, if the coefficient of friction is small, frictional dissipation of W will take longer than if the coefficient of friction is large. But the entropy increase $\Delta S = W/T$ associated with work W being frictionally dissipated at temperature T is the same whether the coefficient of friction is small or large. Even with coefficients of friction in the low range, W will typically be frictionally dissipated immediately or on short time scales.

Of course, efficiency is highest if work W, whether supplied via a heat engine or otherwise, is not frictionally dissipated at all. This would obtain, for example, in perfect (reversible) regenerative braking of an electrically-powered motor vehicle, with the motor operating backward as a generator during braking. It also would obtain, for example, if a noncyclic rocket heat engine's work output is perfectly (reversibly) sequestered as kinetic and gravitational potential energy in the launching of a spacecraft (but typically most of the kinetic energy accelerates the exhaust gases, not the payload) or if a cyclic heat engine's work output is perfectly (reversibly) sequestered as gravitational potential energy in the construction of a building. But of course in practice (as opposed to in principle) *total* avoidance of frictional dissipation of work [and also of additional losses, e.g., due to irreversible heat flows engendered by finite temperature differences (no insulation is perfect)] is not possible.

Although we do not consider them in this chapter, we should note that: (a) There are generalizations of the Curzon-Ahlborn efficiency [13–15] (see also Ref. [3], Section 4-9) at maximum power output both for macroscopic heat engines [18–20] and for microscopic heat engines [21, 22], with irreversible heat flows *not* necessarily directly proportional to temperature differences. (b) There are analyses of maximum heat-engine work output per cycle (as opposed to maximum power output) [23]. Comprehensive discussions concerning the Curzon-Ahlborn efficiency and generalizations thereof are provided in Refs. [24–26]. Some, but not all, such generalized efficiencies [18–26] do not differ greatly from the Curzon-Ahlborn efficiency [13–15] (see also Ref. [3], Section 4-9). In particular, we note that alternative results [26] to the Curzon-Ahlborn efficiency have been derived [26]. But for definiteness and for simplicity, in this chapter, we employ the standard Curzon-Ahlborn efficiency [13–15] (see also Ref. [3], Section 4-9) for cyclic heat engines operating at maximum power output.

A misconception pertaining to the efficiencies of engines (heat engines or otherwise) is discussed and corrected in Section 2.

In Section 3, we review the work outputs, efficiencies, and entropy productions of Carnot (reversible) and Curzon-Ahlborn (endoreversible) heat engines, first without frictional dissipation of a heat engine's work output and then with frictional dissipation thereof into its cold reservoir. In Section 3, we do not consider frictional dissipation of a heat engine's work output at the highest practicable temperature, which we dub as high-temperature recharge (HTR).

In Section 4, we discuss the work outputs, efficiencies, and entropy productions of Carnot and Curzon-Ahlborn heat engines operating with frictional dissipation of a heat engine's work output at the highest practicable temperature—which we dub as high-temperature recharge (HTR)—and the improvements thereof over those obtainable (as per Section 3) without HTR. Cases wherein HTR is practicable include, but are not necessarily limited to, (a) hurricanes, which via HTR are rendered more powerful than they would otherwise be [27–37], (b) thermoelectric generators [38], and (c) heat engines *powered by a cold reservoir*, employing ambient as the *hot* reservoir, for example, heat engines powered by the evaporation of water [39–51] or by liquid nitrogen [52], ocean-thermal-energy-conversion (OTEC) heat engines [53–56], and heat engines powered by the cold of outer space [57].

Concerning (a) in the immediately preceding paragraph, on the one hand, the importance of HTR (dubbed as "dissipative heating") has been confirmed in a study of Hurricane Andrew (1992) [36], and, as one might expect, "dissipative heating appears to be a more important process in intense hurricanes, such as Andrew, than weak ones" [36]. But, on the other hand, more recently it has been contended [37] that, while HTR exists in hurricanes, it is of lesser importance than previously supposed [27–36]. [There are occasional speculations concerning extracting useful energy from hurricanes (with or without help from HTR) and also freshwater. But, of course, except for (strongly built!) windmills and ocean-wave-powered generators for extracting energy and reservoirs for extracting freshwater, this is beyond currently available (and perhaps even currently foreseeable) technology. To the extent that HTR increases wind speeds in hurricanes, it increases the power flux density available to (strongly built!) windmills and ocean-wave-powered generators: wind power flux density is proportional to the cube of the wind speed (and directly proportional to the air density). But, at least for the time being, the main (or perhaps even only) employment of HTR in hurricanes is by the hurricanes themselves, to increase their wind speeds, whether as previously supposed [27–36] or to a lesser degree [37].

To the best knowledge of the author, the concept of HTR was first partially and qualitatively broached by Spanner (see Ref. [6], pp. 11–12, 60–65, and 263–265, especially pp. 263–265) and, later, was first fully and quantitatively expounded and developed by Emanuel [27–34] in the course of his research concerning hurricane science. It was subsequently employed by Apertet et al. [38] for increasing efficiencies of thermoelectric generators. In these works [6, 27–34, 38] and in related works [35–37, 58–62], the concept is not dubbed HTR, but of course it is the concept itself, and not the dubbing it with a name, that is important. To the best knowledge of the author, the concept has not been dubbed HTR (dubbed, if at all, as "dissipative heating") in the previous literature. Heat engines employing it have previously been dubbed "dissipative engines" (see, e.g., Refs. [58–60]).

The increases in efficiency attainable via HTR are not practicable if frictional dissipation of work into other than the cold reservoir is not practicable. Thus they are *never* practicable for noncyclic (necessarily one-time, single-use) heat engines: however the work output of a noncyclic (necessarily one-time, single-use) heat engine might be frictionally dissipated, the heat thereby generated can*not* restore the engine to its initial state. Moreover in many cases the work outputs of noncyclic (necessarily one-time, single-use) heat engines are not frictionally dissipated *at all*, at least not during practicable time scales, for example, a noncyclic rocket heat engine's work output is sequestered essentially permanently as kinetic and gravitational potential energy in the launching of a spacecraft (but typically most of the kinetic energy accelerates the exhaust gases, not the payload). They also are *never* practicable for reverse operation of cyclic heat engines as refrigerators or heat pumps, because for both refrigerators and heat pumps, the *total* energy output (the work W, plus the heat Q_C extracted from a cold reservoir at the expense of W as required by the Second Law of Thermodynamics) *always* is deposited as heat Q_H into the hot reservoir ($Q_H = Q_C + W$): thus there is *never* any *additional* energy to be deposited into the hot reservoir (as there is from frictional dissipation of work done via forward operation of cyclic heat engines). {See Ref. [1], Section 20-3; Ref. [2], Sections 4.3, 4.4, and 4.7 (especially Section 4.7); Ref. [3], Sections 4-4, 4-5, and 4-6 (especially Section 4-6); Ref. [5], Section 5.12 and Problem 5.22; Ref. [7], pp. 233–236 and Problems 1, 2, 4, 6, and 7 of Chapter 8; Ref. [16], Chapter XXI; Ref. [17], Sections 6.7, 6.8, 7.3, and 7.4; and Ref. [54], Sections 5-7-2, 6-2-2, 6-9-2, and 6-9-3, and Chapter 17. [Problem 2 of Chapter 8 in Ref. [7] considers absorption refrigeration, wherein the entire energy output is into an *intermediate*-temperature (most typically ambient-temperature) reservoir, and hence for which HTR is *even more strongly never practicable*.]} They also are not practicable for cyclic heat engines in cases wherein a cyclic heat engine's work output is not frictionally dissipated immediately or on short time scales [16, 17], for example, as gravitational potential energy sequestered for a long time interval in the construction of a building. For a building once erected typically remains standing for a century or longer. Even if, when it is finally torn down, its gravitational potential energy were to be totally frictionally dissipated into a hot reservoir, it is simply impracticable to wait that long. Thus HTR is not practicable in all cases. But in the many cases wherein cyclic heat engines' work outputs are frictionally dissipated immediately or on short time scales [16, 17], practicability obtains: improved conversion—and reconversion—of frictionally dissipated heat into work, and hence improved cyclic heat-engine performance, can then obtain.

Since HTR is *never* practicable for noncyclic (necessarily one-time, single-use) heat engines such as rockets or firearms, or for reverse operation of cyclic heat engines as refrigerators or heat pumps, henceforth we will (except where otherwise mentioned) focus exclusively on forward operation of cyclic heat engines.

Note that the primarily relevant time scale pertaining to "in the many cases wherein cyclic heat engines' work outputs are frictionally dissipated immediately or on short time scales [16, 17]" is (i) the time interval between a cyclic heat engine's work output and frictional dissipation of this work output [16, 17], *not* (ii) the time interval required for frictional dissipation per se. The time interval (ii) is *zero* in all cases wherein work is done against the nonconservative force of friction and hence frictionally dissipated *immediately*. Indeed, in *all* cases of steady-state engine operation against friction (e.g., an automobile traveling at constant speed) *both* time intervals are *zero*. This is by far the most common mode of engine operation. Even work output sequestered when an engine is started or when an automobile accelerates is typically frictionally dissipated only a short time later, when the engine is turned off or when the automobile decelerates. Work output sequestration for a century or longer can obtain (as gravitational potential energy) in the construction of buildings and essentially permanently in the launchings of spacecraft—but these are rare exceptions. Thus we focus on time interval (i): a necessary (but not sufficient) condition for HTR to be practicable is that the time interval (i) be zero or at most short. (This condition is automatically met in steady-state engine operation, wherein work is frictionally dissipated *immediately* and hence *both* time intervals are *zero*.)

But for cyclic heat engines whose work outputs typically are frictionally dissipated immediately or on short time scales [16, 17], HTR often *is* practicable. For cyclic heat engines employing ambient as the *cold* reservoir, the existent hot reservoir is likely already at the practicable upper temperature limit. Hence for these cyclic heat engines, HTR at the temperature of the hot reservoir could increase efficiency, but HTR at a still higher temperature probably would not be practicable. By contrast, consider cyclic heat engines *powered by a cold reservoir*, employing ambient as the *hot* reservoir, for example, cyclic heat engines powered by the evaporation of water [39–51] or by liquid nitrogen [52], ocean thermal-energy-conversion (OTEC) heat engines [53–56], and heat engines powered by the cold of outer space [57]. For *these* cyclic heat engines, HTR at a higher temperature than ambient probably *would* be practicable. For *these* cyclic heat engines, employment of HTR could boost the temperature of the hot reservoir from ambient to the highest practicable temperature for HTR.

Henceforth if HTR is employed we construe the terms "the highest practicable temperature for HTR" and "the hot reservoir" to be synonymous.

Recapitulation and generalization are provided in Section 5. A reply to criticisms [58, 59] of HTR is provided in Section 6 and in references cited therein. Concluding remarks are provided in Section 7.

2. Correcting a misconception pertaining to the efficiencies of engines

The efficiency of any engine in general is its work output (force-times-distance output) divided by its energy input, and the efficiency of a heat engine in particular it is its work output (force-times-distance output) divided by its heat-energy input. What happens to an engine's work output after the work has been done is an entirely different issue.

Work can be done either against a conservative force, in which case it is sequestered, or against the nonconservative force of friction, in which case it is dissipated as heat. To re-emphasize, in either case—whether the opposing force is conservative or nonconservative—the efficiency of any engine in general is its work output (force-times-distance output) divided by its energy input, and the efficiency of a heat engine in particular it is its work output (force-times-distance output) divided by its heat-energy input. What happens to an engine's work output after the work has been done is an entirely different issue.

In this Section 2, we wish to correct a misconception that is sometimes made, according to which an engine's efficiency can exceed zero only if its work output is done against a conservative force. This misconception is erroneous.

In the vast majority of cases, for almost all engines on Earth, work is done against the nonconservative force of friction, and hence instantaneously and continually dissipated as heat. The engines work at steady state, and while working, their internal energy and the internal energy of any equipment they might be operating do not change. Consider, for example, the engine of any automobile, train, ship, submarine, or aircraft traveling at constant speed, any factory or workshop engine such as a power saw operating at constant speed, or any domestic appliance engine such as that of a dishwasher, refrigerator, etc., operating at constant speed. According to the erroneous misconception that an engine's efficiency is zero if its work output is done against the nonconservative force of friction, the efficiency of all of these engines—indeed of almost all engines on Earth—would falsely be evaluated at zero. If their efficiencies were truly zero, they could do zero work against *any* opposing force, conservative or nonconservative, i.e., they could not operate *at all*. A specific example is the following: If the efficiency of an engine (heat engine or otherwise) attempting to maintain an automobile at constant speed was zero, the engine could do zero work against friction, and the automobile's speed would also be zero.

Only in rare cases, such as the construction of buildings and the launchings of spacecraft, is the work done even against conservative forces (e.g., gravity, inertia, etc.) sequestered for any significant lengths of time. Even in most cases wherein work is done against a conservative force, it is frictionally dissipated a short time later. For example, the work done in accelerating an automobile against its own inertia is typically frictionally dissipated as heat a short time later the next time the automobile decelerates. The net effect of the acceleration/deceleration process is frictional dissipation of the automobile's temporarily sequestered kinetic energy, the same as the instantaneous and continual frictional dissipation of its kinetic energy while it operates at constant speed.

In general, when an engine (heat engine or otherwise) is turned on, part of its work output is sequestered as its own kinetic energy and the kinetic energy of any equipment that it might be operating. But this kinetic energy is frictionally dissipated as heat when the engine is turned off, so the net effect of the on/off process is frictional dissipation of this temporarily sequestered kinetic energy, the same as the instantaneous and continual frictional dissipation of the engine's work output while it operates at constant speed between the time it is turned on and the time it is turned off.

3. Carnot and Curzon-Ahlborn efficiencies without high-temperature recharge (HTR)

The standard (without high-temperature recharge or HTR) Carnot efficiency $\epsilon_{\text{Carnot,std}}$ and Curzon-Ahlborn efficiency $\epsilon_{\text{CA,std}}$ corresponding to heat-engine operation between a hot reservoir at temperature T_H and a cold reservoir at temperature T_C are given by the respective well-known formulas:

$$\epsilon_{\text{Carnot,std}} = \frac{W}{Q_H} = \frac{T_H - T_C}{T_H} = 1 - \frac{T_C}{T_H} \equiv 1 - R_T \tag{1}$$

and

$$\epsilon_{\text{CA,std}} = \frac{W}{Q_H} = \frac{T_H^{1/2} - T_C^{1/2}}{T_H^{1/2}} = 1 - \left(\frac{T_C}{T_H}\right)^{1/2} \equiv 1 - R_T^{1/2}. \tag{2}$$

We define the *temperature* ratio between a heat engine's cold and hot reservoirs as $R_T \equiv T_C/T_H$. Obviously in all cases $0 \le R_T \le 1$. The case $R_T = 1$ is of no interest because it corresponds to zero efficiency of any heat engine. The case $R_T = 0$ is unattainable because there is no cold reservoir at absolute zero (0 K) [63], and even if there was [63], it would no longer be at 0 K the instant after a heat engine began operating and exhausting its waste heat into it [see also Ref. [2], Chapter 10 (especially Section 10.4); Ref. [3], Chapter 11 (especially Section 11-3); and Ref. [4], Chapter 14 (especially Sections 14.3–14.5 and 14.7)]. Hence we confine our attention to the range $0 < R_T < 1$.

The *efficiency* ratio $R_{\epsilon 1}$ between these efficiencies is

$$R_{\epsilon 1} = \frac{\epsilon_{\text{Carnot,std}}}{\epsilon_{\text{CA,std}}} = \frac{1 - R_T}{1 - R_T^{1/2}}. \tag{3}$$

$R_{\epsilon 1}$ increases monotonically with increasing R_T. In the limit $R_T \to 0$, it is obvious that $R_{\epsilon 1} \to 1$. In the limit $R_T \to 1$, $R_{\epsilon 1} \to 2$. The latter limit is most easily demonstrated by setting $R_T = 1 - \delta$, letting $\delta \to 0$, and applying the binomial theorem. This yields

$$\lim_{R_T \to 1} R_{\epsilon 1} = \lim_{\delta \to 0} R_{\epsilon 1} = \lim_{\delta \to 0} \frac{1 - (1 - \delta)}{1 - (1 - \delta)^{1/2}} = \frac{\delta}{1 - \left(1 - \frac{1}{2}\delta\right)} = \frac{\delta}{\frac{1}{2}\delta} = 2. \tag{4}$$

By the First and Second Laws of Thermodynamics, for a standard reversible heat engine operating (without HTR) at Carnot efficiency, the heat input Q_H from its hot reservoir, the work output W, the waste heat Q_C exhausted to its cold reservoir, the efficiency $\epsilon_{\text{Carnot,std}}$, the entropy change $\Delta S_{H,\text{Carnot,std}}$ of its hot reservoir, the entropy change $\Delta S_{C,\text{Carnot,std}}$ of its cold reservoir, and the total entropy change $\Delta S_{\text{total,std}} = \Delta S_{H,\text{Carnot,std}} + \Delta S_{C,\text{Carnot,std}}$ are related in accordance with

$$W = Q_H - Q_C = Q_H \epsilon_{\text{Carnot,std}} = Q_H(1 - R_T) \Rightarrow Q_C = R_T Q_H. \tag{5}$$

and

$$\Delta S_{\text{total,Carnot,std}} = \Delta S_{H,\text{Carnot,std}} + \Delta S_{C,\text{Carnot,std}} = -\frac{Q_H}{T_H} + \frac{Q_C}{T_C} = -\frac{Q_H}{T_H} + \frac{R_T Q_H}{R_T T_H} = 0. \tag{6}$$

We note that, in most derivations (in textbooks or elsewhere) of $\epsilon_{\text{Carnot,std}}$, $\Delta S_{\text{total,Carnot,std}} = 0$ is invoked as the initial Second Law assumption or postulate and is employed along with the initial First Law postulate $W = Q_H - Q_C$ [1–7].

Similarly, by the First and Second Laws of Thermodynamics, for a standard endoreversible heat engine operating (without HTR) at Curzon-Ahlborn efficiency, the heat input Q_H from its hot reservoir, the work output W, the waste heat Q_C exhausted to its cold reservoir, the efficiency $\epsilon_{\text{CA,std}}$, the entropy change $\Delta S_{H,\text{CA,std}}$ of its hot reservoir, the entropy change $\Delta S_{C,\text{CA,std}}$ of its cold reservoir, and the total entropy change $\Delta S_{\text{total,CA,std}} = \Delta S_{H,\text{CA,std}} + \Delta S_{C,\text{CA,std}}$ are related in accordance with

$$W = Q_H - Q_C = Q_H \epsilon_{\text{CA,std}} = Q_H\left(1 - R_T^{1/2}\right) \Rightarrow Q_C = R_T^{1/2} Q_H. \tag{7}$$

and

$$\begin{aligned} \Delta S_{\text{total,CA,std}} &= \Delta S_{H,\text{CA,std}} + \Delta S_{C,\text{CA,std}} = -\frac{Q_H}{T_H} + \frac{Q_C}{T_C} = -\frac{Q_H}{T_H} + \frac{R_T^{1/2} Q_H}{R_T T_H} \\ &= \frac{Q_H}{T_H}\left(\frac{1}{R_T^{1/2}} - 1\right) = \frac{Q_H}{T_H}\left(R_T^{-1/2} - 1\right) \\ &> \Delta S_{\text{total,Carnot,std}} = 0. \end{aligned} \tag{8}$$

Note that for any R_T in general and as $R_T \to 0$ in particular, $\Delta S_{\text{total,Carnot,std}} = 0$; by contrast, for any R_T in general $\Delta S_{\text{total,CA,std}} > 0$ ($\Delta S_{\text{total,CA,std}} \to 0$ only in the limit $R_T \to 1$), and as $R_T \to 0$, $\Delta S_{\text{total,CA,std}} \to \infty$.

As we have already noted, heat engines' work outputs are, in almost all cases, totally frictionally dissipated as heat immediately or on short time scales [16, 17]. For example, an automobile heat engine's work output in initially accelerating the automobile is typically frictionally dissipated only a short time later the next time the automobile decelerates; its work output while the automobile travels at constant speed is immediately and continually frictionally dissipated. [Rare exceptions include, for example, a noncyclic rocket heat engine's work output being sequestered essentially permanently as kinetic and gravitational potential energy in the launching of a spacecraft (but typically most of the kinetic energy accelerates the exhaust gases, not the payload) and a cyclic heat engine's work output being sequestered for a long time interval as gravitational potential energy in the construction of a building.] Apart from such rare exceptional cases, in the operation of *any* cyclic heat engine operating at *any* efficiency with*out* HTR—whether reversible at Carnot efficiency, endoreversible at Curzon-Ahlborn efficiency, or otherwise—not only is the waste heat Q_C immediately exhausted into the cold reservoir, but also the heat engine's work output W is frictionally dissipated into the cold reservoir immediately or on short time scales [16, 17]. Thus, apart from such rare exceptional cases, the *ultimate* total entropy increase $\Delta S_{\text{total,ultimate}}$ resulting from the operation of *any* heat engine with*out* HTR at *any* efficiency occurs as soon as the work is done or shortly thereafter and is [1–7, 16, 17]:

$$\begin{aligned}\Delta S_{\text{total,ultimate}} &= -\frac{Q_H}{T_H}+\frac{Q_C}{T_C}+\frac{W}{T_C} = -\frac{Q_H}{T_H}+\frac{Q_C+W}{T_C} = -\frac{Q_H}{T_H}+\frac{Q_H}{T_C}\\ &= -\frac{Q_H}{T_H}+\frac{Q_H}{R_T T_H} = \frac{Q_H}{T_H}\left(\frac{1}{R_T}-1\right) = \frac{Q_H}{T_H}\left(R_T^{-1}-1\right)\\ &> \Delta S_{\text{total,CA,std}} = \frac{Q_H}{T_H}\left(R_T^{-1/2}-1\right)\\ &> \Delta S_{\text{total,Carnot,std}} = 0.\end{aligned} \tag{9}$$

Note that for any $R_T > 0$, the inequality $\Delta S_{\text{total,ultimate}} > 0$ is stronger than the inequality $\Delta S_{\text{total,CA,std}} > 0$. In particular, note that as $R_T \to 1$, $\Delta S_{\text{total,ultimate}} \to 0$ more slowly than $\Delta S_{\text{total,CA,std}} \to 0$, while as $R_T \to 0$, $\Delta S_{\text{total,ultimate}} \to \infty$ faster than $\Delta S_{\text{total,CA,std}} \to \infty$.

4. Carnot and Curzon-Ahlborn efficiencies with high-temperature recharge (HTR)

If, as is almost always the case, a cyclic heat engine's work output W is totally frictionally dissipated as heat immediately or on short time scales [16, 17], the engine's efficiency can be increased if this dissipation is not at the temperature of its cold reservoir but instead at the highest practicable temperature. For cyclic heat engines employing ambient as the *cold* reservoir, the existent hot reservoir is likely already at the practicable upper temperature limit. Hence for these cyclic heat engines, HTR at the temperature of the hot reservoir could increase efficiency, but HTR at a still higher temperature probably would not be practicable. By contrast, for cyclic heat engines *powered by a cold reservoir*, employing ambient as the *hot* reservoir [39–57], frictional dissipation at a higher temperature than ambient probably *would* be practicable. For *these* cyclic heat engines, employment of HTR could boost T_H, the temperature of the hot reservoir, from ambient to the the highest practicable temperature for HTR. In this Section 4, we take T_H, the temperature of the hot reservoir, to be the highest practicable temperature for frictional dissipation of a cyclic heat engine's work output into heat.

Consider first a reversible heat engine operating at Carnot efficiency.
If the engine's work output $W = Q_H(1 - R_T)$ is frictionally dissipated into its hot reservoir (at temperature T_H), then the *net* heat input $Q_{H,\text{net}}$ required from its hot reservoir is reduced from Q_H to $Q_H - W = Q_H - Q_H(1 - R_T) = Q_H - Q_H R_T = Q_C$. Hence, with the help of Eqs. (1) and (5), the efficiency $\epsilon_{\text{Carnot,HTR}}$ of a Carnot engine operating with HTR is

$$\epsilon_{\text{Carnot,HTR}} = \frac{W}{Q_{H,\text{net}}} = \frac{W}{Q_H - W} = \frac{W}{Q_C} = \frac{Q_H - Q_C}{Q_C} = \frac{Q_H}{Q_C} - 1 = \frac{T_H}{T_C} - 1 = R_T^{-1} - 1$$

$$\Rightarrow R_{\epsilon 2} = \frac{\epsilon_{\text{Carnot,HTR}}}{\epsilon_{\text{Carnot,std}}} = \frac{R_T^{-1} - 1}{1 - R_T} = R_T^{-1} > 1. \tag{10}$$

Note that not only is $R_{\epsilon 2} > 1$ in all cases, but also that in some cases also $\epsilon_{\text{Carnot,HTR}} > 1$. The efficiency $\epsilon_{\text{Carnot,HTR}}$ increases monotonically from zero in the limit $R_T \to 1$ to ∞ in the limit $R_T \to 0$. And the efficiency ratio $R_{\epsilon 2}$ increases monotonically from unity in the limit $R_T \to 1$ to ∞ in the limit $R_T \to 0$. If $R_T < \frac{1}{2} \Leftrightarrow R_{\epsilon 2} > 2$, $\epsilon_{\text{Carnot,HTR}} > 1$. Yet the First and Second Laws of Thermodynamics are *not* violated. The First Law is not violated because no energy is created (or destroyed): $\epsilon_{\text{Carnot,HTR}} > 1$ if $R_T < \frac{1}{2} \Leftrightarrow R_{\epsilon 2} > 2$ obtains via recycling and reusing the same energy, not via the creation of new energy. The Second Law is not violated because the change in total entropy is positive:

$$\begin{aligned}
\Delta S_{\text{total,Carnot,HTR}} &= -\frac{Q_H}{T_H} + \frac{W}{T_H} + \frac{Q_C}{T_C} \\
&= \frac{Q_C}{T_C} - \frac{Q_H - W}{T_H} = \frac{Q_C}{T_C} - \frac{Q_C}{T_H} \\
&= Q_C\left(\frac{1}{T_C} - \frac{1}{T_H}\right) = R_T Q_H \left(\frac{1}{R_T T_H} - \frac{1}{T_H}\right) \\
&= \frac{R_T Q_H}{T_H}\left(\frac{1}{R_T} - 1\right) = \frac{R_T Q_H}{T_H}\left(\frac{1 - R_T}{R_T}\right) \\
&= \frac{Q_H}{T_H}(1 - R_T) \\
&> \Delta S_{\text{total,Carnot,std}} = 0.
\end{aligned} \tag{11}$$

In Eq. (11), we applied Eqs. (5) and (6). Yet, also applying Eq. (9), $\Delta S_{\text{total,Carnot,HTR}}$, while positive and hence greater than $\Delta S_{\text{total,Carnot,std}} = 0$, is smaller than $\Delta S_{\text{total,ultimate}}$ by the *entropy* ratio

$$R_{S1} = \frac{\Delta S_{\text{total,Carnot,HTR}}}{\Delta S_{\text{total,ultimate}}} = \frac{1 - R_T}{R_T^{-1} - 1} = R_T. \tag{12}$$

The entropy ratio R_{S1} decreases monotonically from unity in the limit $R_T \to 1$ to zero in the limit $R_T \to 0$.

It may be of interest to note that $\epsilon_{\text{Carnot,HTR}}$ of Eq. (10) is the multiplicative inverse of the Carnot coefficient of performance of a refrigerator, whereas, by contrast, $\epsilon_{\text{Carnot,std}}$ of Eq. (1) is the multiplicative inverse of the Carnot coefficient of performance of a heat pump. These relations may be of interest. But, as we have noted in Section 1, HTR is *never* practicable for reverse operation of cyclic heat engines as refrigerators or heat pumps.

Consider next an endoreversible heat engine operating at Curzon-Ahlborn efficiency. If the engine's work output $W = Q_H\left(1 - R_T^{1/2}\right)$ is frictionally dissipated into its hot reservoir (at temperature T_H), then the *net* heat input $Q_{H,\text{net}}$ required from its hot reservoir is reduced from Q_H to $Q_H - W = Q_H - Q_H\left(1 - R_T^{1/2}\right) = Q_H R_T^{1/2} = Q_C$. Hence, with the help of Eqs. (2) and (7), the efficiency $\epsilon_{\text{CA,HTR}}$ of a Curzon-Ahlborn engine operating with HTR is

$$\begin{aligned}\epsilon_{\text{CA,HTR}} &= \frac{W}{Q_{H,\text{net}}} = \frac{W}{Q_H - W} = \frac{W}{Q_C} = \frac{W}{Q_H}\frac{Q_H}{Q_C} = \frac{Q_H\left(1 - R_T^{1/2}\right)}{Q_H}\frac{Q_H}{Q_C} \\ &= \left(1 - R_T^{1/2}\right)\frac{Q_H}{Q_C} = \left(1 - R_T^{1/2}\right)\frac{Q_H}{R_T^{1/2}Q_H} = \frac{1 - R_T^{1/2}}{R_T^{1/2}} \\ &= R_T^{-1/2} - 1 \\ \Rightarrow R_{\epsilon 3} &= \frac{\epsilon_{\text{CA,HTR}}}{\epsilon_{\text{CA,std}}} = \frac{R_T^{-1/2} - 1}{1 - R_T^{1/2}} = R_T^{-1/2}.\end{aligned} \tag{13}$$

The efficiency $\epsilon_{\text{CA,HTR}}$ increases monotonically from zero in the limit $R_T \to 1$ to ∞ in the limit $R_T \to 0$. And the efficiency ratio $R_{\epsilon 3}$ increases monotonically from unity in the limit $R_T \to 1$ to ∞ in the limit $R_T \to 0$. If $R_T < \frac{1}{4} \Leftrightarrow R_{\epsilon 3} > 2$, $\epsilon_{\text{CA,HTR}} > 1$. Yet the First and Second Laws of Thermodynamics are *not* violated. The First Law is not violated because no energy is created (or destroyed): $\epsilon_{\text{CA,HTR}} > 1$ if $R_T < \frac{1}{4} \Leftrightarrow R_{\epsilon 3} > 2$ obtains via recycling and reusing the same energy, not via the creation of new energy. The Second Law is not violated because the change in total entropy is positive:

$$\begin{aligned}\Delta S_{\text{total,CA,HTR}} &= -\frac{Q_H}{T_H} + \frac{W}{T_H} + \frac{Q_C}{T_C} \\ &= \frac{Q_C}{T_C} - \frac{Q_H - W}{T_H} = \frac{Q_C}{T_C} - \frac{Q_C}{T_H} \\ &= Q_C\left(\frac{1}{T_C} - \frac{1}{T_H}\right) = R_T^{1/2}Q_H\left(\frac{1}{R_T T_H} - \frac{1}{T_H}\right) \\ &= \frac{R_T^{1/2}Q_H}{T_H}\left(\frac{1}{R_T} - 1\right) \\ &= \frac{Q_H}{T_H}\left(R_T^{-1/2} - R_T^{1/2}\right) \\ &> \Delta S_{\text{total,CA,std}} = \frac{Q_H}{T_H}\left(R_T^{-1/2} - 1\right) \\ &> \Delta S_{\text{total,Carnot,std}} = 0.\end{aligned} \tag{14}$$

In Eq. (14), we applied Eqs. (7) and (8). Yet, also applying Eq. (9), $\Delta S_{\text{total,CA,HTR}}$, while not merely positive but greater than $\Delta S_{\text{total,CA,std}}$, is smaller than $\Delta S_{\text{total,ultimate}}$ by the entropy ratio

$$R_{S2} = \frac{\Delta S_{\text{total,CA,HTR}}}{\Delta S_{\text{total,ultimate}}} = \frac{R_T^{-1/2} - R_T^{1/2}}{R_T^{-1} - 1} = R_T^{1/2}. \tag{15}$$

But, as one would expect, $\Delta S_{\text{total,CA,HTR}}$ is larger than $\Delta S_{\text{total,Carnot,HTR}}$. Applying Eqs. (11), (14), and (15), it is larger by the entropy ratio

$$R_{S3} = \frac{\Delta S_{\text{total,CA,HTR}}}{\Delta S_{\text{total,Carnot,HTR}}} = \frac{R_T^{-1/2} - R_T^{1/2}}{1 - R_T} = R_T^{-1/2} = R_{S2}^{-1}. \tag{16}$$

There is one more efficiency ratio that is of interest. Applying Eqs. (10) and (13):

$$R_{\epsilon 4} = \frac{\epsilon_{\text{Carnot,HTR}}}{\epsilon_{\text{CA,HTR}}} = \frac{R_T^{-1} - 1}{R_T^{-1/2} - 1}. \tag{17}$$

The efficiency ratio $R_{\epsilon 4}$ increases monotonically from 2 in the limit $R_T \to 1$ to ∞ in the limit $R_T \to 0$. The latter limit is obvious. The former limit is most easily demonstrated by setting $R_T = 1 - \delta$, letting $\delta \to 0$, and applying the binomial theorem. This yields

$$\lim_{R_T \to 1} R_{\epsilon 4} = \lim_{\delta \to 0} R_{\epsilon 4} = \lim_{\delta \to 0} \frac{(1-\delta)^{-1} - 1}{(1-\delta)^{-1/2} - 1} = \frac{(1+\delta) - 1}{\left(1 + \frac{1}{2}\delta\right) - 1} = \frac{\delta}{\frac{1}{2}\delta} = 2. \tag{18}$$

Note that: (a) Applying Eqs. (3), (4), (17), and (18), in the limit $R_T \to 1$, $R_{\epsilon 1} \to 2$ from below; by contrast $R_{\epsilon 4} \to 2$ from above. (b) Applying Eqs. (3), (4), (17), and (18), in the limit $R_T \to 0$, $R_{\epsilon 1} \to 1$ but $R_{\epsilon 4} \to \infty$. (c) Applying Eqs. (10) and (13), in the limit $R_T \to 0$, $\epsilon_{\text{Carnot,HTR}} \to \infty$ as R_T^{-1} but $\epsilon_{\text{CA,HTR}} \to \infty$ only as $R_T^{-1/2}$; thus, while both approach ∞, $\epsilon_{\text{Carnot,HTR}}$ does so at higher order—hence as $R_T \to 0$, $R_{\epsilon 4} \to \infty$.

As an aside, it may be of interest to note, applying Eqs. (3) and (17), that

$$\frac{R_{\epsilon 4}}{R_{\epsilon 1}} = \frac{\dfrac{R_T^{-1} - 1}{R_T^{-1/2} - 1}}{\dfrac{1 - R_T}{1 - R_T^{1/2}}} = \frac{R_T^{-1} - 1}{1 - R_T} \times \frac{1 - R_T^{1/2}}{R_T^{-1/2} - 1} \tag{19}$$
$$= R_T^{-1} \times R_T^{1/2} = R_T^{-1/2}.$$

5. Recapitulation and generalization

Efficiency is of course highest if work W, whether supplied via a heat engine or otherwise, is not frictionally dissipated at all. This would obtain, for example, in perfect (reversible) regenerative braking of an electrically-powered motor vehicle, with the motor operating backward as a generator during braking. It also would obtain, for example, if a noncyclic (necessarily one-time, single-use) rocket heat engine's work output is perfectly (reversibly) sequestered as kinetic and gravitational potential energy in the launching of a spacecraft (but typically most of the kinetic energy accelerates the exhaust gases, not the payload) or if a cyclic heat engine's work output is perfectly (reversibly) sequestered as gravitational potential energy in the construction of a building. But of course in practice (as opposed to in principle) *total* avoidance of frictional dissipation of work [and also of additional losses, e.g., due to irreversible heat flows engendered by finite temperature differences (no insulation is perfect)] is not possible.

The Second Law of Thermodynamics allows frictional dissipation of work into heat at *any temperature* [1–7] (in Ref. [6], see pp. 11–12, 60–65, and 263–265, especially pp. 263–265). The entropy increase resulting from frictional dissipation of work W at temperature T, namely, $\Delta S = W/T$, decreases monotonically with increasing T but is positive for any finite T—and the Second Law requires only that $\Delta S \geq 0$ [1–7]. The diminution of $\Delta S = W/T$ at higher T is, ultimately, what yields increased efficiency via HTR, within the restriction of frictional dissipation of W being unavoidable. The diminution of $\Delta S = W/T$ via frictional dissipation of W at T_H as opposed to at $T_C = R_T T_H$ is

$$\text{diminution of } \Delta S = \frac{W}{T_C} - \frac{W}{T_H} = W\left(\frac{1}{T_C} - \frac{1}{T_H}\right) = W\left(\frac{1}{R_T T_H} - \frac{1}{T_H}\right) = \frac{W}{T_H}\left(R_T^{-1} - 1\right). \tag{20}$$

And the corresponding saving of exergy or free energy is

$$\Delta X = T_C \times (\text{diminution of } \Delta S) = T_C W\left(\frac{1}{T_C} - \frac{1}{T_H}\right) = W\left(1 - \frac{T_C}{T_H}\right) = W(1 - R_T). \tag{21}$$

Note that in the limit $R_T \to 1$, both (diminution of ΔS) $\to 0$ and $\Delta X \to 0$, while in the limit $R_T \to 0$, (diminution of ΔS) $\to \infty$ and $\Delta X \to W$.

Consider work W from *any* source, heat engine or otherwise, frictionally dissipated into heat at temperature T_H [6]. The temperature T_H could be generated via the friction itself [6]. Thus frictional dissipation of work W from *any* source, heat engine or otherwise, could, in at least in principle, generate an arbitrarily high temperature T_H [6]. In the limit $T_H \to \infty$ (with T_C fixed) or equivalently in the limit $R_T \to 0$, not only the standard (non-HTR) Carnot efficiency $\epsilon_{\text{Carnot,std}}$ but even the standard (non-HTR) Curzon-Ahlborn efficiency $\epsilon_{\text{CA,std}}$ approaches unity, and entropy production even given the standard (non-HTR) Curzon-Ahlborn efficiency $\epsilon_{\text{CA,std}}$ approaches zero. And both HTR efficiencies, $\epsilon_{\text{Carnot,HTR}}$ and $\epsilon_{\text{CA,HTR}}$, approach infinity ($\epsilon_{\text{Carnot,HTR}}$ approaching infinity at higher order), with entropy production corresponding to both approaching zero (that corresponding to $\epsilon_{\text{Carnot,HTR}}$ approaching zero at higher order). Thus in the limit $R_T \to 0$, work frictionally dissipated into heat can completely be reconverted back into work by a heat engine. When this work is, in turn, frictionally dissipated, the process can be repeated over and over again—indefinitely in the limit $R_T \to 0$. We emphasize again: no energy is created (or destroyed)—energy is merely recycled—hence the First Law of Thermodynamics is not violated [6]. No decrease in entropy occurs—$\Delta S = W/T_H > 0$ for any finite T_H—hence the Second Law of Thermodynamics is not violated [6]: as per Eqs. (20) and (21), ΔS is diminished but still remains positive. To re-emphasize, it is the *diminution* of $\Delta S = W/T$ at higher T: $W/T_H < W/T_C$ (notwithstanding that ΔS still remains positive: $W/T_H > 0$) that yields increased efficiency via HTR, within the restriction of frictional dissipation of W being unavoidable.

Consider the following thought experiment. If an automobile travels at constant speed, the work output of its engine is immediately and continually frictionally dissipated, but the work was done and the efficiency was W/Q_H, not zero. In the operation of automobiles at constant speed, W is immediately and continually frictionally dissipated to ambient (the cold reservoir), and hence the consequent entropy increase is $\Delta S = W/T_C$. (Of course, there are in practice other entropy increases accompanying the operation of automobiles, e.g., owing to irreversible heat flows engendered by finite temperature differences.) But if W could instead be frictionally dissipated into the cylinders of an automobile's engine during power strokes (of course this is impracticable), the entropy increase would be diminished from $\Delta S = W/T_C$ to $\Delta S = W/T_H$, and the required heat input would be reduced from Q_H to $Q_H - W = Q_C$. Hence via HTR the efficiency would be increased from W/Q_H to $W/(Q_H - W) = W/Q_C$, and thus also the required fuel consumption would be decreased by the ratio Q_C/Q_H. Even though this is obviously impracticable, given that $\Delta S = W/T_H > 0$ the Second Law allows it, so we can at least do it as a thought experiment.

If a heat engine's work output is frictionally dissipated into its hot reservoir, the net heat input required from the hot reservoir is reduced from Q_H to $Q_H - W$, and hence via HTR the engine's efficiency is increased from W/Q_H to $W/(Q_H - W) = W/Q_C$, which can indeed exceed the Carnot limit—even though the efficiency W/Q_H of the *initial* production of work must be within the Carnot limit. If the temperature T_C of the cold reservoir is only a small fraction of the the temperature T_H of the hot reservoir, W/Q_H can be almost as large as unity or equivalently W can be almost as large as Q_H, and hence $W/(Q_H - W) = W/Q_C$ can *greatly* exceed the Carnot limit.

We note that the temperature of the cosmic background radiation is only 2.7 K, while the most refractory materials remain solid at temperatures slightly exceeding 2700 K. This provides a temperature ratio of $R_T \equiv T_C/T_H \approx 10^{-3}$. Could even smaller values of $R_T \equiv T_C/T_H$ be possible, at least in principle? Perhaps, maybe, if frictional dissipation of work into heat might somehow be possible into a gaseous hot reservoir at temperatures exceeding the melting point or even the critical temperature (the maximum boiling point at any pressure) of even the most refractory material.

While in this chapter we do not challenge the Second Law, we do challenge an *over*statement of the Second Law that is sometimes made: that energy can do work only once. This *over*statement is *false*. Energy can indeed do work more than once, in principle up to an infinite number of times, and even in practice many more times than merely once, before its ability to do work is totally dissipated. Consider these three examples: (i) Energy can do work in an infinite number of times in perfect (reversible) regenerative braking of an electrically-powered motor vehicle, with the motor operating backward as a generator during braking. Even with real-world less-than-perfect (less than completely reversible) regenerative braking, energy can do work many more times than merely once before its ability to do work is totally dissipated. (ii) Energy can do work in an infinite number of times in perfect (reversible) HTR (in the limit $R_T \to 0$). Even with real-world less-than-perfect (less than completely reversible) HTR (finite but small $R_T > 0$), energy can do work many more times than merely once before its ability to do work is totally dissipated. (iii) Energy can do work in an infinite number of times in perfect (reversible) thermal recharge of intermediate heat reservoirs—not to be confused with HTR discussed in this present chapter—see Section VI of Ref. [35] and the improved treatment in another chapter [61] in this book. Even with real-world less-than-perfect (less than completely reversible) thermal recharge of intermediate heat reservoirs, energy can do work more times than merely once before its ability to do work is totally dissipated.

6. Reply to criticisms of HTR

The concept of HTR (without being dubbed HTR) was criticized by Makarieva, Gorshkov, Li, and Nobre [58] and by Bejan [59], as being in conflict with the First and Second Laws of Thermodynamics, especially with the Second Law (see especially Sections 4 and 5 of Ref. [58] and Section 4 of Ref. [59]). These criticisms are addressed directly in Ref. [60]. They are also addressed in works concerning (a) HTR in hurricanes [27–37, 62] and (b) the *experimental verification* of HTR in increasing efficiency of thermoelectric generators [38]. [Concerning (a) immediately above, on the one hand, the importance of HTR (dubbed as "dissipative heating") has been confirmed in a study of Hurricane Andrew (1992) [36], and, as one might expect, "dissipative heating appears to be a more important process in intense hurricanes, such as Andrew, than weak ones" [36]. But, on the other hand, more recently it has been contended [37] that, while HTR exists in hurricanes, it is of lesser importance than previously supposed [27–36].]

Perhaps the simplest and most straightforward reply to these criticisms [58, 59] is that provided by Spanner (see Ref. [6], pp. 11–12, 60–65, and 263–265, especially pp. 263–265): Friction resulting from dissipation of work can in principle generate arbitrarily high temperature T_H without violating the Second Law of Thermodynamics: The entropy increase resulting from frictional dissipation of work W at temperature T_H, namely, $\Delta S = W/T_H$, decreases monotonically with increasing T_H but is positive for any finite T_H—and the Second Law requires only that $\Delta S \geq 0$ [6]. A heat engine operating between this high temperature T_H and a low (cold-reservoir) temperature T_C arbitrarily close to absolute zero (0 K) can in principle recover essentially all of the frictional dissipation as work [6]—and the recycling of energy from work to heat via frictional dissipation and then back to work via the heat engine can in principle then be repeated essentially indefinitely [6]. No energy is created (or destroyed)—energy is merely recycled—hence the First Law of Thermodynamics is not violated [6]. No decrease in entropy occurs—$\Delta S = W/T_H > 0$ for any finite T_H—hence the Second Law of Thermodynamics is not violated [6].

As has been previously emphasized [35], it is only recycling of a heat engine's *waste heat* Q_C into its hot reservoir at T_H instead of rejection thereof into its cold reservoir at T_C—*not* recycling of heat generated by frictional dissipation of its *work output* W back into its hot reservoir at T_H—that would violate the Second Law of Thermodynamics. Recharging W to the hot reservoir does *not* violate the Second Law, because the entropy change $\Delta S = W/T_H$ is positive—albeit less strongly positive than $\Delta S = W/T_C$ that obtains if W is frictionally dissipated into the cold reservoir. Only recharging Q_C to the hot reservoir would violate the Second Law, because the entropy change $\Delta S = Q_C/T_H - Q_C/T_C$ would be negative. And recycling of a heat engine's *waste heat* Q_C into its hot reservoir at T_H instead of its rejection into its cold reservoir at T_C has *never* been claimed [27–38, 60, 62].

There is one caveat: the entropy increase $\Delta S = W/T_H > 0$ owing to frictional dissipation of W at T_H could in principle be employed to pay for pumping a heat engine's waste heat Q_C from T_C to T_H, but no capability to do work would be gained by this procedure. For, even if this procedure could be executed perfectly (reversibly), e.g., via a perfect (reversible) heat pump, we would have [applying Eqs. (1) and (5)]

$$\Delta S_{\text{total}} = \frac{W}{T_H} - \frac{Q_C}{T_C} + \frac{Q_C}{T_H} = \frac{W}{T_H} - Q_C\left(\frac{1}{T_C} - \frac{1}{T_H}\right) = \frac{W}{T_H} - Q_C\frac{T_H - T_C}{T_C T_H} = 0$$

$$\Rightarrow W = Q_C\frac{T_H - T_C}{T_C} = Q_C\left(\frac{T_H}{T_C} - 1\right) = Q_H\frac{T_C}{T_H}\left(\frac{T_H}{T_C} - 1\right) \tag{22}$$

$$= Q_H\left(1 - \frac{T_C}{T_H}\right) = Q_H(1 - R_T) = Q_H\epsilon_{\text{Carnot,std}}.$$

What Eq. (22) brings to light is that the operation of the heat pump, even if perfect (reversible), results merely in the recovery of W. But W is recoverable more simply by avoiding this unnecessary procedure, as per Section 5 and the first three paragraphs of this Section 6.

7. Conclusion

We provided introductory remarks, an overview, and general considerations in Section 1. A misconception pertaining to the efficiencies of engines (heat engines or otherwise) was discussed and corrected in Section 2. Then we discussed the work outputs, efficiencies, and entropy productions of Carnot (reversible) and Curzon-Ahlborn (endoreversible) heat engines. In Section 3, we reviewed the standard (without HTR) work outputs, efficiencies, and entropy productions of Carnot (reversible) and Curzon-Ahlborn (endoreversible) heat engines, first without frictional dissipation of heat engines' work outputs and then with frictional dissipation thereof into their cold reservoirs. In Section 4 we considered them with frictional dissipation of heat engines' work outputs into their hot reservoirs (with HTR). (If HTR is employed, we construe the terms "the highest practicable temperature for HTR" and "the hot reservoir" to be synonymous.) We showed that the efficiencies of both Carnot and Curzon-Ahlborn engines can be increased, indeed in some cases greatly increased, via employing HTR. The increases in efficiencies via employing HTR are minimal in the limit $R_T \equiv T_C/T_H \to 1$ but become arbitrarily large in the limit $R_T \equiv T_C/T_H \to 0$. Efficiencies via employing HTR can exceed unity and can even approach ∞.

We provided recapitulation, as well as generalization, in Section 5. We replied to criticisms [58, 59] of HTR in Section 6.

As we have already noted in Section 1, the increases in efficiency attainable via HTR are not practicable if frictional dissipation of work into other than the cold reservoir is not practicable. Thus they are *never* practicable for noncyclic (necessarily one-time, single-use) heat engines: however the work output of a noncyclic (necessarily one-time, single-use) heat engine might be frictionally dissipated, the heat thereby generated can*not* restore the engine to its initial state. Moreover in many cases the work outputs of noncyclic (necessarily one-time, single-use) heat engines are not frictionally dissipated *at all*, at least not during practicable time scales, for example, a noncyclic rocket heat engine's work output is sequestered essentially permanently as kinetic and gravitational potential energy in the launching of a spacecraft (but typically most of the kinetic energy accelerates the exhaust gases, not the payload). They also are *never* practicable for reverse operation of cyclic heat engines as refrigerators or heat pumps, because for both refrigerators and heat pumps, the *total* energy input (the work W, plus the heat Q_C extracted from a cold reservoir at the expense of W as required by the Second Law of Thermodynamics) *always* is deposited as heat Q_H into a hot reservoir ($Q_H = Q_C + W$): thus there is *never* any *additional* energy to be deposited into the hot reservoir (as there is from frictional dissipation of work done via forward operation of cyclic heat engines). {See Ref. [1], Section 20-3; Ref. [2], Sections 4.3, 4.4, and 4.7 (especially Section 4.7); Ref. [3], Sections 4-4, 4-5, and 4-6 (especially Section 4-6); Ref. [5], Section 5.12 and Problem 5.22; Ref. [7], pp. 233–236 and Problems 1, 2, 4, 6, and 7 of Chapter 8; Ref. [16], Chapter XXI; Ref. [17], Sections 6.7, 6.8, 7.3, and 7.4; and Ref. [54], Sections 5-7-2, 6-2-2, 6-9-2, and 6-9-3, and Chapter 17. [Problem 2 of Chapter 8 in Ref. [7] considers absorption refrigeration, wherein the entire energy output is into an *intermediate*-temperature (most typically ambient-temperature) reservoir, and hence for which HTR is *even more strongly never practicable.*]} They also are not practicable for cyclic heat engines in cases wherein a cyclic heat engine's work output is not frictionally dissipated immediately or on short time scales [16, 17], for example, as gravitational potential energy sequestered for a long time interval in the construction of a building. For a building once erected typically remains standing for a century or longer. Even if, when it is finally torn down, its gravitational potential energy were to be totally frictionally dissipated into a hot reservoir, it is simply impracticable to wait that long. Thus HTR is not practicable in all cases. But in the many cases wherein cyclic heat engines' work outputs are frictionally dissipated immediately or on short time scales [16, 17], practicability obtains: improved conversion—and reconversion—of frictionally dissipated heat into work, and hence improved cyclic heat-engine performance, can then obtain.

We emphasize yet again that First and Second Laws of Thermodynamics are *not* violated. The First Law is not violated because no new energy is created (or destroyed): super-unity efficiencies via employment of HTR obtain via recycling and reusing the same energy, not via the creation of new energy. The Second Law is not violated because the change in total entropy is positive if HTR is employed and frictional dissipation of work as heat is into the hot reservoir, albeit less strongly positive than if HTR is not employed and frictional dissipation of work as heat is into the cold reservoir. The improved heat-engine performance that HTR provides ultimately obtains from this reduction of entropy increase.

While in this chapter we do not challenge the First or Second Laws of Thermodynamics, we should note that there have been many challenges to the Second Law, especially in recent years [64–69]. By contrast, the First Law has been questioned only in cosmological contexts [70–72] and with respect to fleeting violations thereof associated with the energy-time uncertainty principle [73, 74]. But there are contrasting viewpoints [73, 74] concerning the latter issue.

Acknowledgements

I am very grateful to Dr. Donald H. Kobe, Dr. Paolo Grigolini, Dr. Daniel P. Sheehan, Dr. Bruce N. Miller, and Dr. Marlan O. Scully and for many very helpful and thoughtful insights, as well as for very perceptive and valuable discussions and communications, that greatly helped my understanding of thermodynamics and statistical mechanics. Also, I am indebted to them, as well as to Dr. Bright Lowry, Dr. John Banewicz, Dr. Bruno J. Zwolinski, Dr. Roland E. Allen, Dr. Abraham Clearfield, Dr. Russell Larsen, Dr. James H. Cooke, Dr. Wolfgang Rindler, Dr. Richard McFee, Dr. Nolan Massey, and Dr. Stan Czamanski for lectures, discussions, and/or communications from which I learned very much concerning thermodynamics and statistical mechanics. I thank Dr. Stan Czamanski and Dr. S. Mort Zimmerman for very interesting general scientific discussions over many years. I also thank Dan Zimmerman, Dr. Kurt W. Hess, and Robert H. Shelton for very interesting general scientific discussions at times. Additionally, I thank Robert H. Shelton for very helpful advice concerning diction.

Conflicts of interest

The author declares no conflicts of interest.

Author details

Jack Denur
Electric & Gas Technology, Inc., Rowlett, Texas, USA

*Address all correspondence to: jackdenur@my.unt.edu

References

[1] Walker J, Halliday D, Resnick R. Fundamentals of Physics. 11th (Extended) ed. Hoboken, NJ: John Wiley & Sons; 2018, Chapters 18 and 20 (especially Sections 20-2 and 20-3)

[2] Callen HB. Thermodynamics. New York, NY: John Wiley & Sons; 1960, Chapter 4

[3] Callen HB. Thermodynamics and an Introduction to Thermostatistics. 2nd ed. New York, NY: John Wiley & Sons; 1985, Chapter 4

[4] Baierlein R. Thermal Physics. Cambridge, UK: Cambridge University Press; 1999, Chapters 2 and 3

[5] Reif F. Fundamentals of Statistical and Thermal Physics. New York, NY: McGraw-Hill; 1965 (reissued: Long Grove, Illinois: Waveland Press; 2009), Sections 5.11 and 5.12, and Problems 5.22 through 5.26

[6] Spanner DC. Introduction to Thermodynamics. London, UK: Academic Press; 1964, Chapter 6, pp. 213–215, and pp. 263–265

[7] Kittel C, Kroemer H. Thermal Physics. 2nd ed. San Francisco, CA: W. H. Freeman and Company; 1980, Chapter 8

[8] Friction and Friction Coefficients [Online]. Available from: https://www.engineeringtoolbox.com/ [Accessed: 16 March 2020]

[9] Rolling Resistance [Online]. Available from: https://www.engineeringtoolbox.com/ [Accessed: 16 March 2020]

[10] Friction [Online]. Available from: https://www.wikipedia.org/ [Accessed: 16 March 2020]

[11] Rolling Resistance [Online]. Available from: https://www.wikipedia.org/ [Accessed: 16 March 2020]

[12] Tolman RC. Relativity, Thermodynamics, and Cosmology. Oxford, UK: Oxford University Press; 1934 (unabridged and unaltered republication: New York, NY: Dover; 1987), Sections 130 and 169–173

[13] Curzon FL, Ahlborn B. Efficiency of a Carnot engine at maximum power output. American Journal of Physics. 1975;**43**:22-24. DOI: 10.1119/1.10023

[14] Vaudrey A, Lanzetta F, Feidt M. Reitlinger and the origins of the efficiency at maximum power output for heat engines. Journal of Non-Equilibrium Thermodynamics. 2014;**39**: 199-203. DOI: 10.1515/jnet-2014-0018

[15] Endoreversible Thermodynamics [Online]. Available from: https://www.wikipedia.org/ [Accessed: 16 March 2020]

[16] Faries VM. Applied Thermodynamics. Revised Ed. New York, NY: Macmillan; 1949, Chapter VI (especially Sections 54, 60, and 61)

[17] Zemansky MW, Dittman RH. Adapted by Chattopadhyay AK. Heat and Thermodynamics. 8th ed. Chennai, India: McGraw Hill Education (India); 2011 (Seventeenth reprint 2018), Sections 6.9–6.14 and 16.8

[18] De Vos A. Efficiency of some heat engines at maximum-power conditions. American Journal of Physics. 1985;**53**: 570-573. DOI: 10.1119/1.14240

[19] Gordon JM. Maximum power-point characteristics of heat engines as a general thermodynamic problem. American Journal of Physics. 1989;**57**: 1136-1142. DOI: 10.1119/1.16130

[20] Gordon JM. Observations on efficiency of heat engines operating at maximum power. American Journal of Physics. 1990;**58**:370-375. DOI: 10.1119/1.16175

[21] Schmiedl T, Seifert U. Efficiency at maximum power: An analytically solvable model for stochastic heat engines. Europhysics Letters. 2008;**81**:20003. DOI: 10.1209/0295-5075/81/20003

[22] Tu ZC. Efficiency at maximum power of Feynman's ratchet as an engine. Journal of Physics A. 2008;**41**: 312003. DOI: 10.1088/1751-8113/41/31/312003

[23] Leff HS. Thermal efficiency at maximum work output: New results for old heat engines. American Journal of Physics. 1987;**55**:602-610. DOI: 10.1119/1.15071

[24] Ouerdane H, Apertet Y, Goupil C, Lecoeur P. Continuity and boundary conditions in thermodynamics: From Carnot's efficiency to efficiencies at maximum power. European Physical Journal - Special Topics. 2015;**224**: 839-862. DOI: 10.1140/epjst/e2015-02431-x

[25] Parrondo JMR, Ouerdane H, et. al. Debate. Continuity and boundary conditions in thermodynamics: From Carnot's efficiency to efficiencies at maximum power. European Physical Journal - Special Topics. 2015;**224**: 862-864

[26] Apertet Y, Ouerdane H, Goupil C, Lecoeur P. True nature of the Curzon-Ahlborn efficiency. Physical Review E. 2017;**96**:022119. DOI: 10.1103/PhysRevE.96.022119

[27] Emanuel K. Divine Wind. Oxford, UK: Oxford University Press; 2005 (especially Chapter 10)

[28] Emanuel K. Hurricanes: Tempests in a greenhouse. Physics Today. 2006; **59**(8):74-75. DOI: 10.1063/1.2349743

[29] Emanuel K. Tropical cyclones. Annual Review of Earth and Planetary Sciences. 2003;**31**:75-104. DOI: 10.1146/annurev.earth.31.100901.141259

[30] Emanuel K. Thermodynamic control of hurricane intensity. Nature. 1999;**401**:665-669. DOI: 10.1038./44326

[31] Bister M, Emanuel KA. Dissipative heating and hurricane intensity. Meteorology and Atmospheric Physics. 1998;**65**:223-230

[32] Emanuel K. Response of tropical cyclone activity to climate change: Theoretical basis. In: Murmane RJ, Liu K-B, editors. Hurricanes and Typhoons: Past, Present, and Future. New York, NY: Columbia University Press; 2004, pp. 395–407

[33] Emanuel KA, Speer K, Rotunno R, Srivastava R, Molina M. Hypercanes: A possible link in global extinction scenarios. Journal of Geophysical Research: Atmospheres. 1995;**100**: 13755-13765. DOI: 10.1029/95JD01368

[34] Emanuel K, Callagham J, Otto PA. A hypothesis for redevelopment of warm-core cyclones over northern Australia. Monthly Weather Review. 2008;**136**: 3863-3872. DOI: 10.1175/2008MWR2409.1

[35] Denur J. The apparent " super-Carnot" efficiency of hurricanes: Nature's steam engine versus the steam locomotive. American Journal of Physics. 2011;**79**:631-643. DOI: 10.1119/1.3534841

[36] Zhang D-L, Altshuler E. The effects of dissipative heating on hurricane intensity. Monthly Weather Review. 1999;**127**:3032-3038

[37] Kieu C. Revisiting dissipative heating in tropical cyclone maximum potential intensity. Quarterly Journal of the Royal Meteorological Society. 2015; **141**:2497-2504. DOI: 10.1002/qj.2534

[38] Apertet Y, Ouerdane H, Goupil C, Lecoeur P. Efficiency at maximum power of thermally coupled heat engines. Physical Review E. 2012;**85**. DOI: 041144, 10.1103/PhysRevE. 85041144

[39] Bachhuber C. Energy from the evaporation of water. American Journal of Physics. 1983;**51**:259-264. DOI: 10.1119/1.13297

[40] Güémez J, Valiente B, Fiolhais C, Fiolhais M. Experiments with the drinking bird. American Journal of Physics. 2003;**71**:1257-1264. DOI: 10.1119/1.1603272

[41] Abraham N, Paliffy-Muhoray P. A drinking bird of the second kind. American Journal of Physics. 2004;**72**: 782-785. DOI: 10.1119/1.1703543

[42] Lorenz R. Finite-time thermodynamics of an instrumented drinking-bird toy. American Journal of Physics. 2006;**74**:677-682. DOI: 10.1119/ 1.2190688

[43] Murrow RB. A Simple Heat Engine of Possible Utility in Primitive Environments [Online]. 1966. Available from: https://www.rand.org/content/dam/rand/pubs/papers/2008/P3367.pdf [Accessed: 16 March 2020]

[44] Lear J. New Waterbird for Egypt: A robot Shadoof. Saturday Review. 1967; **34**:49-50

[45] Murrow RB. The research frontier. Saturday Review. 1967;**34**:51-53

[46] Craig RJ. Modeling of a Thermodynamically Driven Heat Engine with Application Intended for Water Pumping. Master's Thesis. Stellenbosch, South Africa: Stellenbosch University; December 2014

[47] Chen X, Goodnight D, Gao Z, Cavusoglu AH, Sabharwal N, DeLay M, Driks A, Sahin O. Scaling up nanoscale water-driven energy conversion into evaporation-driven engines and generators. Nature Communications. 16 June 2015; 6: Article Number 7346, 7 pages. DOI: 10.1038/ncomms8346. Summaries of the work discussed in this article are provided in References [48] and [49]

[48] Temming M. Water, water everywhere. Scientific American. September 2015;**313**(3):26

[49] Ornes S. Spore Power. Discover. April 2016; 37(3): 14. Available from: www.DiscoverMagazine.com/HYDRA provided in this article, which redirects to http://blogs.discovermagazine.com/d-brief/2015/06/16/evaporation-powered-engine/ [the latter website can also be accessed directly, both websites: [Accessed: 16 March 2020]

[50] Uechi ST, Uechi H. The Analysis of Thermomechanical Periodic Motions of a Drinking Bird [Online]. 2019. Available from: https://www.researchgate.net/publication/333049178 [Accessed: 16 March 2020]

[51] Drinking Bird [Online]. 2019. Available from: https://www.wikipedia.org/ [Accessed: 16 March 2020]

[52] Liquid Nitrogen Engine [Online]. 2019. Available from: https://www.wikipedia.org/ [Accessed: 16 March 2020]

[53] Takahashi P, Trenka A. Ocean Thermal Energy Conversion. Hoboken, NJ: John Wiley & Sons; 1996

[54] Wark K, Richards DE. Thermodynamics, 6th ed. Boston, MA: WCB/McGraw-Hill; 1999, Section 16-6-3, and Problems 16-116 and 16-117E

[55] Etemadi A, Emdadi A, AsefAfshar O, Emami Y. Electricity generation by ocean thermal energy. Energy Procedia. 2011;**12**:936-943

[56] Ocean Thermal Energy Conversion [Online]. Available online: https://www.wikipedia.org/ [Accessed: 16 March 2020]

[57] Temming M. Nighttime panel makes electricity. Science News. 2019; **196**(7):20

[58] Makarieva AM, Gorshkov VC, Li B-L, Nobre AD. A critique of some modern applications of the Carnot heat engine concept: The dissipative engine cannot exist. Proceedings of the Royal Society A. 2010;**466**:1893-1902. DOI: 10.1098/rspa.2009.0581

[59] Bejan A. Thermodynamics of heating. Proceedings of the Royal Society A. 2019;**475**:20180820. DOI: 10.1098/rspa.2018.0820

[60] Bister M, Renno N, Pauluis O, Emanuel K. Comment on Makarieva *et. al.* 'A critique of some modern applications of the Carnot heat engine concept: The dissipative engine cannot exist'. Proceedings of the Royal Society A. 2011;**467**:1-6. DOI: 10.1098/rspa.2010.0087

[61] Denur J. Heat-engine operation employing multiple heat reservoirs. In: Petrică Vizureanu, Academic Editor. Applied Thermodynamics and Energy Engineering. London, UK: IntechOpen; 2019

[62] Ozawa H, Shimokawa S. Thermodynamics of a tropical cyclone: Generation and dissipation of mechanical energy in a self-driven convection system. Tellus A. 2015;**67**: 24216. DOI: 10.3402/tellusa.v67.24216

[63] Baierlein R. Atoms and Information Theory. San Francisco, CA: W. H. Freeman; 1971, Sections 7.5, 7.6, 10.5, 11.4, and 11.5 (especially Section 7.5 and pp. 400–403)

[64] Sheehan DP, editor. Quantum Limits to the Second Law, AIP Conference Proceedings Volume 643. Melville, NY: American Institute of Physics; 2002

[65] Nikulov AV, Sheehan DP, editors. Special issue: Quantum limits to the second law of Thermodynamics. Entropy March 2004;**6**(1)

[66] Čápek V, Sheehan DP. Challenges to the Second Law of Thermodynamics: Theory and Experiment. Dordrecht, The Netherlands: Springer; 2005

[67] Sheehan DP, editor. Special issue: The second law of thermodynamics: Foundations and status. Foundations of Physics. December 2007;**37**(12)

[68] Sheehan DP, editor. Second Law of Thermodynamics: Status and Challenges, AIP Conference Proceedings Volume 1411. Melville, NY: American Institute of Physics; 2011

[69] Sheehan DP, editor. Special issue: Limits to the second law of thermodynamics: Experiment and theory. Entropy 2017;**19**

[70] Harrison ER. Mining energy in an expanding universe. The Astrophysical Journal. 1995;**446**:63-66

[71] Sheehan DP, Kriss VG. Energy Emission by Quantum Systems in an Expanding FRW Metric [Online]. Available from: arXiv:astro-ph/0411299v1 [Accessed: 16 March 2020]

[72] Parry R. Extracting Energy from the Expanding Universe: Can we Avoid the Heat Death? Honours Physics 2015. Sydney, Australia: Sydney Institute for Astronomy, School of Physics, The University of Sydney; 2015

[73] Griffiths DJ, Schroeter DF. Introduction to Quantum Mechanics. 3rd ed. Cambridge, UK: Cambridge University Press; 2018, Section 3.5.3

[74] Hagmann MJ. Distribution of times for barrier traversal caused by energy fluctuations. Journal of Applied Physics. 1993;**74**:7302-7305

Chapter 7

Improving Heat-Engine Performance by Employing Multiple Heat Reservoirs

Jack Denur

Abstract

The efficiencies of heat-engine operation employing various numbers (≥ 2) of heat reservoirs are investigated. Operation with the work output of the heat engines sequestered, as well as with it being totally frictionally dissipated, is discussed. We consider mainly heat engines whose efficiencies depend on ratios of a higher and lower temperature or on simple functions of such ratios but also provide brief comments concerning more general cases. We show that, if a hot reservoir supplies a heat engine whose waste heat is discharged *and* whose work output is totally frictionally dissipated into a cooler reservoir, which in turn supplies heat-engine operation that discharges waste heat into a still cooler reservoir, the total work output can exceed the heat input from the initial hot reservoir. This extra work output increases with increasing numbers (≥ 3) of reservoirs. We also show that this obtains within the restrictions of the First and Second Laws of Thermodynamics.

Keywords: First Law of Thermodynamics, Second Law of Thermodynamics, heat engines, work, heat, entropy, multiple heat reservoirs

1. Introduction

The efficiencies of heat-engine operation employing various numbers (≥ 2) of heat reservoirs are investigated. In Section 2, we discuss heat-engine operation with the work output of the heat engines sequestered. In Section 3, we discuss heat-engine operation with the work output of the heat engines being totally frictionally dissipated. We consider mainly heat engines whose efficiencies depend on ratios of a higher and lower temperature or on simple functions of such ratios. Examples include heat engines operating not only via the Carnot cycle [1–9] but also via the Ericsson, Stirling, air-standard Otto, and air-standard Brayton cycles [2–9], and endoreversible heat engines operating at maximum power output assuming Curzon-Ahlborn efficiency [10–12] (see also Ref. [4], Section 4-9). But we also provide brief comments concerning more general cases. Endoreversible heat-engine operation assumes irreversible heat flows directly proportional to temperature differences but otherwise reversible operation [10–12]. Although we do not employ them in this chapter, we note that generalizations of the Curzon-Ahlborn efficiency, and also various related efficiencies, have also been investigated [13–21]. In particular, we note that alternative results [21] to the Curzon-Ahlborn efficiency [10–12] (see also Ref. [4], Section 4-9) have been derived [21]. But for definiteness and for simplicity, in this chapter, we employ the standard Curzon-Ahlborn efficiency [10–12] (see also Ref. [4], Section 4-9) for cyclic heat engines operating at maximum power output.

We show that, if a hot reservoir supplies a heat engine whose waste heat is discharged *and* whose work output is totally frictionally dissipated into a cooler reservoir, which in turn supplies heat-engine operation that discharges waste heat into a still cooler reservoir, the total work output can exceed the heat input from the initial hot reservoir. This extra work output increases with increasing numbers (≥ 3) of reservoirs. We also show that this obtains within the restrictions of the First and Second Laws of Thermodynamics.

We fill in details and correct a few mistakes in an earlier, briefer, consideration of the efficiencies of heat-engine operation employing various numbers (≥ 3) of heat reservoirs [22]. We note that heat-engine operation employing various numbers (≥ 3) of heat reservoirs [22] should not be confused with recycling heat engines' frictionally dissipated work outputs into the hottest available reservoir [22–37], which is a *different* process that has been thoroughly investigated and discussed previously [22–37], and which we further investigate in another chapter [38] in this book.

We consider only cyclic heat engines. Noncyclic (necessarily one-time, single-use) heat engines are not limited by the Carnot bound and can in principle operate at unit (100%) efficiency. A simple example is the one-time expansion of a gas pushing a piston. Other examples include rockets: the piston (payload) is launched into space by a one-time power stroke (but typically most of the work output accelerates the exhaust gases, not the payload) and firearms: the piston (bullet) is accelerated by a one-time power stroke and then discarded (but some, typically less than with rockets, of the work output accelerates the exhaust gases resulting from combustion of the propellant). Even if the work output of a noncyclic engine could be frictionally dissipated and the resulting heat returned to the system, there would be, at best, restoration of temperature to its initial value but not restoration of the piston to its initial position. Hence the method investigated in this chapter is useless with respect to noncyclic heat engines.

General remarks, especially concerning entropy, are provided in Section 4. Concluding remarks are provided in Section 5.

2. Multiple-reservoir heat-engine efficiencies with work output sequestered

We designate the temperatures of the heat reservoirs via subscripts, with T_1 being the temperature of the initial, hottest, reservoir, T_2 the temperature of the second-hottest reservoir, T_3 the temperature of the third-hottest reservoir, etc., and T_n the temperature of the nth, coldest, reservoir.

Let a heat engine operate between two reservoirs, extracting heat Q_1 from a hot reservoir at temperature T_1 and rejecting waste heat to a cold reservoir at temperature T_2. If its efficiency is $\epsilon_{1\to2}$, its work output is

$$W_{1\to2} = Q_1\epsilon_{1\to2}. \tag{1}$$

It rejects waste heat $Q_1 - W_{1\to2} = Q_1(1 - \epsilon_{1\to2})$ to the reservoir at temperature T_2. If there is a third reservoir at temperature T_3 and $W_{1\to2}$ is sequestered, that is, not frictionally dissipated, and if the efficiency of heat-engine operation between the second and third reservoirs is $\epsilon_{2\to3}$, a heat engine can then perform additional work

$$W_{2\to3} = Q_1(1 - \epsilon_{1\to2})\epsilon_{2\to3} \tag{2}$$

by employing the reservoir at temperature T_2 as a hot reservoir and the reservoir at temperature T_3 as a cold reservoir. All told it can do work:

$$W_{1\to2} + W_{2\to3} = Q_1\epsilon_{1\to2} + Q_1(1-\epsilon_{1\to2})\epsilon_{2\to3} \\ = Q_1(\epsilon_{1\to2} + \epsilon_{2\to3} - \epsilon_{1\to2}\epsilon_{2\to3}). \tag{3}$$

By contrast, if the heat engine operates in a single step at efficiency $\epsilon_{1\to3}$, employing the reservoir at temperature T_1 as a hot reservoir and the reservoir at temperature T_3 as a cold reservoir, it can do work

$$W_{1\to3} = Q_1\epsilon_{1\to3}. \tag{4}$$

Anticipating that we will eventually deal with n heat reservoirs, let us consider efficiencies of the form

$$\epsilon_{i\to j} = 1 - \left(\frac{T_i}{T_j}\right)^x, \tag{5}$$

where i and j are positive integers in the respective ranges $1 \le i \le n-1$ and $i < j \le n$ and where x is a positive real number in the range $0 < x \le 1$. Applying Eqs. (3) and (5), $W_{1\to3} = W_{1\to2} + W_{2\to3}$, as we will now show. We have

$$\begin{aligned} W_{1\to2} + W_{2\to3} &= Q_1\left\{\begin{matrix} \left[1-\left(\frac{T_2}{T_1}\right)^x\right] + \left[1-\left(\frac{T_3}{T_2}\right)^x\right] \\ -\left[1-\left(\frac{T_2}{T_1}\right)^x\right]\left[1-\left(\frac{T_3}{T_2}\right)^x\right] \end{matrix}\right\} \\ &= Q_1\left\{\begin{matrix} 2-\left(\frac{T_2}{T_1}\right)^x - \left(\frac{T_3}{T_2}\right)^x \\ -\left[1-\left(\frac{T_2}{T_1}\right)^x - \left(\frac{T_3}{T_2}\right)^x + \left(\frac{T_2}{T_1}\right)^x\left(\frac{T_3}{T_2}\right)^x\right] \end{matrix}\right\} \\ &= Q_1\left[1-\left(\frac{T_2}{T_1}\right)^x\left(\frac{T_3}{T_2}\right)^x\right] \\ &= Q_1\left[1-\left(\frac{T_3}{T_1}\right)^x\right] = W_{1\to3}. \end{aligned} \tag{6}$$

We note that $x = 1$ for the Carnot, Ericsson, Stirling, air-standard Otto, and air-standard Brayton cycles [1–9] and $x = 1/2$ for endoreversible heat engines operating at Curzon-Ahlborn efficiency [10–12] (see also Ref. [4], Section 4-9). For all of these cycles, the temperature in the numerator is that of the coldest available reservoir for a given cycle [1–12]. For the Carnot, Ericsson, and Stirling cycles, and for endoreversible heat engines operating at Curzon-Ahlborn efficiency, the temperature in the denominator is that of the hottest available reservoir for a given cycle [1–12]. For the air-standard Otto and air-standard Brayton cycles, the temperature in the denominator is that at the end of the adiabatic-compression process but before the addition of heat from the hottest available reservoir (substituting, in air-standard cycles, for combustion of fuel) [2–9] in a given cycle. The Second Law of Thermodynamics forbids $x > 1$ if the temperature in the numerator is that of the coldest available reservoir for a given cycle and the temperature in the denominator is that of the hottest available reservoir for a given cycle, because then the Carnot efficiency would be exceeded. Since for the aforementioned heat engines, and indeed for any heat engine for which Eq. (5) is applicable, $W_{1\to3} = W_{1\to2} + W_{2\to3}$, this additivity of W obtains for any number of steps, that is, we have

$$W_{1\to n} = W_{1\to2} + W_{2\to3} + \ldots + W_{n-1\to n} = \sum_{j=1}^{n-1} W_{j\to j+1}. \tag{7}$$

For more complex efficiencies than those of Eq. (5), for example, those of the Diesel and dual cycles, which are functions of more than two temperatures, and also for some more complex efficiencies that are functions of two temperatures, the equality of Eq. (7) may not always obtain [3–9, 13–19]. But whether or not the equality of Eq. (7) obtains, the Second Law of Thermodynamics requires that, whichever reservoirs are employed, the efficiency with all work outputs sequestered, whether $W_{j\to j+1}/Q_j$ $(1 \leq j \leq n-1)$, $W_{j\to j+k}/Q_j$ $(1 \leq j \leq n-1$ and $1 \leq k \leq n-j)$, or $W_{1\to n}/Q_1$, cannot exceed the Carnot limit.

3. Multiple-reservoir heat-engine efficiencies with work output totally frictionally dissipated

Let a heat engine operate between two reservoirs, extracting heat Q_1 from a hot reservoir at temperature T_1 and rejecting waste heat to a cold reservoir at temperature T_2. If its efficiency is $\epsilon_{1\to 2}$, its work output is

$$W_{1\to 2} = Q_1 \epsilon_{1\to 2}. \tag{8}$$

It rejects waste heat $Q_1 - W_{1\to 2} = Q_1(1 - \epsilon_{1\to 2})$ to a reservoir at temperature T_2. But now, in addition, we let the work output $W^D_{1\to 2} = Q_1 \epsilon_{1\to 2}$ be totally frictionally dissipated and rejected into the reservoir at temperature T_2 (indicated via a superscript D). This is in fact by far the most common mode of heat-engine operation. With rare exceptions (e.g., a heat engine's work output being sequestered for a long time interval as gravitational potential energy in the construction of a building, or essentially permanently in the launching of a spacecraft), heat engines' work outputs are typically totally frictionally dissipated immediately or on short time scales (see Ref. [6], Chapter VI (especially Sections 54, 60, and 61); and Ref. [7], Sections 6.9–6.14 and 16.8). Indeed, this is true of almost all engines, heat engines or otherwise. The work outputs of all engines of vehicles (automobiles, trains, ships, submarines, aircraft, etc.) operating at constant speed, and of all factory and appliance engines operating at constant speed, are immediately and continually frictionally dissipated. The work output temporarily sequestered as kinetic energy when a vehicle accelerates, or when a factory or appliance engine is turned on, is frictionally dissipated a short time later when the vehicle decelerates, or when the factory or appliance engine is turned off.

If both the waste heat $Q_1 - W^D_{1\to 2} = Q_1(1 - \epsilon_{1\to 2})$ has been rejected and the work output $W^D_{1\to 2} = Q_1 \epsilon_{1\to 2}$ has been totally frictionally dissipated into the reservoir at temperature T_2, and there is a third reservoir at temperature T_3, a heat engine operating at efficiency $\epsilon_{2\to 3}$ can then perform additional work

$$W_{2\to 3} = Q_1 \epsilon_{2\to 3} \tag{9}$$

by employing the reservoir at temperature T_2 as a hot reservoir and the reservoir at temperature T_3 as a cold reservoir. ($W^D_{2\to 3}$ may or may not be frictionally dissipated, so it only optionally carries the superscript D.) All told the total work output is

$$\begin{aligned} W^D_{1\to 3} &= W^D_{1\to 2} + W^D_{2\to 3} = Q_1 \epsilon_{1\to 2} + Q_1 \epsilon_{2\to 3} \\ &= Q_1(\epsilon_{1\to 2} + \epsilon_{2\to 3}). \end{aligned} \tag{10}$$

If $\epsilon_{i\to j} = 1 - (T_i/T_j)^x$, where i and j are positive integers in the respective ranges $1 \leq i \leq n-1$ and $i < j \leq n$, and where x is a positive real number in the range $0 < x \leq 1$, applying Eqs. (5) and (10), we have:

$$W_{1\to3}^{D} = W_{1\to2}^{D} + W_{2\to3}^{D} = Q_1\left\{\left[1-\left(\frac{T_2}{T_1}\right)^x\right] + \left[1-\left(\frac{T_3}{T_2}\right)^x\right]\right\}$$
$$= Q_1\left[2-\left(\frac{T_2}{T_1}\right)^x - \left(\frac{T_3}{T_2}\right)^x\right]. \tag{11}$$

We now maximize $W_{1\to3}^{D}$ with respect to T_2:

$$\frac{dW_{1\to3}^{D}}{dT_2} = 0 \Rightarrow \frac{d}{dT_2}\left[2-\left(\frac{T_2}{T_1}\right)^x - \left(\frac{T_3}{T_2}\right)^x\right] = 0$$
$$\Rightarrow \frac{d}{dT_2}\left(\frac{T_2}{T_1}+\frac{T_3}{T_2}\right) = 0$$
$$\Rightarrow \frac{1}{T_1} - \frac{T_3}{T_2^2} = 0$$
$$\Rightarrow T_{2,\text{opt}} = (T_1T_3)^{1/2}. \tag{12}$$

Thus, the optimum value T_2,opt of T_2, which maximizes $W_{1\to3}^{D}$, is the geometric mean of T_1 and T_3. Applying Eqs. (11) and (12), the maximum value $W_{1\to3,\max}^{D}$ of $W_{1\to3}^{D}$ is

$$W_{1\to3,\max}^{D} = Q_1\left\{2-\left[\frac{(T_1T_3)^{1/2}}{T_1}\right]^x - \left[\frac{T_3}{(T_1T_3)^{1/2}}\right]^x\right\}$$
$$= Q_1\left[2-\left(\frac{T_3}{T_1}\right)^{x/2} - \left(\frac{T_3}{T_1}\right)^{x/2}\right]$$
$$= Q_1\left[2-2\left(\frac{T_3}{T_1}\right)^{x/2}\right]$$
$$= 2Q_1\left[1-\left(\frac{T_3}{T_1}\right)^{x/2}\right]. \tag{13}$$

Note that

$$W_{1\to3,\max}^{D} > Q_1 \text{ if } \left(\frac{T_3}{T_1}\right)^{x/2} < \frac{1}{2} \Leftrightarrow \frac{T_3}{T_1} < \frac{1}{2^{2/x}}. \tag{14}$$

This obtains if $T_3/T_1 < 1/4$ for $x = 1$ and if $T_3/T_1 < 1/16$ for $x = 1/2$. Also, comparing the last line of Eq. (6) with Eq. (13), we find for the maximum extra work $W_{1\to3,\max}^{D,\text{extra}}$:

$$W_{1\to3,\max}^{D,\text{extra}} = W_{1\to3,\max}^{D} - W_{1\to3}$$
$$= 2Q_1\left[1-\left(\frac{T_3}{T_1}\right)^{x/2}\right] - Q_1\left[1-\left(\frac{T_3}{T_1}\right)^x\right]$$
$$= Q_1\left\{2\left[1-\left(\frac{T_3}{T_1}\right)^{x/2}\right] - \left[1-\left(\frac{T_3}{T_1}\right)^x\right]\right\}$$
$$= Q_1\left[2-2\left(\frac{T_3}{T_1}\right)^{x/2} - 1 + \left(\frac{T_3}{T_1}\right)^x\right]$$
$$= Q_1\left[1+\left(\frac{T_3}{T_1}\right)^x - 2\left(\frac{T_3}{T_1}\right)^{x/2}\right] \geq 0. \tag{15}$$

It is easily shown that $W^{D,\text{extra}}_{1\to3,\max} \geq 0$, with the equality obtaining if and only if $\frac{T_3}{T_1} = 1 \Rightarrow W^{D}_{1\to3,\max} = W_{1\to3} = 0 \Rightarrow W^{D}_{1\to3,\max} - W_{1\to3} = W^{D,\text{extra}}_{1\to3,\max} = 0$. For, denoting the ratio $\left(\frac{T_3}{T_1}\right)^{x/2}$ as r and setting $dW^{D,\text{extra}}_{1\to3,\max}/dr = 0$ yields

$$\begin{aligned}\frac{dW^{D,\text{extra}}_{1\to3,\max}}{dr} &= 0 \Rightarrow \frac{d}{dr}\left(r^2 - 2r\right) = 0\\ &\Rightarrow 2r - 2 = 0\\ &\Rightarrow r = 1.\end{aligned} \tag{16}$$

Thus $W^{D}_{1\to3,\max}$ is minimized at 0 if $r = \left(\frac{T_3}{T_1}\right)^{x/2} = 1 \Rightarrow \frac{T_3}{T_1} = 1$. For all $\frac{T_3}{T_1} < 1,\ W^{D}_{1\to3,\text{extra}} > 0$. Moreover, applying Eqs. (5), (13), and (15), note that

$$\begin{aligned}&\lim_{T_3/T_1\to0} W^{D}_{1\to3,\max} = 2Q_1 = 2\lim_{T_3/T_1\to0} W_{1\to3}\\ \Rightarrow\ &\lim_{T_3/T_1\to0} W^{D,\text{extra}}_{1\to3,\max} = 2Q_1 - Q_1 = Q_1 = \lim_{T_3/T_1\to0} W_{1\to3}.\end{aligned} \tag{17}$$

Now consider heat-engine operation employing four heat reservoirs, with all work totally frictionally dissipated (except possibly at the last step; thus, $W^{D}_{3\to4}$ only optionally carries the superscript D). Thus we have

$$\begin{aligned}W^{D}_{1\to4} &= W^{D}_{1\to2} + W^{D}_{2\to3} + W^{D}_{3\to4} = Q_1\epsilon_{1\to2} + Q_1\epsilon_{2\to3} + Q_1\epsilon_{3\to4}\\ &= Q_1\left(\epsilon_{1\to2} + \epsilon_{2\to3} + \epsilon_{3\to4}\right).\end{aligned} \tag{18}$$

If $\epsilon_{i\to j} = 1 - \left(T_i/T_j\right)^x$, where i and j are positive integers in the respective ranges $1 \leq i \leq n-1$ and $i < j \leq n$, and where x is a positive real number in the range $0 < x \leq 1$, applying Eqs. (5) and (18), we have:

$$\begin{aligned}W^{D}_{1\to4} &= W^{D}_{1\to2} + W^{D}_{2\to3} + W^{D}_{3\to4}\\ &= Q_1\left\{\left[1 - \left(\frac{T_2}{T_1}\right)^x\right] + \left[1 - \left(\frac{T_3}{T_2}\right)^x\right] + \left[1 - \left(\frac{T_4}{T_3}\right)^x\right]\right\}\\ &= Q_1\left[3 - \left(\frac{T_2}{T_1}\right)^x - \left(\frac{T_3}{T_2}\right)^x - \left(\frac{T_4}{T_3}\right)^x\right].\end{aligned} \tag{19}$$

We wish to maximize $W^{D}_{1\to4}$. Based on Eq. (12) and the associated discussions, the optimum value $T_{j,\text{opt}}$ of T_j of reservoir j $(1 < j < n \Leftrightarrow 2 \leq j \leq n-1)$, which maximizes $W^{D}_{j-1\to j+1}$, is the geometric mean of T_{j-1} and T_{j+1}. Thus we have

$$T_{2,\text{opt}} = \left(T_1 T_{3,\text{opt}}\right)^{1/2} \tag{20}$$

and

$$T_{3,\text{opt}} = \left(T_{2,\text{opt}} T_4\right)^{1/2}. \tag{21}$$

Applying Eqs. (20) and (21), we obtain

$$\frac{T_{2,\text{opt}}}{T_1} = \frac{\left(T_1 T_{3,\text{opt}}\right)^{1/2}}{T_1} = \left(\frac{T_{3,\text{opt}}}{T_1}\right)^{1/2} \tag{22}$$

and

$$\frac{T_4}{T_{3,\text{opt}}} = \frac{T_4}{\left(T_{2,\text{opt}} T_4\right)^{1/2}} = \left(\frac{T_4}{T_{2,\text{opt}}}\right)^{1/2}. \tag{23}$$

Applying Eqs. (20)–(23), we obtain

$$\begin{aligned} \frac{T_{3,\text{opt}}}{T_{2,\text{opt}}} &= \frac{T_{3,\text{opt}}}{\left(T_1 T_{3,\text{opt}}\right)^{1/2}} = \left(\frac{T_{3,\text{opt}}}{T_1}\right)^{1/2} \\ &= \frac{\left(T_{2,\text{opt}} T_4\right)^{1/2}}{T_{2,\text{opt}}} = \left(\frac{T_4}{T_{2,\text{opt}}}\right)^{1/2} \\ &\Rightarrow \left(\frac{T_{3,\text{opt}}}{T_1}\right)^{1/2} = \left(\frac{T_4}{T_{2,\text{opt}}}\right)^{1/2} \\ &\Rightarrow \frac{T_{2,\text{opt}}}{T_1} = \frac{T_{3,\text{opt}}}{T_{2,\text{opt}}} = \frac{T_4}{T_{3,\text{opt}}}. \end{aligned} \tag{24}$$

Applying Eqs. (22)–(24), we obtain

$$\begin{aligned} \frac{T_4}{T_1} &= \frac{T_2}{T_1}\frac{T_3}{T_2}\frac{T_4}{T_3} \text{ in general} \\ &= \frac{T_{2,\text{opt}}}{T_1}\frac{T_{3,\text{opt}}}{T_{2,\text{opt}}}\frac{T_4}{T_{3,\text{opt}}} \text{ in particular} \\ &= \left(\frac{T_{2,\text{opt}}}{T_1}\right)^3 = \left(\frac{T_{3,\text{opt}}}{T_{2,\text{opt}}}\right)^3 = \left(\frac{T_4}{T_{3,\text{opt}}}\right)^3 \\ &\Rightarrow \frac{T_{2,\text{opt}}}{T_1} = \frac{T_{3,\text{opt}}}{T_{2,\text{opt}}} = \frac{T_4}{T_{3,\text{opt}}} = \left(\frac{T_4}{T_1}\right)^{1/3}. \end{aligned} \tag{25}$$

Applying Eqs. (19) and (25), we obtain

$$\begin{aligned} W^D_{1\to 4,\max} &= Q_1\left[3 - 3\left(\frac{T_4}{T_1}\right)^{x/3}\right] \\ &= 3Q_1\left[1 - \left(\frac{T_4}{T_1}\right)^{x/3}\right]. \end{aligned} \tag{26}$$

We now slightly modify Eqs. (14)–(17) to apply for our four-reservoir system. We obtain

$$W^D_{1\to 4,\max} > Q_1 \text{ if } \left(\frac{T_4}{T_1}\right)^{x/3} < \frac{2}{3} \Leftrightarrow \frac{T_4}{T_1} < \left(\frac{2}{3}\right)^{3/x}. \tag{27}$$

This obtains if $T_4/T_1 < (2/3)^3 = 8/27$ for $x = 1$ and if $T_4/T_1 < (2/3)^6 = 64/729$ for $x = 1/2$. Also, applying Eqs. (5) and (26),

$$\begin{aligned} W^{D,\text{extra}}_{1\to4,\max} &= W^{D}_{1\to4,\max} - W_{1\to4} \\ &= 3Q_1\left[1-\left(\frac{T_4}{T_1}\right)^{x/3}\right] - Q_1\left[1-\left(\frac{T_4}{T_1}\right)^{x}\right] \\ &= Q_1\left\{3\left[1-\left(\frac{T_4}{T_1}\right)^{x/3}\right] - \left[1-\left(\frac{T_4}{T_1}\right)^{x}\right]\right\} \\ &= Q_1\left[3-3\left(\frac{T_4}{T_1}\right)^{x/3} - 1 + \left(\frac{T_4}{T_1}\right)^{x}\right] \\ &= Q_1\left[2+\left(\frac{T_4}{T_1}\right)^{x} - 3\left(\frac{T_4}{T_1}\right)^{x/3}\right] \geq 0. \end{aligned} \tag{28}$$

It is easily shown that $W^{D,\text{extra}}_{1\to4,\max} \geq 0$, with the equality obtaining if and only if $\frac{T_4}{T_1} = 1 \Rightarrow W^{D}_{1\to4,\max} = W_{1\to4} = 0 \Rightarrow W^{D}_{1\to4,\max} - W_{1\to4} = W^{D,\text{extra}}_{1\to4,\max} = 0$. For, denoting the ratio $\left(\frac{T_4}{T_1}\right)^{x/3}$ as r and setting $dW^{D,\text{extra}}_{1\to4,\max}/dr = 0$ yields

$$\begin{aligned} \frac{dW^{D,\text{extra}}_{1\to4,\max}}{dr} = 0 &\Rightarrow \frac{d}{dr}\left(r^3 - 3r\right) = 0 \\ &\Rightarrow 3r^2 - 3 = 0 \\ &\Rightarrow r^2 = 1 \\ &\Rightarrow r = 1. \end{aligned} \tag{29}$$

Thus $W^{D,\text{extra}}_{1\to4,\max}$ is minimized at 0 if $r = \left(\frac{T_4}{T_1}\right)^{x/3} = 1 \Rightarrow \frac{T_4}{T_1} = 1$. For all $\frac{T_4}{T_1} < 1$, $W^{D,\text{extra}}_{1\to4,\max} > 0$. Moreover, applying Eqs. (5), (26), and (28), note that

$$\begin{aligned} &\lim_{T_4/T_1\to0} W^{D}_{1\to4,\max} = 3Q_1 = 3\lim_{T_4/T_1\to0} W_{1\to4} \\ \Rightarrow\ &\lim_{T_4/T_1\to0} W^{D,\text{extra}}_{1\to4,\max} = 3Q_1 - Q_1 = 2Q_1 = 2\lim_{T_4/T_1\to0} W_{1\to4}. \end{aligned} \tag{30}$$

Comparing Eqs. (13)–(17) with Eqs. (26)–(30), note the larger values in Eqs. (26), (28), and (30) than in Eqs. (13), (15), and (17), respectively, and the easier fulfillment of the inequality in Eq. (27) than in Eq. (14) (concerning the latter point: $8/27 > 1/4$ and $64/729 > 1/16$).

Generalizing Eqs. (20)–(30) for an n-reservoir system (n = any positive integer ≥ 4), we obtain:

$$T_{j+1} = \left(T_j T_{j+2}\right)^{1/2}, \tag{31}$$

where j is any positive integer in the range $1 \leq j \leq n-2$ and

$$T_{j+2} = \left(T_{j+1} T_{j+3}\right)^{1/2}, \tag{32}$$

where j is any positive integer in the range $1 \leq j \leq n-3$. The respective temperatures T_1 and T_n of the extreme (hottest and coldest) reservoirs are assumed to be fixed. The temperatures T_2 through T_{n-1} of all intermediate reservoirs are all assumed to be optimized in accordance with Eqs. (31) and (32). With that understood, for brevity and to avoid using different subscripts for the extreme and intermediate reservoirs, the subscript "opt" is omitted in Eqs. (31)–(35). Applying Eqs. (31) and (32), we obtain:

$$\frac{T_{j+1}}{T_j} = \frac{\left(T_j T_{j+2}\right)^{1/2}}{T_j} = \left(\frac{T_{j+2}}{T_j}\right)^{1/2} \tag{33}$$

and

$$\frac{T_{j+2}}{T_{j+1}} = \frac{T_{j+2}}{\left(T_j T_{j+2}\right)^{1/2}} = \left(\frac{T_{j+2}}{T_j}\right)^{1/2}. \tag{34}$$

Applying Eqs. (33) and (34), and recognizing that Eqs. (33) and (34) obtain for *all* values of j such that j is any positive integer in the range $1 \le j \le n-2$, we obtain:

$$\begin{aligned} &\frac{T_{j+2}}{T_{j+1}} = \frac{T_{j+1}}{T_j} \\ &\Rightarrow \frac{T_{j+2}}{T_j} = \frac{T_{j+1}}{T_j}\frac{T_{j+2}}{T_{j+1}} = \left(\frac{T_{j+1}}{T_j}\right)^2 \Leftrightarrow \frac{T_{j+1}}{T_j} = \left(\frac{T_{j+2}}{T_j}\right)^{1/2} \\ &\Leftrightarrow \frac{T_n}{T_1} = \left(\frac{T_{j+1}}{T_j}\right)^{n-1} \Leftrightarrow \frac{T_{j+1}}{T_j} = \left(\frac{T_n}{T_1}\right)^{1/(n-1)}. \end{aligned} \tag{35}$$

The first two lines of Eq. (35) obtain for all values of j such that j is any positive integer in the range $1 \le j \le n-2$, and the third line of Eq. (35) obtain for all values of j such that j is any positive integer in the range $1 \le j \le n-1$. The first two lines of Eq. (35) pertain to any three adjacent heat reservoirs, and hence 2 appears in the exponents of the second line thereof; the third line of Eq. (35) pertains to all n heat reservoirs, and hence $n-1$ appears in the exponents thereof. The second and third lines of Eq. (35) mutually justify each other: the third line of Eq. (35) *must* obtain because the second line thereof obtains for *all* values of j; and, conversely, given that the third line of Eq. (35) obtains, the second line thereof *must* obtain for *all* values of j.

If, as per Eq. (5), $\epsilon_{i\to j} = 1 - \left(T_i/T_j\right)^x$, where i and j are positive integers in the respective ranges $1 \le i \le n-1$ and $i < j \le n$, and where x is a positive real number in the range $0 < x \le 1$, then, applying Eqs. (5) and (31)–(35), we now generalize Eqs. (13)–(17) and (26)–(30), as well as the associated discussions, to apply for our n-reservoir system. We obtain:

$$W^D_{1\to n,\max} = (n-1)Q_1\left[1 - \left(\frac{T_n}{T_1}\right)^{x/(n-1)}\right], \tag{36}$$

$$W^D_{1\to n,\max} > Q_1 \text{ if } \left(\frac{T_n}{T_1}\right)^{x/(n-1)} < \frac{n-2}{n-1} \Leftrightarrow \frac{T_n}{T_1} < \left(\frac{n-2}{n-1}\right)^{(n-1)/x}, \tag{37}$$

and

$$\begin{aligned}W^{D,\text{extra}}_{1\to n,\max} &= W^{D}_{1\to n,\max} - W_{1\to n}\\ &= (n-1)Q_1\left[1-\left(\frac{T_n}{T_1}\right)^{x/(n-1)}\right] - Q_1\left[1-\left(\frac{T_n}{T_1}\right)^{x}\right]\\ &= Q_1\left\{(n-1)\left[1-\left(\frac{T_n}{T_1}\right)^{x/(n-1)}\right] - \left[1-\left(\frac{T_n}{T_1}\right)^{x}\right]\right\}\\ &= Q_1\left[n-1-(n-1)\left(\frac{T_n}{T_1}\right)^{x/(n-1)} - 1 + \left(\frac{T_n}{T_1}\right)^{x}\right]\\ &= Q_1\left[n-2+\left(\frac{T_n}{T_1}\right)^{x} - (n-1)\left(\frac{T_n}{T_1}\right)^{x/(n-1)}\right] \ge 0.\end{aligned} \tag{38}$$

It is easily shown that $W^{D,\text{extra}}_{1\to n,\max} \ge 0$, with the equality obtaining if and only if $\frac{T_n}{T_1} = 1 \Rightarrow W^{D}_{1\to n,\max} = W_{1\to n} = 0 \Rightarrow W^{D}_{1\to n,\max} - W_{1\to n} = W^{D,\text{extra}}_{1\to n,\max} = 0$. For, denoting the ratio $\left(\frac{T_n}{T_1}\right)^{x/(n-1)}$ as r and setting $dW^{D,\text{extra}}_{1\to n,\max}/dr = 0$ yields

$$\begin{aligned}\frac{dW^{D,\text{extra}}_{1\to n,\max}}{dr} = 0 &\Rightarrow \frac{d}{dr}\left[r^{n-1} - (n-1)r\right] = 0\\ &\Rightarrow (n-1)r^{n-2} - (n-1) = 0\\ &\Rightarrow r^{n-2} = 1\\ &\Rightarrow r = 1.\end{aligned} \tag{39}$$

Thus $W^{D,\text{extra}}_{1\to n,\max}$ is minimized at 0 if $r = \left(\frac{T_n}{T_1}\right)^{x/(n-1)} = 1 \Rightarrow \frac{T_n}{T_1} = 1$. For all $\frac{T_n}{T_1} < 1$, $W^{D,\text{extra}}_{1\to n,\max} > 0$. Moreover, applying Eqs. (5), (36), and (38), note that

$$\begin{aligned}&\lim_{T_n/T_1\to 0,\, n \text{ fixed}} W^{D}_{1\to n,\max} = (n-1)Q_1 = (n-1)\lim_{T_n/T_1\to 0,\, n \text{ fixed}} W_{1\to n}\\ &\Rightarrow \lim_{T_n/T_1\to 0,\, n \text{ fixed}} W^{D,\text{extra}}_{1\to n,\max} = \lim_{T_n/T_1\to 0,\, n \text{ fixed}} \left(W^{D}_{1\to n,\max} - W_{1\to n}\right)\\ &= (n-1)Q_1 - Q_1 = (n-2)Q_1 = (n-2)\lim_{T_n/T_1\to 0,\, n \text{ fixed}} W_{1\to n}.\end{aligned} \tag{40}$$

Note that the values in Eqs. (36), (38), and (40) increase monotonically with increasing n and that the fulfillment of the inequality in Eq. (37) becomes monotonically easier with increasing n. Equation (40) is valid not only for Carnot efficiency ($x = 1$) but even for Curzon-Ahlborn efficiency ($x = 1/2$), indeed for any x finitely greater than 0 in the range $0 < x \le 1$, because $\left(\frac{T_n}{T_1}\right)^{x/(n-1)} \to 0 \Leftrightarrow 1 - \left(\frac{T_n}{T_1}\right)^{x/(n-1)} \to 1$ in the limit $T_n/T_1 \to 0$, albeit ever more slowly with decreasing x.

By contrast, even granting Carnot efficiency ($x = 1$) [22]:

$$\lim_{n\to\infty,\, T_n/T_1 \text{ fixed}} W^{D}_{1\to n,\max} = Q_1 \ln\frac{T_1}{T_n} = \left(\lim_{T_n/T_1\to 0,\, n \text{ fixed}} W_{1\to n}\right)\ln\frac{T_1}{T_n}. \tag{41}$$

Note the *linear* divergence of $W^D_{1\to n,\max}$ in the limit $T_n/T_1 \to 0$ with n fixed as per Eq. (40) even *not* assuming Carnot efficiency, as contrasted with the paltry *logarithmic* divergence of $W^D_{1\to n,\max}$ in the limit $n \to \infty$ with T_n/T_1 fixed even granting Carnot efficiency as per the derivation [22] of Eq. (41).

But we note that the temperature of the cosmic background radiation is only 2.7 K, while the most refractory materials remain solid at temperatures slightly exceeding 2700 K. This provides a temperature ratio of $T_1/T_n \approx 10^3 \Leftrightarrow T_n/T_1 \approx 10^{-3}$. Could even larger values of T_1/T_n be possible, at least in principle? Perhaps, maybe, if frictional dissipation of work into heat might somehow be possible into a gaseous hot reservoir at temperatures exceeding the melting point or even the critical temperature (the maximum boiling point at any pressure) of even the most refractory material. Yet even with the paltry *logarithmic* divergence of $W^D_{1\to n,\max}$ in the limit $n \to \infty$ with T_1/T_n fixed as per Eq. (41) and even with a temperature ratio of $T_1/T_n \approx 10^3 \Leftrightarrow T_n/T_1 \approx 10^{-3}$, assuming Carnot efficiency by Eq. (41) $W^D_{1\to n,\max}/Q_1 \approx \ln 10^3 \approx 7$. Hence by Eq. (41) an advanced civilization employing 7 concentric Dyson spheres [39, 40] can procure 7 times as much work output (to the nearest whole number) as its host star's total energy output. Actually the limit $n \to \infty$ with T_1/T_n fixed is not sufficiently closely approached to apply Eq. (41): we should instead apply Eq. (36). Applying Eq. (36) and assuming Carnot efficiency with $T_1/T_n \approx 10^3 \Leftrightarrow T_n/T_1 \approx 10^{-3}$, $W^D_{1\to n,\max}/Q_1 \approx 4$. Hence by Eq. (36) an advanced civilization employing 4 concentric Dyson spheres [39, 40] can procure 4 times as much work output (to the nearest whole number) as its host star's total energy output.

4. General remarks, especially concerning entropy

It is important to emphasize that the super-unity cyclic-heat-engine efficiencies $W^D_{1\to n,\max}/Q_1$ that can obtain with work output totally frictionally dissipated (if $n \geq 3$) are consistent with both the First and Second Laws of Thermodynamics. The two laws are *not* violated because, if the work output of a heat engine is frictionally dissipated as heat into a cooler reservoir, both laws allow this heat to be partially converted to work again if another, still cooler, reservoir is available.

In this Section 4 we do not restrict heat-engine efficiencies to the form given by Equation (5), nor necessarily assume efficiencies of the same form at each step $j \to j+1$ or $j \to j+k$ $(1 \leq k \leq n-j)$. The validity of this Section 4 requires only that the efficiency with all work sequestered, or at any one given step $j \to j+1$ whether work is sequestered or not, be within the Carnot limit, in accordance with the Second Law.

The extra work that is made available via frictional dissipation into cooler reservoirs is paid for by an extra increase in entropy. Consider the work available via heat-engine operation between reservoir j at temperature T_j and reservoir $j+2$ at temperature T_{j+2} without versus with frictional dissipation into reservoir $j+1$ at temperature $T_{j+1}(T_j > T_{j+1} > T_{j+2})$. Without frictional dissipation a heat engine performs work

$$W_{j\to j+1} = Q_j \epsilon_{j\to j+1} \tag{42}$$

by employing the reservoir at temperature T_j as a hot reservoir and the reservoir at temperature T_{j+1} as a cold reservoir. It rejects waste heat $Q_j - W_{j\to j+1} = Q_j(1 - \epsilon_{j\to j+1})$ to the reservoir at temperature T_{j+1}. If a third reservoir at temperature T_{j+2} and $W_{j\to j+1}$ is sequestered, that is, not frictionally dissipated, a heat engine can then perform additional work:

$$W_{j+1\to j+2} = Q_j(1 - \epsilon_{j\to j+1})\epsilon_{j+1\to j+2} \tag{43}$$

by employing the reservoir at temperature T_{j+1} as a hot reservoir and the reservoir at temperature T_{j+2} as a cold reservoir. All told it can do work:

$$\begin{aligned} W_{j\to j+2} &= W_{j\to j+1} + W_{j+1\to j+2} = Q_j\epsilon_{j\to j+1} + Q_j(1 - \epsilon_{j\to j+1})\epsilon_{j+1\to j+2} \\ &= Q_j(\epsilon_{j\to j+1} + \epsilon_{j+1\to j+2} - \epsilon_{j\to j+1}\epsilon_{j+1\to j+2}). \end{aligned} \tag{44}$$

With total frictional dissipation of $W_{j\to j+1}$ into reservoir $j+1$ at temperature T_{j+1}, we still have

$$W^D_{j\to j+1} = W_{j\to j+1} = Q_1\epsilon_{j\to j+1}. \tag{45}$$

But now we let the work output $W^D_{j\to j=1} = Q_1\epsilon_{j+1\to j+2}$ be totally frictionally dissipated into the reservoir at temperature T_{j+1} (indicated via a superscript D). If there is a third reservoir at temperature T_{j+2}, a heat engine can then perform additional work:

$$W^D_{j+1\to j+2} = Q_1\epsilon_{j+1\to j+2}. \tag{46}$$

All told it can do work:

$$\begin{aligned} W^D_{j\to j+2} &= W^D_{j\to j+1} + W^D_{j+1\to j+2} = Q_j\epsilon_{j\to j+1} + Q_j\epsilon_{j+1\to j+2} \\ &= Q_j(\epsilon_{j\to j+1} + \epsilon_{j+1\to j+2}). \end{aligned} \tag{47}$$

The extra work

$$\begin{aligned} W^D_{\text{extra}} &= W^D_{j+1\to j+2} \\ &= Q_j\epsilon_{j\to j+1}\epsilon_{j+1\to j+2} \\ &= W_{j\to j+1}\epsilon_{j+1\to j+2} \\ &= W^D_{j\to j+1}\epsilon_{j+1\to j+2} \end{aligned} \tag{48}$$

is paid for by the extra increase in entropy owing to frictional dissipation into extra heat Q^D_{extra} of the work output as per Eqs. (42) and (45)

$$Q^D_{\text{extra}} = W_{j\to j+1} = W^D_{j\to j+1} = Q_j\epsilon_{j\to j+1} \tag{49}$$

into reservoir $j+1$ at temperature T_{j+1}. This extra increase in entropy is

$$\Delta S^D_{\text{extra}} = \frac{Q^D_{\text{extra}}}{T_{j+1}} = \frac{Q_j\epsilon_{j\to j+1}}{T_{j+1}} = \frac{W_{j\to j+1}}{T_{j+1}} = \frac{W^D_{j\to j+1}}{T_{j+1}} = \frac{W^D_{\text{extra}}}{\epsilon_{j+1\to j+2}T_{j+1}}. \tag{50}$$

[In the last four steps of Eq. (50), we applied Eqs. (42), (45), (48), and (49).] Thus

$$W^D_{\text{extra}} = T_{j+1}\Delta S^D_{\text{extra}}\epsilon_{j+1\to j+2}. \tag{51}$$

In no case do we assume an efficiency with all work sequestered, or at any one given step $j \to j+1$ whether work is sequestered or not, exceeding the Carnot efficiency, and hence we are within the restrictions of the Second Law. (The First Law, of course, puts no restrictions whatsoever on the recycling of energy, except that it is conserved—and we *never* violate conservation of energy.)

We note that, while frictional dissipation of work into intermediate reservoirs can yield extra work W^D_{extra} in heat-engine operation (albeit at the expense of $\Delta S^D_{\text{extra}}$), it seems to be of no help in reverse, that is, refrigerator or heat pump, operation. For, in refrigerator or heat pump operation, with an intermediate reservoir $j+1$ at temperature T_{j+1}, $Q_{j+2} + W_{j+2\to j+1} = Q_{j+1}$, $Q_{j+1} + W_{j+1\to j} = Q_j$, hence $Q_{j+2} + W_{j+2\to j+1} + W_{j+1\to j} = Q_{j+2} + W_{j+2\to j} = Q_j$. Without an intermediate reservoir $j+1$ at temperature T_{j+1}, $Q_{j+2} + W_{j+2\to j} = Q_j$. The bottom line $Q_{j+2} + W_{j+2\to j} = Q_j$ is identical with or without an intermediate reservoir $j+1$ at temperature T_{j+1}. With or without the intermediate reservoir $j+1$ at temperature T_{j+1}, *all* of the energy must end up as Q_j; thus, there is *none* left over to be frictionally dissipated. Hence the presence or absence of this intermediate reservoir makes no difference with respect to reverse, that is, refrigerator or heat pump, operation: See Ref. [1], Section 20-3; Ref. [2], Section 5.12 and Problem 5.22; Ref. [3], Sections 4.3, 4.4, and 4.7 (especially Section 4.7); Ref. [4], Sections 4-4, 4-5, and 4-6 (especially Section 4-6); Ref. [5], Sections 5-7-2, 6-2-2, 6-9-2, and 6-9-3, and Chapter 17; Ref. [6], Chapter XXI; Ref. [7], Sections 6.7, 6.8, 7.3, and 7.4); and Ref. [9], pp. 233–236 and Problems 1, 2, 4, 6, and 7 of Chapter 8. [Problem 2 of Chapter 8 in Ref. [9] considers absorption refrigeration, wherein the entire energy output is into an intermediate-temperature (most typically ambient-temperature) reservoir, and hence for which also there is *no* energy left over to be frictionally dissipated.]

5. Conclusion

We investigated the increased heat-engine efficiencies obtained via operation employing increasing numbers (≥ 3) of heat reservoirs and with work output totally frictionally dissipated into all reservoirs except the first, hottest, one at temperature T_1 and (possibly) also the last, coldest, one at temperature T_n. We emphasize again that our results are consistent with both the First and Second Laws of Thermodynamics. The two laws are *not* violated because, if the work output of a heat engine is frictionally dissipated as heat into a cooler reservoir, both laws allow this heat to be partially converted to work again if another, still cooler, reservoir is available.

We do, however, challenge an *over*statement of the Second Law that is sometimes made, namely, that energy can do work only once. Energy can indeed do work more than once, because the Second Law does not forbid recycling of energy, so long as total entropy does not decrease as a result. This criterion of non-decrease of total entropy *is* obeyed, as per Section 4. In no case do we assume an efficiency with all work sequestered, or at any one given step $j \to j+1$ whether work is sequestered or not, exceeding the Carnot efficiency, and hence we are within the restrictions of the Second Law. (The First Law, of course, puts no restrictions whatsoever on the recycling of energy, except that it is conserved—and we *never* violate conservation of energy).

While in this chapter we do not challenge the First or Second Laws of Thermodynamics, we should note that there have been many challenges to the Second Law, especially in recent years [41–46]. By contrast, the First Law has been questioned only in cosmological contexts [47–49] and with respect to fleeting violations thereof associated with the energy-time uncertainty principle [50, 51]. But there are contrasting viewpoints [50, 51] concerning the latter issue.

Acknowledgements

I am very grateful to Dr. Donald H. Kobe, Dr. Paolo Grigolini, Dr. Daniel P. Sheehan, Dr. Bruce N. Miller, and Dr. Marlan O. Scully and for many very helpful and thoughtful insights, as well as for very perceptive and valuable discussions and communications, which greatly helped my understanding of thermodynamics and statistical mechanics. Also, I am indebted to them, as well as to Dr. Bright Lowry, Dr. John Banewicz, Dr. Bruno J. Zwolinski, Dr. Roland E. Allen, Dr. Abraham Clearfield, Dr. Russell Larsen, Dr. James H. Cooke, Dr. Wolfgang Rindler, Dr. Richard McFee, Dr. Nolan Massey, and Dr. Stan Czamanski for lectures, discussions, and/or communications from which I learned very much concerning thermodynamics and statistical mechanics. I thank Dr. Stan Czamanski and Dr. S. Mort Zimmerman for the very interesting general scientific discussions over many years. I also thank Dan Zimmerman, Dr. Kurt W. Hess, and Robert H. Shelton for the very interesting general scientific discussions at times. Additionally, I thank Robert H. Shelton for very helpful advice concerning diction.

Conflict of interest

The author declares no conflict of interest.

Author details

Jack Denur
Electric & Gas Technology, Inc., Rowlett, Texas, USA

*Address all correspondence to: jackdenur@my.unt.edu

References

[1] Walker J, Halliday D, Resnick R. Fundamentals of Physics. 11th extended ed. Hoboken, NJ: John Wiley & Sons; 2018, Chapters 18 and 20 (especially Sections 20-2 and 20-3)

[2] Reif F. Fundamentals of Statistical and Thermal Physics. New York: McGraw-Hill; 1965. (reissued: Long Grove, IL: Waveland Press; 2009), Sections 5-11 ad 5-12, and Problems 5.22 through 5.26

[3] Callen HC. Thermodynamics: New York: John Wiley & Sons; 1960, Chapter 4

[4] Callen HC. Thermodynamics and an Introduction to Thermostatistics. 2nd ed. New York: John Wiley & Sons; 1985, Chapter 4

[5] Wark K, Richards DE. Thermodynamics. 6th ed. Boston, MA: WCB/McGraw-Hill; 1999, Chapters 6, 8, 9, and 15–17

[6] Faries VM. Applied Thermodynamics. Revised Ed. New York, NY: MacMillan; 1949, Chapters V–VIII, XVII, and XIX

[7] Zemansky, MW, Dittman RH. Adapted by Chattopadhyay AK. Heat and Thermodynamics, 8th ed. Chennai, India: McGraw Hill Education (India); 2011. (Seventeenth reprint 2018), Chapters 6 and 7

[8] Baierlein R. Thermal Physics. Cambridge, UK: Cambridge University Press; 1999, Chapters 2 and 3

[9] Kittel C, Kroemer H. Thermal Physics. 2nd ed. San Francisco, CA: W. H. Freeman and Company; 1980, Chapter 8

[10] Curzon FL, Ahlborn B. Efficiency of a Carnot engine at maximum power output. American Journal of Physics. 1975;**43**:22-24. DOI: 10.1119/1.10023

[11] Vaudrey A, Lanzetta F, Feidt M. Reitlinger and the origins of the efficiency at maximum power output for heat engines. Journal of Non-Equilibrium Thermodynamics. 2014;**39**: 199-203. DOI: 10.1515/jnet-2014-0018

[12] Endoreversible thermodynamics [Online]. Available from: https://www.wikipedia.org/ [Accessed: 16 March 2020]

[13] De Vos A. Efficiency of some heat engines at maximum-power conditions. American Journal of Physics. 1985;**53**: 570-573. DOI: 10.1119/1.14240

[14] Gordon JM. Maximum power-point characteristics of heat engines as a general thermodynamic problem. American Journal of Physics. 1989;**57**: 1136-1142. DOI: 10.1119/1.16130

[15] Gordon JM. Observations on efficiency of heat engines operating at maximum power. American Journal of Physics. 1990;**58**:370-375. DOI: 10.1119/1.16175

[16] Schmiedl T, Seifert U. Efficiency at maximum power: An analytically solvable model for stochastic heat engines. Europhysics Letters. 2008;**81**: 20003. DOI: 10.1209/0295-5075/81/20003

[17] Tu ZC. Efficiency at maximum power of Feynman's ratchet as an engine. Journal of Physics A. 2008;**41**: 312003. DOI: 10.1088/1751-8113/41/31/312003

[18] Leff HS. Thermal efficiency at maximum work output: New results for old heat engines. American Journal of Physics. 1987;**55**:602-610. DOI: 10.1119/1.15071

[19] Ouerdane H, Apertet Y, Goupil C, Lecoeur P. Continuity and boundary conditions in thermodynamics: From Carnot's efficiency to efficiencies at maximum power. European Physical

Journal - Special Topics. 2015;**55**: 839-862. DOI: 10.1140/epjst/e2015-02431-x

[20] Parrando JMR, Ouerdane H, et al. Debate. Continuity and boundary conditions in thermodynamics: From Carnot's efficiency to efficiencies at maximum power. European Physical Journal - Special Topics. 2015;**224**: 862-864

[21] Apertet Y, Ouerdane H, Goupil C, Lecoeur Ph. True nature of the Curzon-Ahlborn efficiency. Physical Review E. 2017;**96**:022119. DOI: 10.1103/PhysRevE.96.022119

[22] Denur J. The apparent "super-Carnot" efficiency of hurricanes: Nature's steam engine versus the steam locomotive. American Journal of Physics. 2011;**79**:631-643. DOI: 10.1119/13534841 (especially Section VI)

[23] Emanuel K. Divine Wind. Oxford, UK: Oxford University Press; 2005 (especially Chapter 10)

[24] Emanuel K. Hurricanes: Tempests in a greenhouse. Physics Today. 2006; **59**(8):74-75. DOI: 10.1063/1.2349743

[25] Emanuel K. Tropical cyclones. Annual Review of Earth and Planetary Sciences. 2003;**31**:75-104. DOI: 10.1146/annurev.earth.31.100901.141259

[26] Emanuel K. Thermodynamic control of hurricane intensity. Nature. 1999;**401**:665-669. DOI: 10.1038./44326

[27] Bister M, Emanuel KA. Dissipative heating and hurricane intensity. Meteorology and Atmospheric Physics. 1998;**65**:223-230

[28] Zhang DL, Altshuler E. The effects of dissipative heating on hurricane intensity. Monthly Weather Review. 1999;**127**:3032-3038

[29] Emanuel K. Response of tropical cyclone activity to climate change: Theoretical basis. In: Murmane RJ, Liu K-B, editors. Hurricanes and Typhoons: Past, Present, and Future. New York: Columbia University Press; 2004, pp. 395–407

[30] Emanuel KA, Speer K, Rotunno R, Srivastava R, Molina M. Hypercanes: A possible link in global extinction scenarios. Journal of Geophysical Research-Atmospheres. 1995;**100**: 13755-13765. DOI: 10.1029/95JD01368

[31] Emanuel K, Callagham J, Otto PA. A hypothesis for redevelopment of warm-core cyclones over northern Australia. Monthly Weather Review. 2008;**136**: 3863-3872. DOI: 10.1175/2008MWR2409.1

[32] Kieu C. Revisiting dissipative heating in tropical cyclone maximum potential intensity. Quarterly Journal of the Royal Meteorological Society. 2015; **141**:2497-2504. DOI: 10.1002/qj.2534

[33] Apertet Y, Ouerdane H, Goupil C, Lecoeur P. Efficiency at maximum power of thermally coupled heat engines. Physical Review E. 2012;**85**: 041144. DOI: 10.1103/PhysRevE.85041144

[34] Makarieva AM, Gorshkov VC, Li B-L, Nobre AD. A critique of some modern applications of the Carnot heat engine concept: The dissipative engine cannot exist. Proceedings of the Royal Society A. 2010;**466**:1893-1902. DOI: 10.1098/rspa.2009.0581

[35] Bejan A. Thermodynamics of heating. Proceedings of the Royal Society A. 2019;**475**:20180820. DOI: 10.1098/rspa.2018.0820

[36] Bister M, Renno N, Pauluis O, Emanuel K. Comment on Makarieva et al. 'A critique of some modern applications of the Carnot heat engine concept: The dissipative engine cannot exist'. Proceedings of the Royal Society A. 2011;**467**:1-6. DOI: 10.1098/rspa.2010.0087

[37] Ozawa H, Shimokawa S. Thermodynamics of a tropical cyclone: Generation and dissipation of mechanical energy in a self-driven convection system. Tellus A. 2015;**67**: 24216. DOI: 10.3402/tellusa.v67.24216. 15 pages

[38] Denur J. Improving heat-engine performance via high-temperature recharge. In: Vizureanu P, Academic editor. Applied Thermodynamics and Energy Engineering. London, UK: IntechOpen; 2019

[39] Dyson Sphere. Available from: https://www.wikipedia.org/ [Accessed: 16 March 2020]

[40] Dyson Spheres in Popular Culture. Available from: https://www.wikipedia.org/ [Accessed: 16 March 2020]

[41] Sheehan DP, editor. Quantum limits to the second law. In: AIP Conference Proceedings Volume 643; Melville, NY: American Institute of Physics; 2002

[42] Nikulov AV, Sheehan DP, editors. Special issue: Quantum limits to the second law of thermodynamics. Entropy 2004;**6**(1)

[43] Čápek V, Sheehan DP. Challenges to the Second Law of Thermodynamics: Theory and Experiment. Dordrecht, The Netherlands: Springer; 2005

[44] Sheehan DP, editor. Special issue: The second law of thermodynamics: Foundations and status. Foundations of Physics. 2007;**37**(12)

[45] Sheehan DP, editor. Second law of thermodynamics: Status and challenges. In AIP Conference Proceedings Volume 1411; Melville, NY: American Institute of Physics; 2011

[46] Sheehan DP, editor. Special issue: Limits to the second law of thermodynamics: Experiment and theory. Entropy. 2017;**19**

[47] Harrison ER. Mining energy in an expanding universe. The Astrophysical Journal. 1995;**446**:63-66

[48] Sheehan DP, Kriss VG. Energy Emission by Quantum Systems in an Expanding FRW Metric [Online]. Available from: arXiv:astroph/0411299v1 [Accessed: 16 March 2020]

[49] Parry R. Extracting Energy from the Expanding Universe: Can we Avoid the Heat Death? Honours Physics 2015. Sydney, Australia: Sydney Institute for Astronomy, School of Physics, The University of Sydney; 2015

[50] Griffiths DJ, Schroeter DF. Introduction to Quantum Mechanics. 3rd ed. Cambridge, UK: Cambridge University Press; 2018, Section 3.5.3

[51] Hagmann MJ. Distribution of times for barrier traversal caused by energy fluctuations. Journal of Applied Physics. 1993;**74**:7302-7305

Chapter 8

Energy Storage in PCM Wall Used in Buildings' Application: Opportunity and Perspective

Majdi Hazami, Farah Mehdaoui, Hichem Taghouti, Marco Noro, Renato Lazzarin and AmenAllah Guizani

Abstract

This chapter deals with the investigation of the effect of a PCM wall on building indoor thermal comfort. To achieve this objective, an experimental framework was installed in the laboratory of thermal processes in Borj Cedria, Tunisia, which is essentially composed of a test cell having the dimension (0.5, 0.5, 0.5 m^3) conceived with a new structure of wallboards. One of the sides of the test cell is a cavity filled with PCM-27, which represents the PCM wall. A numerical investigation by using specific FORTRAN program was also achieved to solve the energy and the exergy mathematic relations to evaluate the PCM wall performances. TRNSYS simulation program was also achieved to simulate the behavior of the integration of the PCM wall in a typical modern house according to Tunisian scenario. It is found that during the hottest period of the day, the temperature of the tested room with PCM wall achieves 25°C, while that without PCM wall exceeds 27°C. During the night, the temperature of the tested room, with PCM wall, decreases in the value of 20°C. It was also found that during the night, the kid's room with PCM wall is reduced by 8°C.

Keywords: PCM wall, building, thermal storage, FORTRAN, TRNSYS

1. Overview

Nowadays, the building sector has become the main consumer of energy in the developed countries. Taking the EU as an example, the building sector accounts for around 40% of the total CO_2 emissions. In Tunisia, the building sector consumes about 30% of the total final energy and is consumed by domestic water heating systems and air conditioning equipments (**Figure 1**) [1]. In order to reduce the energy consumption in buildings and to improve the thermal comfort of occupants, many researchers are focusing on storing of the thermal energy excess as latent heat by using specific phase change material (PCM). PCM may be integrated into the construction material in three ways: by direct incorporation, impregnation or encapsulation. It has been shown that the incorporation of the microencapsulated PCM into the building material is a particularly attractive technology. Recently, many experimental investigations have been conducted about the incorporation of the microencapsulated PCMs into different building elements, such as into plaster, cement, concrete walls or concrete floors. These investigations were aimed to assess

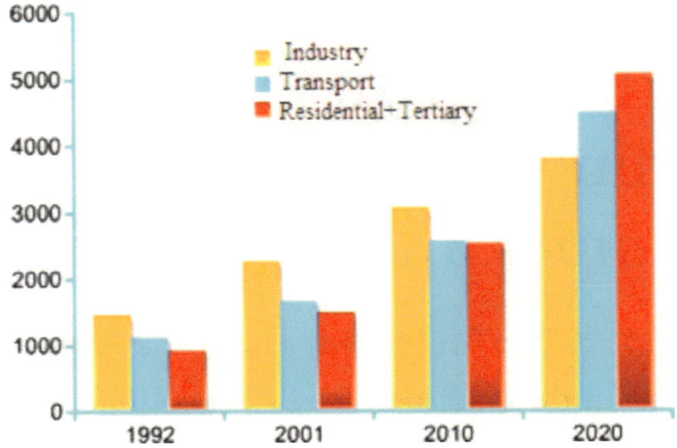

Figure 1.
The rates of energy expenditure in Tunisia (Mehdaoui et al. [1]).

the potential of PCM integration in walls and/or building envelopes to increase their thermal inertia to improve their energy performance [2–7]. In this context, Soares et al. [8] proposed the study of the incorporation of PCM drywalls in lightweight steel-framed building envelop. The authors evaluated the impact of PCM drywalls in the annual and monthly heating and cooling thermal performances and energy savings. It was seen that the energy savings due to PCM drywall incorporation range from 46 to 62%. Navarro et al. [9] studied the incorporation of the PCM inside the concrete core slab for cooling purposes. In this context, a prefabricated concrete slab incorporating PCM was used as internal separation inside the building. The results show that the energy savings in building were registered between 30 and 55%. Solgi et al. [2] presented that PCMs have a great influence on enhancing the performance of night purge ventilation and cooling load reduction of buildings in hot-arid climate. It was found that paraffin with 27°C melting point permits the reduction of about 47% in cooling energy. A performance of a collector storage wall system using PCMs was investigated by Zhou et al. [10]. PCM slabs were integrated in the gap-side wall surface to enhance the heat storage. The test was carried out for a whole day with charging period of 6.5 h and discharging period of 17.5 h. They investigated the variations of surface temperature as well as the indoor temperatures. It was found that the indoor temperature was about 22°C during the whole discharging period under given conditions. Barzin et al. [11] presented an experimental study dealing with the building's space cooling by using PCM energy storage in combination with night ventilation. Hence, two experimental tests were achieved: one with PCM-impregnated gypsum boards and the other with normal gypsum board. The result of the experimental investigation shows that substantial electricity saving is about to 73%. Sajjadian et al. [12] presented the study of the potential of using PCMs to reduce domestic cooling energy loads for current and future UK climates. The study used simulations of a high performance detached house model with a near Passivhaus Standard in London, where the impact of climate change effect is predicted to be significant. It was shown that appropriate levels of PCM, with a suitable incorporation mechanism into the building construction, have significant advantages for residential buildings in terms of reducing total discomfort hours. In this context, Royon et al. [13] studied the optimization of PCM implanted in a floor panel envelope of buildings. The study is mainly based on numerical investigation. The numerical results were confronted to experimental ones with the same boundary conditions in order to validate the model. Łukasz et al. [14] presented a parametric study of the thermal performance characteristics of thermal energy storage unit based on PCM integrated in building structure. In order to perform the analysis of the storage unit, a simulation program was developed. Using the program, many computer simulations were performed. In their work, Łukasz et al. presented important conclusions regarding the selection of PCM, and mainly its melting temperature range was formulated. Xiaoming et al. [15] studied the potential of exploiting ventilation systems with thermal energy storage (TES) and by using phase change materials (PCMs) for space cooling in air conditioned buildings during the summer. A dynamic computational model was achieved in order to simulate the indoor thermal environment and energy consumption of the room. The results showed that the electricity energy saving ratio (ESR) by using the TES system over the base case ranges between 16.9 and 50.8%, while considering the conventional NV system, the ESR ranges between 9.2 and 33.6%. Stropnik et al. [16] presented a study a system assuring self-sufficient heating and cooling of building from solar energy and interconnection between PV, electrical storage, heat pump, thermal energy storage and building energy management system. They showed that with such a smart energy system the almost zero-energy buildings can be reached in residential sector. The results show that thermal energy storage unit with integrated PCM modules supplies desired quantity of water temperature for longer period of time. Pushpendra et al. [17] presented a detailed review of various approaches to integrate the PCM in the building envelope. They showed that this method not only improves the indoor thermal behavior of the buildings but also reduces the cooling load without or little compromise with the mechanical strength of the building structure. They studied also the effect of the PCM integration on indoor thermal behavior and reduction in cooling load. They presented also an investigation of various materials used for making containers for encapsulation and it was also investigated. From the studied technologies, a great attention was given to investigate the effects of design parameters on thermal performances of PCM radiant floor heating system integrated in buildings. In this context, Li [18] proposed a numerical investigation aiming at the evaluation of the thermal performance of different kinds of roofs with and without PCM installed in Northeast China. They showed that the effect of transition temperature and latent heat of PCM on the thermal performance of roofs is relatively weak, compared with the roof slope, PCM layer thickness and absorption coefficients of external roof surface. In 2015, Joulin et al. [19] proposed an experimental and a numerical investigation of a PCM-27 conditioned in a rectangular container located between two heat exchangers. It was found that the PCM needs about 1.48 h to melt during the charging process. The evaluation of the effect of the integration of PCM inside a building was also studied with experimental and simulation methods by Huang et al. [20]. They showed that the PCM floor is able to supply about 37.7 MJ heat for 16 h in a building. In the same context, Prieto et al. [21] concluded that the integration of solar collectors holding PCM as storage material provided about 18–23% of total daily thermal energy needs of the building. Krese et al. [22] present also an experimental study of a small-scale wall composite containing PCM-27. The result of the investigation indicates that the heat recovery throughout the night is about 25 W/m^2.

2. Experimental methodology and framework

The aim of the research presented in this chapter is to evaluate the effectiveness of a PCM wall used as a storage medium in reducing the building's air temperature and in improving the occupant's thermal comfort in a Tunisian real house. A specific experimental framework was presented to characterize the PCM wall behavior during the storage and the discharging process (**Figure 2**). In this experimental

Figure 2.
PCM wall filled with Paraffin-27.

investigation, a PCM wall was installed in a test cell (**Figure 3**). A framework was installed to follow the indoor air temperatures of the various rooms of the house (with and without PCM). Thirteen T-type copper-constantan thermocouples firstly calibrated with a measurement inaccuracy of $\pm 0.2°C$ were incorporated to PCM wall at the front surface, inside PCM, the back surface and inside the test cell as shown in **Figure 4**. Between the exchanging plates, T-type thermocouples are inserted in both sides of the Plexiglas container to measure the temperature fields on each side of the PCM wall. T-type thermocouples were previously calibrated by using the comparative method. **Figure 4** shows the positions of all thermocouples inside the test cell.

The measurement test was continued for 14 consecutive days during February and March 2016 (from 25/02/2016 to 13/03/2016) to evaluate PCM wall performance. The temperature was measured at front surfaces of PCM wall, at back surfaces of PCM wall, inside the PCM and inside the test cell. The method adopted in the investigation consisted of imposing heating flux (lamp of 120 W) on the exposed PCM wall of the test cell. The simultaneous measurements of the temperature evolution and the heat flux exchanged during charging and the discharging process were accomplished to evaluate the PCM wall thermal performances. A primary experimental test was conducted in the laboratory to determine the characteristics of the phase change material (PCM) during the melting phase. PCM is initially in the solid state at the ambient temperature of the room, 22°C. Then, the right side of the PCM wall was heated by using a special lamp of 120 W held at a distance of 10 cm to ensure the uniformity of heat over the PCM wall surface. By

Figure 3.
The test cell.

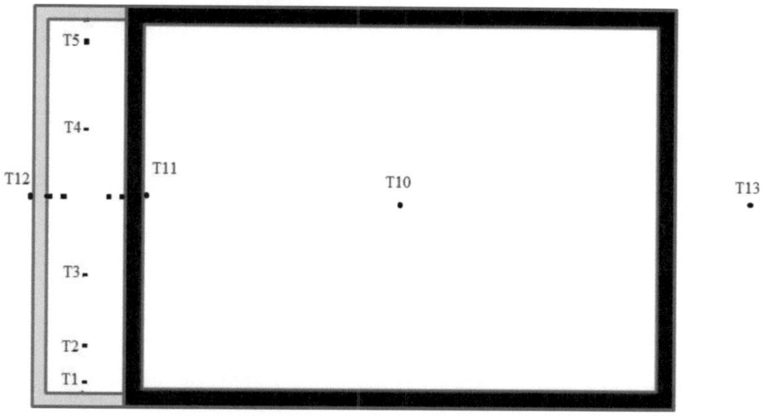

Figure 4.
The position of the 13 thermocouples inside the test cell.

using the data acquisition system, the temperature change at the interior of PCM wall during the charging process is followed. In the second test, the lamp was extinguished and then the PCM temperature fields were followed to evaluate the heat exchanged through the PCM wall during the discharge process.

3. PCM wall numerical characterization

A numerical investigation by using specific FORTRAN program was achieved to solve the energy and the exergy mathematic relations to evaluate the PCM wall performances by determining the melting phase proprieties (velocity, isotherm, melting front evolution, etc.).

The proposed numeric investigations describe the heat transfer phenomena inside the PCM wall and evaluate its thermal behavior and effects on test cell ambiance. It allows also the appraisal of the energy and exergy stored during the charging process and to evaluate the thermal characteristics. The considered assumptions are the flow is two-dimensional and laminar, the expansion of the PCM is negligible and the phase change is isothermal. The PCM wall is subjected to an imposed temperature superior to the melting temperature of PCM-27 (**Figure 5**).

The other walls are maintained adiabatic (**Figure 5**). Considering the mentioned assumptions:

- The continuity equation is given by [1, 23]:

$$\frac{\partial u}{\partial x}+\frac{\partial v}{\partial y}=0 \tag{1}$$

- The quantity of movement relations are given by [1, 23]:

$$\rho_l\frac{\partial u}{\partial t}+\rho_l\left(\frac{\partial uu}{\partial x}+\frac{\partial uv}{\partial y}\right)=-\frac{\partial p}{\partial x}+\mu_l\left(\frac{\partial^2 u}{\partial x^2}+\frac{\partial^2 u}{\partial y^2}\right)+Bu \tag{2}$$

$$\rho_l\frac{\partial v}{\partial t}+\rho_l\left(\frac{\partial uv}{\partial x}+\frac{\partial vv}{\partial y}\right)=-\frac{\partial p}{\partial y}+\rho_l g+\mu_l\left(\frac{\partial^2 v}{\partial x^2}+\frac{\partial^2 v}{\partial y^2}\right)+Bv \tag{3}$$

In the Boussinesq approximation, the source terms B_u and B_v that appear in the momentum Eqs. (2) and (3) are used to account for this buoyancy force when the PCM is solid. The technique used to cancel the velocity introduces a Darcy term [1].

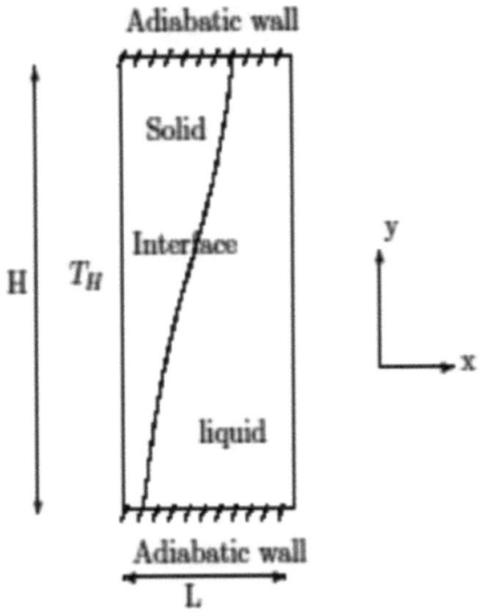

Figure 5.
Boundary conditions through the PCM wall.

$$B = -C\frac{(1-H^2)}{(H^3+b)} \quad (4)$$

The constant C is chosen so that it is high enough to cancel the velocities in the solid region and a low number b is introduced to avoid a division by zero in the case of a zero liquid fraction: where b = 0.001 is a small computational constant used to avoid division by zero, and C is a constant reflecting the morphology of the melting front. A value of C = 10^5 has been used in the literature [1]. The liquid fraction (H) is given by [1, 24].

$$H = \begin{cases} 0 & if \quad T < T_{melting} \\ ]0,1[& if \quad T = T_{melting} \\ 1 & if \quad T > T_{melting} \end{cases} \quad (5)$$

The energy equation is written as [25, 26].

$$Cp_{eq}\frac{\partial T}{\partial t} = \lambda_{eq}\left(\frac{\partial^2 T}{\partial x^2} + \frac{\partial^2 T}{\partial y^2}\right) + Cp_{eq}\left(u\frac{\partial T}{\partial x} + v\frac{\partial T}{\partial y}\right) \quad (6)$$

Cp_{eq} and λ_{eq} that appear in Eq. (8), respectively, represent the equivalent volume capacity (*J/K*) and the equivalent thermal conductivity (*W/m K*) of the two solid and liquid phases of PCM, such as:

$$Cp_{eq} = \sum \theta_i \rho_i (Cp)_i \quad (7)$$

$$\lambda_{eq} = \sum \theta_i \lambda_i \quad (8)$$

The term θ_i that appears in Eqs. (9) and (10) is a quantity related to the H value, which provides information on the state-owned PCM whether liquid or solid. The thermal and dynamic initial conditions are, *respectively,* $u = v = 0$ *(once* $t = 0$*) and* $T = T_0 = 288.15$ *K.*

- In order to account for the phase change process happening when the PCM is melting, the energy balance was applied at the interface as follows [27, 28]:

$$k_s \vec{\nabla} T_s - k_l \vec{\nabla} T_l = \rho L \frac{dX}{dt} \tag{9}$$

- Since the phase change of pure substances occur at a single temperature, the temperatures of liquid and solid at the interface is given by [1]:

$$T_l = T_s = T_i \tag{10}$$

where the subscripts s and l stand for the solid and liquid phase, L is the latent heat (enthalpy) of fusion, and X is the position of the melting interface.

- The Plexiglas cavity was partially filled with PCM-27 heated by a lamp placed at 0.1 m. The lamp imposes the uniformity of heat and temperature, T_H, on the directly exposed PCM wall surface. The condition of adhesion was used to express the velocity fields. The thermal and the dynamic bounder conditions in the PCM vertical enclosure are given by the following expressions:

- The left vertical side of area ($x = L, y, t$) is maintained to a temperature T_H upper to the PCM melting temperature:

$$u(x = L, y, t) = v(x = L, y, t) = 0 \tag{11}$$

$$T(x = L, y, t) = T_H \tag{12}$$

- The right vertical side of the PCM wall ($x = 0, y, t$) of the area is maintained adiabatic:

$$u(x = 0, y, t) = v = (x = 0, y, t) = 0 \text{ and } \partial t/\partial x = 0 \tag{13}$$

- The horizontal walls of the domain are maintained adiabatic ($x, y = 0, t$) and ($x, y = H, t$):

$$u(x, y = 0, t) = v(x, y = 0, t) = 0, \ \frac{\partial T(x, y = 0, t)}{\partial y} = 0 \tag{14}$$

$$u(x, y = H, t) = v(x, y = H, t) = 0, \ \frac{\partial T(x, y = H, t)}{\partial y} = 0 \tag{15}$$

3.1 Energy and exergy analysis

- The energy stored, E_S, in the PCM wall is given by [25]:

$$E_S = M_{PCM-27}.C_{PCM-27}.(T_{PCM,F} - T_{PCM,1}) + M_{PCM-27}.L \tag{16}$$

where $M_{PCM\text{-}27}$ (kg) represents the PCM-27 mass, $C_{PCM\text{-}27}$ (kJ/kg°C) represents the specific heat of PCM-27, $T_{PCM,\ I}$ (°C) is the initial temperature of PCM, $T_{PCM\text{-}27,\ F}$ (°C) is the final temperature of PCM, L (kJ/kg) is the latent heat of fusion, $A_{PCM\ wall}$ (m^2) is the aperture area of the PCM wall, T_{cell} (°C) is the test

cell temperature, and $h_{PCM\text{-}27}$ ($W/m^{2}{}^{\circ}C$) is the heat transfer coefficient of PCM-27.
Energy input during charging is given by:

$$E_{ic} = I.A_{PCM\,wall} \tag{17}$$

where I (W/m^2) is the irradiance intensity of the lamp and $A_{PCM\ wall}$ (m^2) is the PCM wall area.

- The energy efficiency of the PCM wall during the charging and the discharging processes are given by:

$$\eta_c = \frac{E_0}{E_{ic}} \text{ and } \eta_d = \frac{E_S}{E_0} \tag{18}$$

where E_0 is the energy transferred to the PCM wall.

- The overall exergy transferred to the PCM wall is given by [29, 30]:

$$EX_0 = M_{PCM-27}.C_{PCM-27}.(T_f - T_i) + M_{PCM-27}.L.\left(1 - \frac{T_a}{T_s}\right) - M_{PCM-27}.T_a.C_p.\ln\left(\frac{T_f}{T_i}\right) \tag{19}$$

where $T_a(K)$ is the ambient temperature and $T_s(K)$ is the temperature of sun.

- Exergy input during the thermal storage is given by [30]:

$$EX_{ic} = I.A_{PCM\,wall}.\left(1 - \frac{T_a}{T_s}\right) \tag{20}$$

- Exergy efficiency of the PCM wall during the thermal storage and the thermal discharging are respectively given by:

$$\Psi_C = \frac{EX_0}{EX_{ic}} \text{ and } \Psi_d = \frac{E_S X}{EX_0} \tag{21}$$

4. Numerical results

4.1 Validation of the numerical model

This section is devoted to the obtained experimental results and their comparison with experimental data. The validation of the numerical model used in this study was performed by following the temperature changes inside the PCM wall and the melting front states during the charging process. The numerical results were compared with the experimental temperature data recorded in the laboratory during the same period between February 25 and March 13, 2016. **Figure 6** shows the temperature evolution in the vertical plane x = L/2 vs. local time for the low position (a) (y = 1 cm) and the high position (b) (y = 7.5 cm) of the PCM wall. **Figure 6a** shows that the numerical results are quite similar to the experimental measurement. Indeed, the difference between the simulated and the measured values of the temperature at the bottom of PCM is about 0–4°C. It is also seen that in a higher position of PCM wall (**Figure 6b**) the experimental and numerical results of

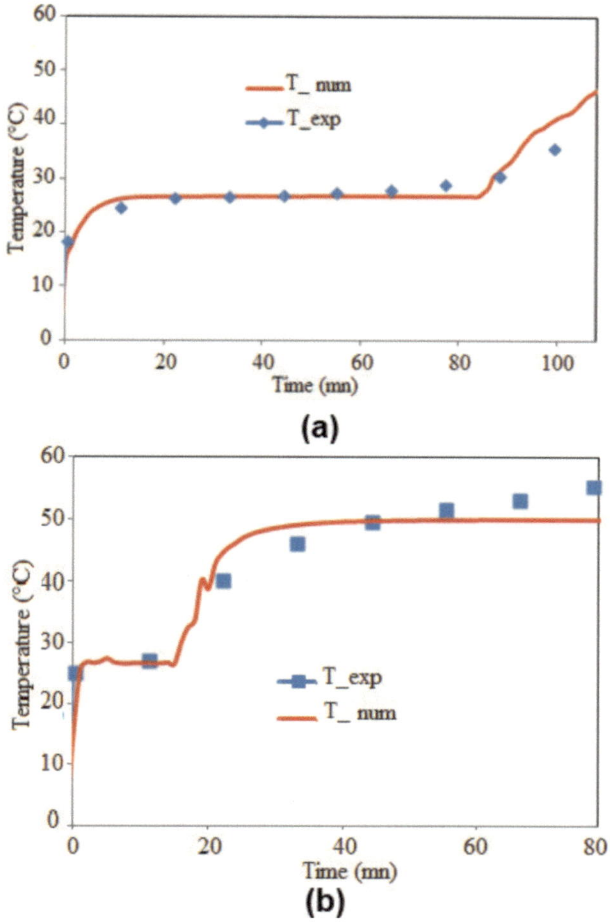

Figure 6.
The simulated and the experimental temperature profile inside the PCM wall for two different times: (a) 4000 and (b) 6000 s.

temperature obtained show an acceptable agreement, of about 0–5°C. It is concluded that the numerical model permits the simulation of the PCM wall thermal behavior with an acceptable accuracy.

4.2 Exploitation of the numerical model

Figure 7 shows the variation of the energy and the exergy stored in the PCM wall during storage process. It is found that the recovered energy incessantly increases vs. charging time. It ranges between 95 and 780 W. This variation takes roughly 130 min and then the energy stored reaches a maximum, which value is due to the fact that the test cell temperature also fluctuates that is in the range of 22–24°C (**Figure 7**). On the other hand, it is found that the exergy stored grows with the charging time. However, it is seen that the exergy is lesser than the stored energy. It varies between 50 and 460 W.

Figure 8 shows the variation of the energy and exergy efficiencies of PCM wall during the charging process. It is seen that the PCM wall performance increases gradually from 10 to 95%. **Figure 9** shows the variation of the energy and exergy efficiencies of PCM wall during the discharging process. It is seen that the PCM wall performance decreases regularly from 100 to 10%. It is found that the energy and exergy efficiencies are more important than the charging process. It is also seen that

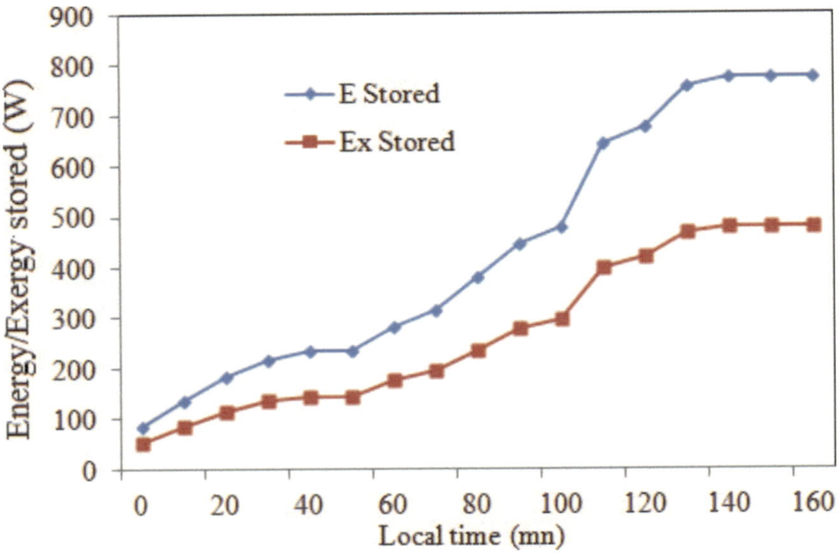

Figure 7.
Thermal energy and exergy changes during the charging process.

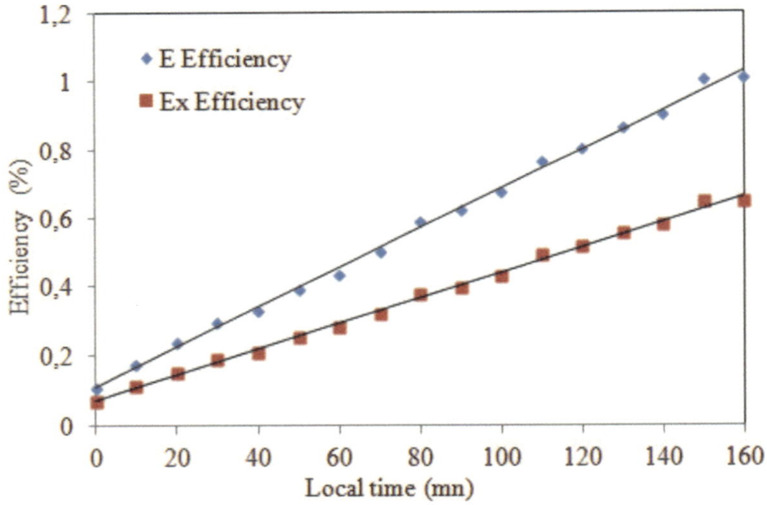

Figure 8.
Thermal energy and exergy efficiency changes during the storage stage.

the exergy efficiency is always found to be lower than the energy efficiency, which is due to the consideration of the losses/irreversibility during exergy analysis, which ultimately gives the information about the quality of energy or available energy. On the other hand, the energy efficiency is all about the quantity of energy rather than quality as it does not consider the losses/irreversibility in the analysis. Both the efficiencies are found to be decreasing with increasing loads. This is due to the fact that backup time is inversely proportional to the increase of the heating load.

In **Figure 10**, the evolution of the melting front inside the vertical enclosure for two different instances (4000 and 6000 s) is represented. At the beginning of the heating process, the PCM-27 inside the vertical enclosure was in solid phase. Then, we detected the presence of two distinct phases: a liquid phase and a solid phase separated by melting front. It was seen that the ending of the melting process was observed after 6000 s. It is also seen that the liquid in the vicinity of the directly

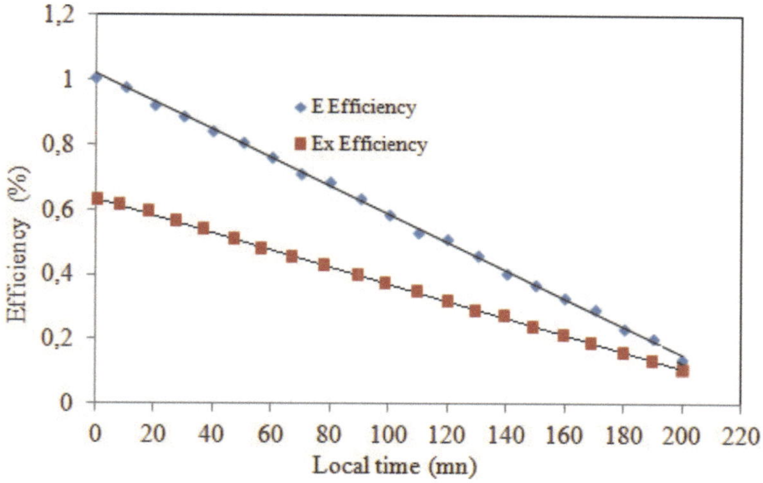

Figure 9.
Thermal energy and exergy efficiency changes during the discharging phase.

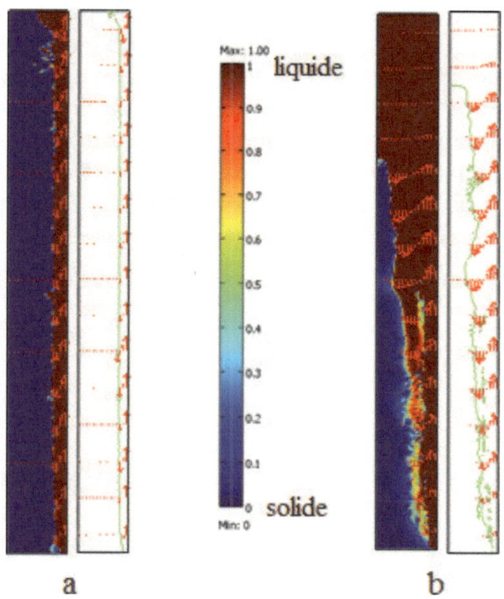

Figure 10.
Evolution of the melting front and the velocity fields inside the PCM wall for two different instances ((a) 4000 and (b) 6000 s).

heated side of the PCM wall acquires heat, causing the temperature increases of the PCM-27. Consequently, a decrease of PCM-27 density was noted, which ascends along the heated PCM wall. At the top of the test cell, the velocity of the fluid is very important, so the liquid descends along the solid-liquid interface. During its descent, it loses its heat to the cold interface. At the bottom of the interface, the fluid is cold, and the temperature of the melting rate gradients is low. A blocking of the thermal transfers leading to the slowdown of the interface movement occurs in the latter region. In the liquid phase, PCM-27, which is at the top of the field, has a slightly higher temperature than the bottom of the cavity temperature. It is noted that the interface movement forms a contour from the bottom of the cavity, along the heated side to descend on the other side of the PCM wall. It is seen that as convection increases, the melting rate increases in the upper part of the PCM wall.

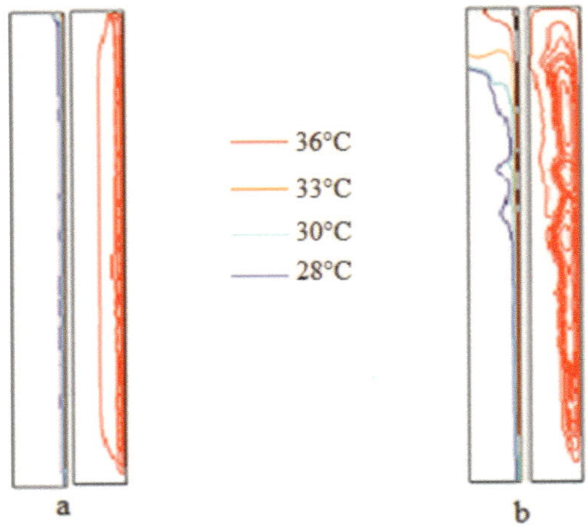

Figure 11.
Isothermal and current lines inside the PCM wall for two different instances ((a) 4000 and (b) 6000 s).

This is explained by the fact that the paraffin in the vicinity of the PCM wall heated side acquires heat, which causes the climb of the upper part with a high speed and then the liquid paraffin descends along the solid-liquid interface.

Figure 11 shows the thermal and the dynamic behavior of the PCM-27 for two different instances (4000 and 6000 s). It was seen that at the beginning of the melting process, the interface is almost vertical, indicating the predominance of heat transfer by conduction mode. Isotherms remain vertical and parallel. Gradually as the convection increases, the melting rate increases in the upper portion of the interface. Therefore, the PCM-27 in the upper area has a higher temperature than the bottom of the cavity and the isotherms do not remain parallel. It is noted also that these movements are not made of the plate toward the solid–liquid interface. They form an outline by gravity from the bottom of the cavity, along the plate to descend on the other side on the solid-liquid interface.

5. Experimental investigation according to Tunisian scenario

An experimental framework and procedure were accomplished in the laboratory in order to evaluate the PCM wall thermal performances, in particular its capacity to store the heat and to moderate internal test cell temperature. The experimental framework considered for the investigation of the thermal performances of the PCM wall comprises essentially a test cell with the dimensions $0.5 \times 0.5 \times 0.5\ m^3$ and managed to imitate a test room. Each sides of the test cell have the dimensions: 0.22 m length, 0.22 m width and 0.026 m thickness. One side of the conceived test cell is fixed with a Plexiglas parallelepiped-shaped container with a size of $22 \times 22 \times 2.6\ mm^3$. The sides of the PCM wall are fixed with the epoxy resin to form a strong bond. Then, the Plexiglas container was field with paraffin-27, which melts at 27°C with a high latent heat storage capacity (about 110 J/g). Upon PCM solidification, a $7\ 10^{-2}$ m free space was left from the top of Plexiglas container to accommodate volume changes and release trapped air during successive melting and solidification phases. A 120-W incandescent lamp is used for heating the exposed side of the PCM wall. The thermophysical properties of the PCM-27 are given in **Table 1**.

Data acquisition is achieved by an autonomous acquisition device controlled by a Lab VIEW program adapted to measure temperature fluctuations during melting

Thermal conductivity (W/m·K)		Heat capacity (kJ/kg·K)		Density (kg/m³)		Enthalpy of fusion (kJ/kg)	Melting temperature (K)
Solid	Liquid	Solid	Liquid	Solid	Liquid		
1.05	0.58	1.42	2.22	1530	1710	172.42	300.15

Table 1.
Thermophysical properties PCM-27 [30].

processes. It acquires the output signals from the thermocouples, digitizes, treats them and then saves the results. Data are directly collected with the help of an interface network between the measurement station and a computer equipped with application software. Storing data is performed with a regular pitch, equal to 60 s, in the form of text file. **Figure 12** shows the entire device and framework. Thirteen T-type copper-constantan thermocouples firstly calibrated with a measurement inaccuracy of ±0.2°C were incorporated to PCM wall at the front surface, inside PCM, the back surface and inside the test chamber as shown in **Figure 10**. Between the exchanging plates, thermocouples (T-type) are inserted on both sides of the sample to measure the temperature fields on each side of the PCM wall. The thermocouples were fixed at the front and back surface with strong white tape and were shielded from direct irradiation.

The experimental investigation was conducted for 14 consecutive days in September 2015 (from 23/09/2015 to 07/10/2015) to evaluate PCM wall performance. The temperature was measured at front surfaces of PCM wall, at back surfaces of PCM wall, inside the PCM and inside the test cell. The method adopted in the investigation consisted of imposing heating flux (lamp of 120 W) on the exposed PCM wall of the test cell. Simultaneous measurements of the temperature variations and heat flux exchanged during charging and discharging processes were accomplished to evaluate the PCM wall thermophysical properties [29, 30]. The experimental test conducted in the laboratory aims to determine the PCM characteristics during the storage phase. The test starts at 9:00 am and continues until the total melting of PCM-27. PCM is initially in the solid state at the ambient temperature of the room, which is about 22°C. Then, the right side of the PCM wall was heated by using a special lamp with a thermal power of about 120 W. The lamp was held at a distance of 10 10^{-2} m to guarantee the uniformity of heat over the entire PCM wall area. By using the data acquisition system, the temperature variation at the interior of PCM wall during the storage stage was followed. To evaluate the heat

Figure 12.
Experimental device and framework.

exchanged through the PCM wall during the discharge process, the lamp was omitted and then the PCM temperature fields were tracked.

The result of the experimental tests permits the characterization of the PCM wall and the description of the test cell thermal behavior. **Figure 13** shows the temperature evolution of PCM-27 inside the wall during the charging and discharging processes along the vertical axis ($x = 10^{-2}$ m) (T1, T2, T3 and T4) and along the horizontal axis ($y = 7.5\ 10^{-2}$ m) (T5, T6, T7, T8 and T9). It is found that the PCM solidification process during the discharging phase takes more time than the storage stage. This is explained by the formation of a solid layer of paraffin in contact with the PCM wall sides, which make a thermal isolation and consequently slow down the crystallization in the other parts of the wall. It is seen that by the launch of the melting process all temperature profiles grow linearly up to 27°C. This phase corresponds to sensible heat storage in the solid PCM. Then, it is noted that during about 20 min of heating process with the lamp of 120 W all positions confess symmetrical temperature profiles, around 27°C. It is noted that PCM-27 temperature rises rapidly especially in the first 20 min of the melting process, which corresponds to the sensible storage process inside the PCM wall. Then, the PCM-27 temperature varies slowly between 27 and 29°C during the latent heat storage process. After about 40 min of heating by the lamp, the PCM-27 temperature increases to reach 50°C. **Figure 13** shows also that the temperature of thermocouples that are close to the heated side increases rapidly than those of the other sides. During this phase, the storage of heat is achieved by sensible process inside the melt PCM. After the melting process, it is noted that the different vertical positions inside the PCM wall ($y = 2, 8, 12$ and $18\ 10^{-3}$ m) presented a dissimilarity in the measured temperature. This delay is explained by the trajectory of the melting front of the solid–liquid interface, which merges firstly from the positions located above the PCM wall to the bottom. The duration of this phase varies between 110 and 40 minutes from one position to another according to the thermocouple on-axis position ($x = 10^{-2}$ m).

Figure 14 shows the variation of test cell air temperatures with and without PCM-27. The test was accomplished by exposing the PCM wall to the lamp of 120 W used as a heating load source. It is seen that during the charging phase the temperature profile inside the test cell with PCM wall is almost stable at 29°C. Indeed, the PCM wall permits the storage of the excess of heat supplied by the lamp. It is also

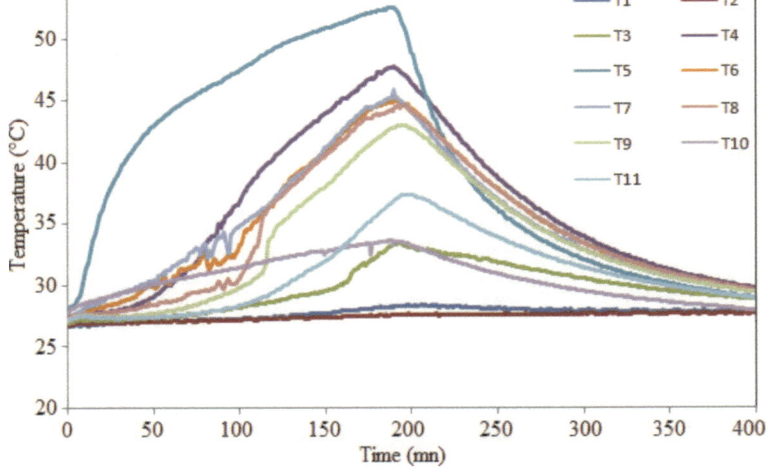

Figure 13.
Evolution of PCM temperature during charging and discharging phases.

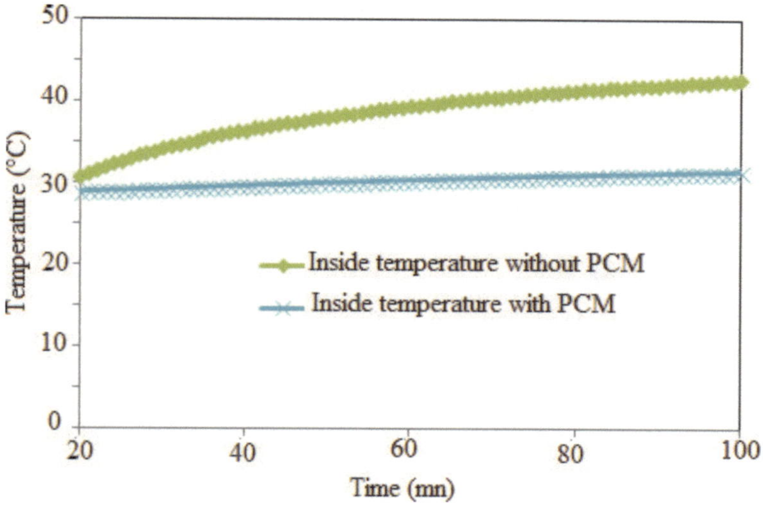

Figure 14.
Evolution of the temperature inside the test cell during the storage phase.

found that once the PCM wall is replaced by simple wooden wall the internal temperature of the test cell grows seriously from 29 to 40°C.

Figure 15 shows the temperature variation of the air inside the test cell with and without PCM wall during the discharging phase (cooling). As can be seen, the air temperature inside the test cell is stabilized around 21°C, which represents the ambient temperature, whereas with PCM wall, the air temperature inside the test cell decreases continuously with respect to time from 27 to 22°C. It is also noted that the temperature of the test cell with PCM is always higher than the temperature of the test cell without PCM. This shows the significance of using PCM inside the room and the temperature in the thermal comfort range can be maintained for a long time even with the heating load. Consequently, intruding of the PCM wall in the test cell presents a good potential to be applied for space conditioning.

To generalize the investigation, a simple room (kid's room) inside a typical modern house (**Figure 16**) in Tunis, Tunisia, is considered. The building is composed of five rooms with a floor area of 128 m^2. The technical specifications of the selected room are presented in **Table 2**. This scenario is archived by considering a

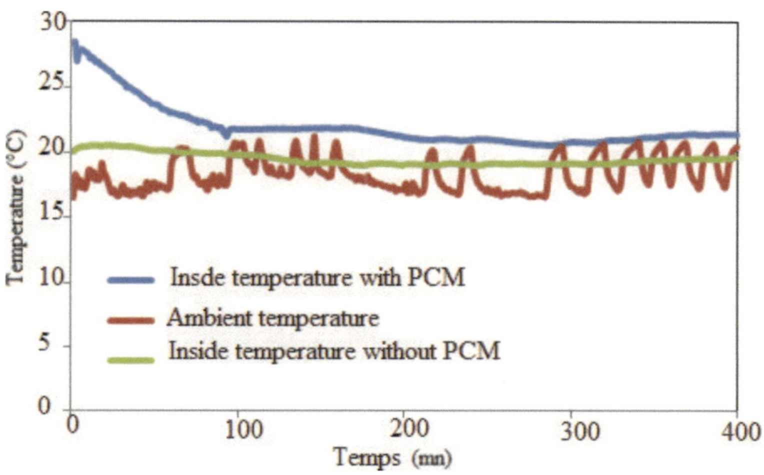

Figure 15.
Evolution of the temperature inside the test cell during the discharging phase.

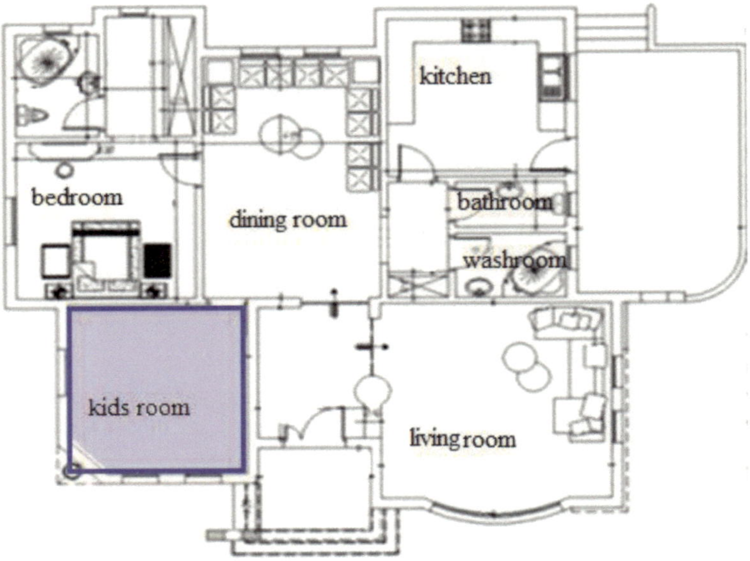

Figure 16.
The typical house plan used in this study.

Type	Layer	Thickness (m)	Conductivity (kJ/(h m K))	Density (kg/m^3)	Specific heat (kJ/kg K)
Wall	Brick	0.15	3.2	1800	1
	Concrete	0.15	7.56	2400	0.8
	Gypsum	0.05	0.75	1200	1
Ground	Concretes	0.06	4.06	1400	1
	Insulation	0.05	0.15	40	0.8
Roof	Concrete	0.24	7.56	2400	0.8

Table 2.
Structure and physical properties of the building structure.

TRNSYS program. The component of the TRNSYS model is the flat-plate solar collector (type73) used as a heat source, storage tank (type4c) and the building (type 56a). In this section, the integration of the PCM wall in the envelope of a typical Tunisian building is simulated by using TRNSYS. The main TRNSYS component used is Type 399, which models a PCM wall. Type 399 is designed to interact with Type 56a and can simulate a PCM wall located in any position within the tested room. It should be also noted that a new Type 399 models a pure PCM-27 that is assumed to go through its freeze/thaw process at constant temperature, to have a constant specific heat in the liquid phase and to have a constant specific heat in the solid phase. The basic architectural specification of the selected room used in Tunisian scenario is given in **Table 2**.

Figure 17 illustrates the evolution of the energy stored and destocked during four days, from January 1 to 4, 2015. During the day, the PCM wall stores a rate of heat brought by the solar collector, and this phase of storage is characterized by positive values of the energy. The latter forms a peak dependent on the period of sunshine and that can reach 1200 kJ m^{-2}. The heat accumulated during the day by the PCM wall will be restored to be used at the end of the day and during the night.

Figure 18 illustrates the variation of the temperature inside the selected room with and without integration of PCM wall during the period from January 1 to 4,

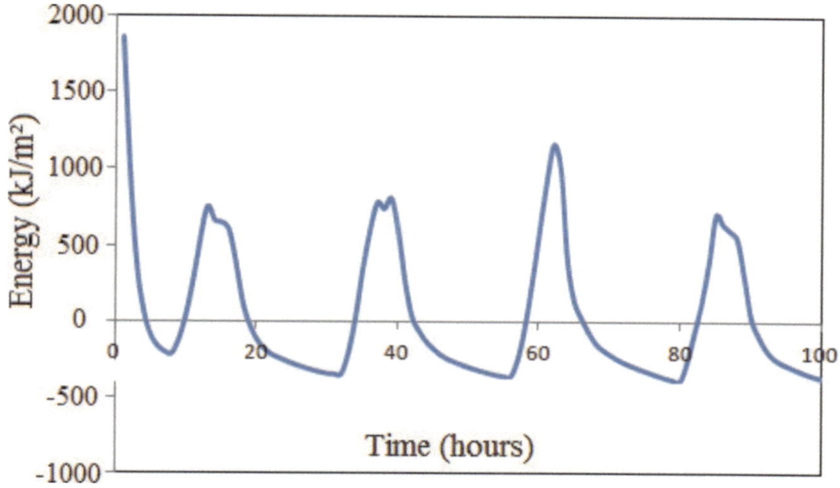

Figure 17.
Evolution of the energy stored in the PCM wall during the discharging phase.

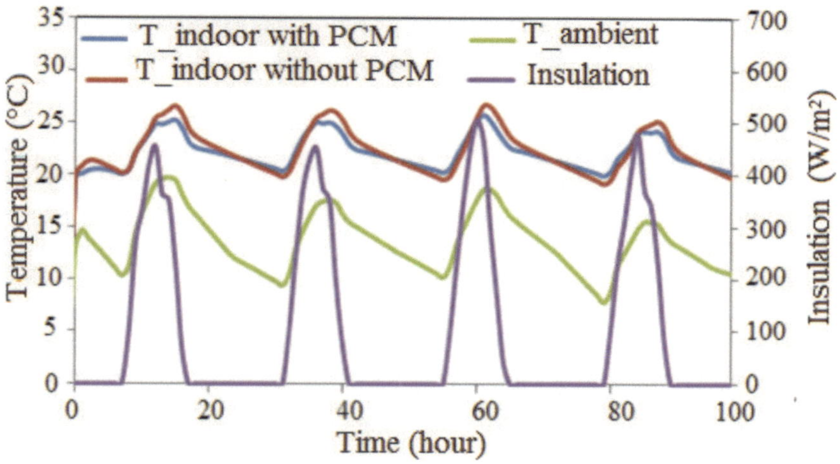

Figure 18.
Evolution of the ambient temperature inside the kid's room with and without PCM wall application.

2015. The outside temperature and the solar irradiation were also evaluated during the same period. During the hottest period of the day, the temperature of the tested room with PCM wall achieves 25°C, while that without PCM wall exceeds 27°C. During the night, the temperature of the tested room, with PCM wall, decreases in the value of 20°C. During the night, the temperatures for the system using the PCM wall become more marked, more of 1°C compared with that of the system without PCM wall. It is seen that the PCM wall performs its function of thermal shock absorber.

6. Conclusion

The objective of this chapter is to show the importance of using PCM as storage material in buildings and to study the thermal potential offered by the integration of a PCM wall to enhance the thermal comfort of the occupant by reducing the thermal fluctuation and by improving the thermal inertia of the buildings' envelop. Accordingly, an experimental prototype represented by a small-scale home

(0.5 × 0.5 × 0.5 m3) was conceived in our laboratory. The test cell was equipped with a PCM-27 vertical enclosure placed at one side of the test cell. Several tests were carried out with an experimental setup designed for testing the viability of using PCM wall integrated in building structure. The experimental study was carried out by measuring temperature through the PCM wall. The test cell indoor temperature was also evaluated to appraise the thermal inertia of the wall envelope. During the heating 22 phase, the temperature inside PCM shelter appears constant at about 28°C. But it varied between 29 and 40°C inside the test room without PCM wall.

A numerical simulation based on FORTRAN program was also carried out to interpret the experimental data. The numerical simulation was achieved to solve the energy and the exergy mathematic relations to evaluate the PCM wall performances by determining the melting phase proprieties during the charging and the discharging processes. The numerical study constitutes a preliminary step before construction of cells equipped with such wallboards in order to obtain a certain indoor passive air conditioning and especially to avoid overheating of buildings during summer. The test of the numerical model shows that there is a good agreement between experimental and numerical results. The numerical model was then exploited to evaluate the PCM wall thermal behavior. It was found that the following of the evolution of the melting front, the velocity fields, the isothermal and the current lines shows that the paraffin-27 melting process is more significant in the upper part of the PCM wall.

To generalize the investigation for a typical modern house composed of five rooms with a floor area of 128 m^2, a TRNSYS program simulation was proposed by considering the Tunisian scenario. The results were presented for a single room equipped with PCM wall. It was seen that during the day, the PCM wall stores a rate of heat that can reach 1200 kJ m^{-2}. It was found that during the hottest period of the day the temperature of the tested room with PCM wall achieves 25°C, while that without PCM wall exceeds 27°C. During the night, the temperature of the tested room, with PCM wall, is about 20°C. It is seen that the PCM wall performs its function of thermal shock absorber. The investigation showed that the efficiency of PCM wall is remarkable in the control and the reduction of the indoor temperature amplitude in the building.

The results obtained by this investigation are exploited in another new experimental work, which is in progress to optimize the geometric and the physical parameters of the PCM wall according to Tunisian buildings' specificity.

Acknowledgements

The authors would like to thank the Laboratoire des Procédés Thermiques (LPT) and the Centre de Recherches et des Technologies de l'Energie (CRTEn), Tunis, Tunisia, for financially supporting the project and for supplying useful data.

Nomenclature

A	PCM wall area (m^2)
C_{PCM}	heat capacity (kJ/kg·K)
K	thermal conductivity (W/m·K)
g	Gravitational acceleration (m/s^2)
H	Liquid fraction
L_m	enthalpy of fusion (kJ/kg)
P	pressure (Pa)

T	temperature (K)
T_i	inside air temperature (K)
T_o	outside air temperature (K)
T_{wi}	inside wall temperature (K)
T_{wo}	outside wall temperature (K)
h	Enthalpy
t	Time (s)
(x, y)	Cartesian coordinates (m)
L	cavity width
(u, v)	velocity components (m/ss^{-1})

Greek symbols

B	coefficient of thermal expansion
μ	dynamic viscosity (kg/m sm^{-1} s^{-1})
α	Thermal diffusivity coefficient
ρ	density (kg/m^3)
ν	kinematic viscosity (m^2/s)

Indices

s	Solid
l	Liquid
fus	Fusion
eq	Equivalent

Author details

Majdi Hazami[1*], Farah Mehdaoui[1], Hichem Taghouti[2], Marco Noro[3], Renato Lazzarin[3] and AmenAllah Guizani[1]

1 Laboratory of Thermal Processes, Research Center for Energy Technologies, Technopole Borj Cedria, Tunisia

2 Department of Electrical Engineering, Laboratory of Analysis, Design and Systems Control, National Engineering School of Tunis, Tunisia

3 Department of Management and Engineering, University of Padua, Vicenza, Italy

*Address all correspondence to: hazamdi321@yahoo.fr

References

[1] Mehdaoui F, Hazami M, Messaouda A, Taghouti H, Guizani A. Thermal testing and numerical simulation of PCM wall integrated inside a test cell on a small scale and subjected to the thermal stresses. Renewable Energy. 2019;**135**:597-607

[2] Solgi E, Fayaz R, Kari BM. Cooling load reduction in office buildings of hotarid climate, combining phase change materials and night purge ventilation. Renewable Energy. 2016;**85**: 725-731

[3] Lizana J, Chacartegui R, Barrios-Padura A, Valverde JM. Advances in thermal energy storage materials and their applications towards zero energy buildings: A critical review. Applied Energy. 2017;**203**:219-239

[4] Lizana J, Chacartegui R, Barrios-Padura A, Ortiz C. Advanced low-carbonenergy measures based on thermal energy storage in buildings: A review. Renewable and Sustainable Energy Reviews. 2018;**82**:3705-3749

[5] Zeinelabdein R, Omer S, Gan G. Critical review of latent heat storage systems for free cooling in buildings. Renewable and Sustainable Energy Reviews. 2018;**82**:2843-2868

[6] Reddy KS, Mudgal V, Mallick TK. Review of latent heat thermal energy storage for improved material stability and effective load management. Journal of Energy Storage. 2018;**15**:205-227

[7] Weinläder H, Klinker F, Yasin M. PCM cooling ceilings in the energy efficiency center – Passive cooling potential of two different system designs. Energy and Buildings. 2016;**119**: 93-100

[8] Soares N, Gaspar AR, Santos P, Costa JJ. Multi-dimensional optimization of the incorporation of PCM-drywalls in lightweight steel-framed residential buildings in different climates. Energy and Buildings. 2014;**70**: 411-421

[9] Navarro L, De Gracia A, Castell A, Cabeza LF. Experimental evaluation of a concrete core slab with phase change materials for cooling purposes. Energy and Buildings. 2016;**116**:411-419

[10] Zhou G, Mengmeng PM. Experimental investigations on the performance of a collector–storage wall system using phase change materials. Energy Conversion and Management. 2015;**105**:178-188

[11] Barzin R, Chen J, Young B, Farid M. Application of PCM energy storage in combination with night ventilation for space cooling. Applied Energy. 2015;**158**: 412-421

[12] Sajjadian SM, Lewis J, Sharples S. The potential of phase change materials to reduce domestic cooling energy loads for current and for future UK climates. Energy and Buildings. 2015;**93**:83-89

[13] Royon L, Karim L, Bontemp A. Optimization of PCM embedded in a floor panel developed for thermal management of the lightweight envelope of buildings. Energy and Buildings. 2014;**82**:385-390

[14] Łukasz W, Maciej J. Computer simulations of heat transfer in a building integrated heat storage unit made of PCM composite. Thermal Science and Engineering Progress. 2017;**2**:109-118

[15] Xiaoming C, Quan Z, Zhiqiang JZ, Xiaowei M. Potential of ventilation systems with thermal energ y storage using PCMs applied to air conditioned buildings. Renewable Energy. 2019; **138**:39-53

[16] Stropnik R, Koželj R, Zavrl E, Uro S. Improved thermal energy storage for nearly zero energy buildings with PCM integration. Solar Energy. 2019;**190**: 420-426

[17] Pushpendra K, Singh R, Shailendra KS. Potential of macroencapsulated pcm for thermal energy storage in buildings: A comprehensive review. Construction and Building Materials. 2019;**225**: 723-744

[18] Li D, Zheng Y, Liu C, Wu G. Numerical analysis on thermal performance of roof contained PCM of a single residential building. Energy Conversion and Management. 2015;**100**: 147-156

[19] Joulin A, Younsi Z, Zalewski L, Lassue S, Rousse DR, Cavrot JP. Experimental and numerical investigation of a phase change material: Thermal-energy storage and release. Applied Energy. 2011;**88**:2454-2462

[20] Huang K, Feng G, Zhang J. Experimental and numerical study on phase change material floor in solar water heating system with a new design. Solar Energy. 2014;**105**:126-138

[21] Prieto C, Cooper P, Fernández AI, Cabeza LF. Review of technology: Thermochemical energy storage for concentrated solar power plants. Renewable and Sustainable Energy Reviews. 2016;**60**:909-929

[22] Krese G, Butala V, Stritih U. Thermochemical seasonal solar energy storage for heating and cooling of buildings. Energy and Buildings. 2018; **164**:239-253

[23] Tasnim SH, Hossain R, Mahmud S, Dutta A. Convection effect on the melting process of nano-PCM inside porous enclosure. International Journal of Heat and Mass Transfer. 2015;**85**: 206-210

[24] Park JH, Lee J, Wi S, Jeon J, Chang SJ, Chang JD, et al. Optimization of phase change materials to improve energy performance within thermal comfort range in the South Korean climate. Energy and Buildings. 2019; **185**:12-25

[25] Fořt J, Trník A, Pavlíková M, Pavlík Z, Černy R. Fabrication of dodecanol/diatomite shape-stabilized PCM and its utilization in interior plaster. International Journal of Thermophysics. 2018;**39**(12):137

[26] Zhang C, Zhang Z, Ye R, Gao X, Ling Z. Characterization of $MgCl_2$ $6H_2O$ based eutectic/expanded perlite composite phase change material with low thermal conductivity. Materials. 2018;**11**(12):2369

[27] Ye R, Zhang C, Sun W, Fang X, Zhang Z. Novel wall panels containing $CaCl_2$ $6H_2O$-Mg $(NO_3)_2$ $6H_2O$/ expanded graphite composites with different phase change temperatures for building energy savings. Energy and Buildings. 2018;**176**:407-417

[28] Ramakrishnan S, Wang X, Sanjayan J. Thermal enhancement of paraffin/hydrophobic expanded perlite granular phase change composite using graphene nanoplatelets. Energy and Buildings. 2018;**169**:206-215

[29] Sarı A, Bicer A, Al-Ahmed A, Al-Sulaiman FA, Zahir MH, Mohamed SA. Silica fume/capric acid-palmitic acid composite phase change material doped with CNTs for thermal energy storage. Solar Energy Materials & Solar Cells. 2018;**179**:353-361

[30] Aadmi M, Karkri M, El Hammouti M. Heat transfer characteristics of thermal energy storage for PCM (phase change material) melting in horizontal tube: Numerical and experimental investigations. Energy. 2015;**85**:339-352

Chapter 9

Water Desalination Using PCM to Store Solar Energy

Paritosh Kulkarni

Abstract

The rising water pollution levels and depleting freshwater sources have formed a delirious inverse proportionality for which the cause is human and effect is also on humanity. A possible solution to this problem is harnessing solar energy to engender thermal energy for solar distillation. Thus solar distillation is one of the potential solutions to asses both the ever-increasing demands for clean water and the inquisition for finding eco-friendly techniques to yield the water. This analysis was undertaken to discover the possible utilization of phase change material on solar distillation in double-slope solar still. The equipment that performs distillation is called "Solar Still." A phase change material (PCM) is a substance that either releases or absorbs energy comparable to the sensible heat during its phase transition to provide useful heat/cooling. Examples of PCM include phenol, paraffin, salt hydrate, and fatty acid. The experiment includes a blank distillate output without PCM, followed by possible optimization on the designed solar still. Solar distillation was performed in the improved solar still with varying types of PCMs. A theoretical model discerning the above phenomena and experiments were performed on the solar still. It was reported that for the yield maximum of water distillate with PCM (Phenol—5 cm) is 370 mL and without PCM is 345 mL, showing a 7.2% increase.

Keywords: solar still, phase change material, phenol, sustainability, water-desalination, renewable energy

1. Introduction

In recent years the energy shortage and water pollution have been rising around the world. With the rapid increase in the level of greenhouse gas emissions, the discovery of alternative sources of energy is increasingly gaining importance for the development of a sustainable world. The rising oil price and environmental regulations have dramatically increased the demand for utilizing alternative power sources [1]. In 2015 WHO and UNICEF reported that around 663 million people still use non-potable drinking water. During the emergency conditions, the need for developing efficient and portable techniques to obtain a clean source of water is paramount. One such method is solar distillation using phase change material.

In the twenty-first century, the impact of energy and water on the socioeconomic development of developed and developing nations is significant [1]. Solar energy is one of many renewable energy sources to obtain stable thermal energy future generations [2]. The process of distillation can be used to get fresh water from brackish or contaminated water. Water is available in different forms, such as seawater, underground water, surface water, and atmospheric water. Clean water is essential for good health. These current conditions serve as a motivational factor for the research conducted, to effectively use Phase Change Material for optimum solar distillation to desalinate the water, abundant in situ around Udupi (near the Indian Ocean) [3].

Single/double-slope solar still is a popular solar device used for converting available brackish or wastewater into potable water. Solar still absorbs the thermal energy solar radiation to distillate polluted water into potable water in an enclosed space—still. The principles of heat transfer and energy balance were the governing equations for the operation of single-slope solar still. Because of its lower productivity, it is not popularly used. Numbers of works are undertaken to improve the productivity and efficiency of the solar still [2, 4].

Several PCMs melt and solidify at an array of temperatures, thus creating a focus on various possible applications. These PCMs are applied for numerous thermal storage systems utilizing latent heat, applications in heat pumps, engineering using solar radiation, and space travel. PCMs have been used for heating and cooling for many years, and the study in this regard has been attracting attention since the past decade. The pragmatic results reckoned in the field of water distillation process with the help of solar energy in the presence of energy storage materials like water and MOFs [2, 5].

Solar still is a latent heat storage system, which uses phase change materials (PCMs). Using PCM is an impactful way of storing thermal energy and has benefits in terms of high-energy storage density and the isothermal nature of the storage process. PCMs have been widely used in latent heat thermal-storage systems for heat pumps, solar engineering, and spacecraft thermal control applications. There are large numbers of PCMs that melt and solidify at a wide range of temperatures, making them attractive in a number of applications [6].

The desalination can provide a 24-hour supply of heat and water in greenhouse-based agricultural projects [7]. In another unsteady state modeling and simulation approach, El-Sebaii and his co-authors presented the transient mathematical models for a single-slope single-basin solar still with and without phase change material under the basin liner [3, 8]. They used stearic acid as PCM and used a computer-based simulation procedure to obtain a better insight of temperatures of the still elements and the PCM. The data were correlated using summer and winter day's temperature data in Jeddah, Saudi Arabia. It was observed that during phase change (liquid to stable) of PCM, the convective heat transfer coefficient from the basin liner to basin water is doubled; thus, the evaporative heat transfer coefficient is increased by 27% upon using 3.3-cm layer of stearic acid beneath the basin liner. Dashtban [9] used paraffin wax as PCM in their theoretical study of PCM-based weir-type cascade solar still. It was expected to obtain enhanced productivity by using PCM, which helps in keeping the temperature of basin high enough to produce the distilled water without interruption, especially after sunset [10]. In this study, water desalination and hot water production using solar still involving PCM are theoretically investigated. The numerical approach is presented to study the performance of desalination units—with and—without phase change materials. The effect of the PCM on the productivity expressed as the amount of water produced is theoretically studied. The following parameters and their effects were theoretically investigated: the type of the PCM, melting point of PCM, solar irradiation. It is hoped to determine the optimum parameter that will result in higher unit productivity. The purpose of a solar distillation system is to clean or purify water within the permissible limit [11, 12]. Besides the problem of water shortage, process energy constitutes another problem area. Due to the high cost of conventional energy sources, which are also environmentally harmful, renewable energy sources have gained more attraction since their use in distillation plants will save conventional energy for other applications, reduce environmental pollution and provide a free, continuous, and low maintenance energy source [13].

The objective of this thesis is to study how solar distillation is used by nature to produce rain, which is the primary source of freshwater supply and replicate the process using knowledge of engineering. Solar radiation falling on the surface of the sea is absorbed as heat and causes evaporation of the water. The vapor rises above the surface and is moved by winds [7]. When this vapor cools down to its dew point, condensation occurs, and freshwater precipitates as rain. All available artificial distillation systems are small-scale duplications of this natural process.

Solar distillation differs from other forms of desalination that are more energy-intensive, such as methods such as reverse osmosis, or simply boiling water due to its use of free energy. A very common and, by far, the most significant example of solar distillation is the natural water cycle that the Earth experiences [4].

The novelty of the present study is an in-depth and multifaceted comparison of two solar distillation methods, that is, one utilizing PCM to discover an effective way to bring about solar distillation and another technique reflecting the natural phenomenon of distillation. Two methods can accompany the distillation of water using solar energy. The first method utilizes the so-called greenhouse effect to evaporate saltwater enclosed in a simple solar still (direct collection). A typical basin type containing the saline water is covered with a transparent airtight top. The top is mounted sloping downward toward sweet water collecting troughs. Solar energy is absorbed by the basin, causing the water to evaporate and condense on the inside of the cover and slides down into the collecting trough. The second method or indirect collection system often involves more than one subsystem, one for collecting and another for energy storage and the third system for energy utilization in the distillation process, multi-stage flash evaporation mar offer good potential when utilized in a solar distillation system [6].

To emphasize the scope of this research, consider rural areas in and around Udupi district and several such underdeveloped regions around the tropic of cancer. In these areas, the solar distillation process yields drinking water tantamount to the process of rain generation in the water cycle. The solar radiation causes water to evaporate, segregating the water vapor from salt or impurities. The vapor formed from the process of evaporation condenses on the still for collection [14]. The fundamental operation is evaporation. As the temperature of water increases, vaporization starts at the surface of the liquid. The vapor then rises from the surface of the water and gets condensed on the top cover. The condensed vapor is free from minerals and impurities and thus separated through some distillate channel [15]. The application of finding from this thesis encompasses solar energy applications, primarily supported by government initiatives in the countries along tropic of cancer, including India, Mexico, and UAE.

We are choosing two heights—5 and 7 cm— of PCM to understand and relate this distillation over real application in salt pans. The average height of human-made salt pans near the coast of Karnataka is in the range of 25 cm. Since the height of PCM is less than the height of the salt pan, its implementation is possible. If implemented, this method of distillation can obtain not only potable water but also residue salt [16]. The 2-cm increment is to study the effect.

Figure 1 below represents the solar still to be designed as intended, upon which experimental work has been performed to yield the results as presented below.

Figure 2 represents the schematic of that implementation. The angle of Glass Cover is kept at 32° following the laws of reflection and refraction, to explain that let us consider Snell's law of refraction and law of reflection.

Figure 1 schematic is a focused view of glass cover in **Figure 2**, say $\mu_2 = 1.003$ and $\mu_1 = 1.517$ are refractive indices of air and glass cover [17];

we have,

$$\mu_1 \times Sin\ \theta_2 = \mu_2 \times Sin\ \theta_1 \quad (1)$$

And so assuming all rays are incoming perpendicular, thus,

$$\text{Sin}\,\theta_1 = 1,$$

we get

$$\theta_2 = \text{Sin}^{-1}(\mu_2/\mu_1) = 41.14°, \quad (2)$$

which is the ultimate critical angle of the glass cover, and the angle we chose is 32° signifying maximum refraction [4].

Giving as of **Figure 1**, say we have $\theta_1 = 32°$ and $\theta_2 = 58°$

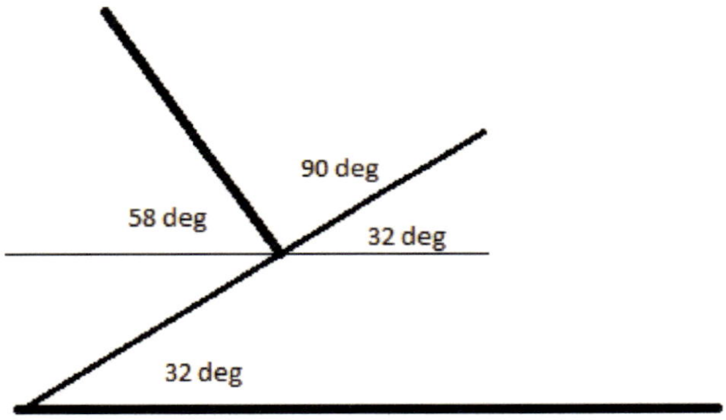

Figure 1.
Schematic of glass cover and sunrays.

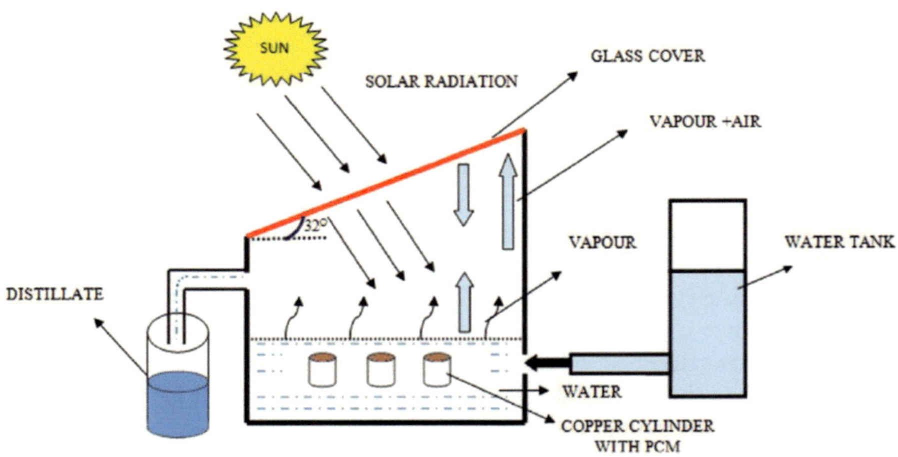

Figure 2.
Solar distillation still with PCM: schematic [4].

2. Methodology

2.1 Materials

Figure 3 shows PCM classified according to their commonalities as per the melting point and the enthalpy of fusion. It follows that the two vital characteristics of phase change material, relating to their semantics "phase" and "change," are derivates of temperature and heat released during the phenomenon of phase change [18, 19].

In this pragmatic study, the experimental setup is similar to the one described in **Figure 1**, and the following phase change material was used:

- Water and water solutions with eutectic compositions are used below the triple point [5].

- Phenol.

As we can observe in **Figure 2**, both of these compounds are on the left corner of the graph with melting temperature near zero or below zero and enthalpy of fusion around 300 MJ/m^3. Hence, the comparison is rather challenging owning to the similarities between the two compounds [20].

As far as the solar distillation goes, the following **Figure 4** summarizes the various substances used for an optimized solar distillation setup. Each material novel and being researched upon [14].

2.2 Experimental synopsis

2.2.1 Experimental setup

Double-slope solar still shown schematically in **Figure 5** was used to conduct the experiments [21]. Concerning **Figures 1** and **2**, we have designed this schematic

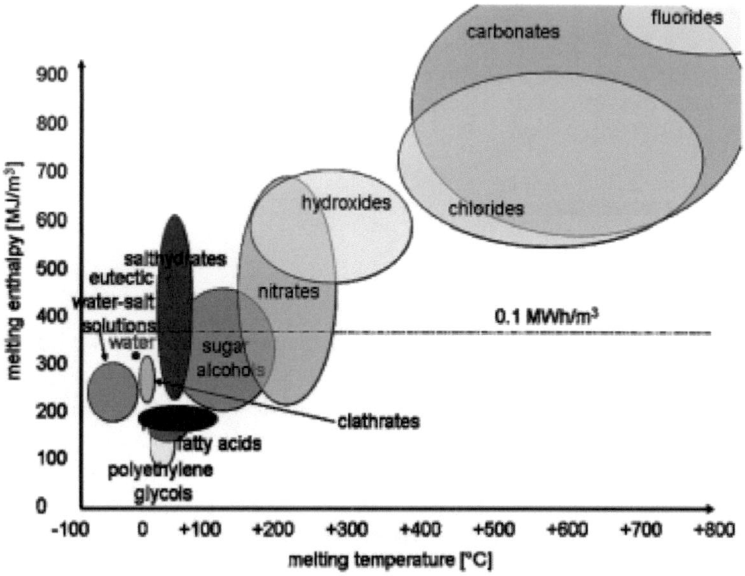

Figure 3.
Classes of materials that can be used as PCM and their typical range of melting temperature and melting enthalpy [15].

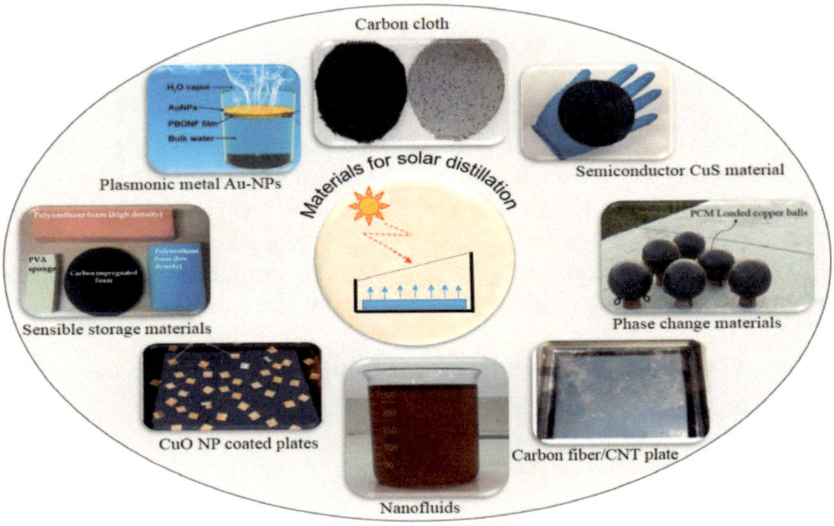

Figure 4.
Various materials used for solar distillation [3].

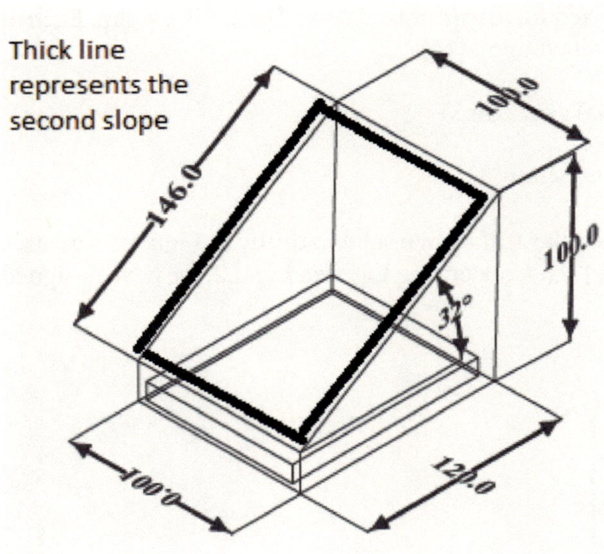

Figure 5.
Schematic of double-slope solar still measurements in cm.

with lines of varying strokes and measurements of solar still that we implemented for real-time experiments.

As shown in **Figure 5**, the base or basin of double-slope solar still was made using an 18-mm-thick waterproof plywood obtained from a local vendor marking the instance of the in situ experimental setup. The side walls were constructed using the same 6-mm thick plywood. The solar basin had an approximate active area (A) of 0.9 m^2. The inside of still was coated with waterproofing M-Seal an epoxy compound with a resin and a hardener. The compound prevents leakage through joints of sidewalls and the base of the still [22].

Since solar radiation has three components for the receiving surface namely, absorption, reflection, or transmission.

To account for these characteristics, we introduce additional properties:

- Absorptivity, **α**, as the fraction of incident radiation absorbed.
- Reflectivity, **ρ**, the fraction of incident radiation reflected.
- Transmissivity, **τ**, the fraction of incident radiation transmitted.

We see, from conservation of energy:

$$\alpha + \rho + \tau = 1 \qquad (3)$$

Since the solar still includes opaque surfaces, as we are painted the walls with black,

$\tau = 0$**, so that:** $\alpha + \rho = 1$

The basin of the still was also painted black. Owning the height (h) of 0.2 m and area (A) of 0.9 m^2 basin has the capacity of

$$V = h \times A \qquad (4)$$

$$0.900 \times 0.2 = 0.0180 \ \mathbf{m^3} \ \textit{or} \ 180 \ \mathbf{L}$$

Through the sidewalls, the distillate was collected via streamline channels. In an enclosed basin tank that was subjected to the solar radiation was filled with tap water via the inlet valve. Temperatures of water, glass cover, and water-vapor mixture were noted every hour using thermocouples of k-type, which have an accuracy of ±0.2°C and a least count of 0.1°C [23–25].

Distillate collected was also measured during temperature recording. Solar intensity falling on solar still was taken from the reading measured by Pyranometer. We had statistical knowledge from SynergyEnvio-Engineerings that around Udupi, Karnataka (*Latitude: 13.35, Longitude: 74.75*). Annual average of solar irradiation (***E***) is 5.44 kWh/m^2/day.

With an area of 0.9m^2 *translates to* **A ∗ E = 5.44 ∗ 0.9 = 4.895 kWh/day** *on a bright sunny day*.

For approximately 90 bright sunny days in summer, this translates to (90 × 4.895) or 440.55 kWh.

440.55 kWh energy per still, considering a rooftop has room for 10 stills (maximum space needed is 13 m^2), this energy is equivalent to *4405.5 kWh* [18].

2.2.2 Solar distillation without PCM

Water was the sole element in the still influencing the heat released and the rate of interphase mass transfer. So we studied the water up to two depths 0.05 and 0.07 m and calculated relevant parameters [20].

2.2.3 Experimental investigation to find the effect of PCM on solar distillation

The PCM material was evenly distributed and covered by a 5-mm thick metal plate. The sides of the metal plate were sealed using M-Seal chemical to avoid leakage or contact of PCM and water. The same solar distillation experiments were conducted with a fixed amount of different PCM, and the distillate collected was recorded on an hourly basis; 5 kg of phenol as PCM was used to perform the experiments at 0.05 and 0.07 m.

3. Results

We have used single-basin, double-slope solar distillation still for the study and hence the results have signified the need for multi-slope and multi-basin for improved performance. Observations were made for water temperature ($\mathbf{T}_w$), top condensing surface temperature ($\mathbf{T}_g$), and distillate output in mL. The temperature of water and glass was measured using two k-type thermocouples. The vapor temperature ($\mathbf{T}_v$) was taken as the average water temperature and glass temperature.

$$\mathbf{T}_v = (\mathbf{T}_w + \mathbf{T}_g)/2 \tag{5}$$

3.1 Solar distillation readings at various heights

Tables 1 and **2** describes the variation of water temperature recorded from 8:30 am until 5:30 pm in 9 hours, each with different heights of water—5 and 7 cm respectively. Water and condensing surface temperature are averaged and recorded in a separate column, which is correlated with distillate collected, **Figures 6** and 7 show variation of temperatures $\mathbf{T}_w$, $\mathbf{T}_g$, and $\mathbf{T}_v$ with increasing time in hours. As the day progresses till noon, the water temperature increases faster as compared to the condensate temperature due to exposure of the glass surface to the ambient atmosphere. Alteration in condensate and water temperature can be attributed to unstable climatic conditions. However, in all cases, the pattern followed by the hourly variation shows a constant rise and fall in all line plots (**Figure 8**).

It can be inferred from the above implications and **Figures** 7 and **9** that the yield of distillate and the pattern of rising and falling of the curve is similar indicating no variation upon water depth change. The temperature of the condensing surface temperature is rising since morning till past noon and decreases after maxima. We can notice that the temperature variation of water basin coupled with PCM, as shown in **Figures 10** and **11** show similar and broader variation when compared to the curves of water in **Figure 8** and **9** [26, 27]. In the early hours of the day, the

S. No.	Time (h)	T_w (°C)	T_g (°C)	T_v (°C)	Distillate (mL)
1	9	30.3	28.3	30.01	0
2	10	39.3	32.7	36.3	23
3	11	45.6	37.2	421.4	106
4	12	52.2	44.1	47.15	204
5	13	54.3	52.7	52.5	287
6	14	56.1	53.3	55.7	348
7	15	55.2	48.3	53.55	273
8	16	49.1	45.7	48.2	217
9	17	48.7	38.5	42.65	149

Water depth = 0.05 m.

Table 1.
Experimental results without PCM.

S. No.	Time (h)	T_w (°C)	T_g (°C)	T_v (°C)	Distillate (mL)
1	9	30	27.8	29.95	0
2	10	35.3	32.8	36	5
3	11	41.2	37.6	41.4	9
4	12	48.7	43.8	48.15	109
5	13	51.3	47.4	53.5	201
6	14	54.8	50.8	54.7	330
7	15	52.3	45.3	52.55	292
8	16	48.7	39.1	47.2	215
9	17	45	34.2	41.65	157

Water depth (d) = 0.07 m.

Table 2.
Experimental results without PCM.

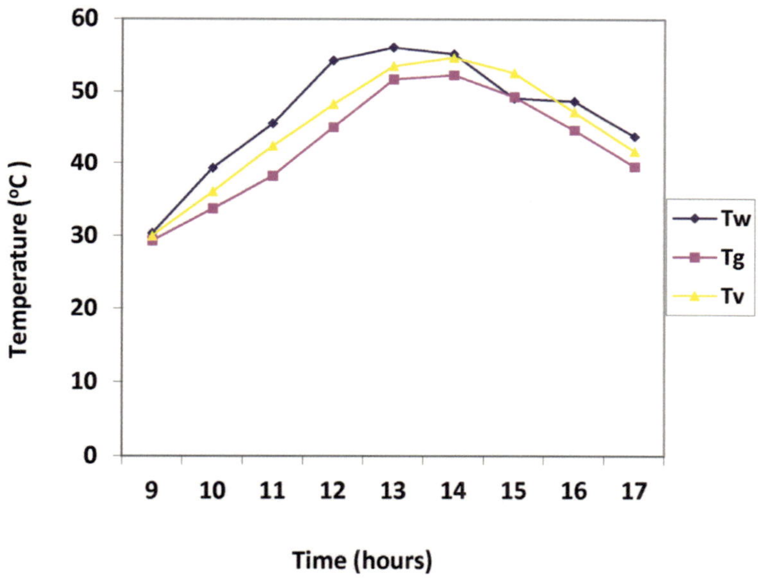

Figure 6.
Hourly variation of temperature at water depth-1.

inner glass temperature is close to that of water basin temperature. However, as the day, progresses the difference broadens because water can absorb some of the incident solar energy, whereas glass transmits most of the incident solar intensity. From figures about PCM phenol, we can see how phenol is retaining the solar energy, which decreases the slope of the line, which indicates declining temperature with an increase in hours.

3.2 Solar distillation using PCM

Five kilograms of phenol was covering the 5-mm metal plate in the solar still basin while the experiment was being conducted, there was no mixing of water and PCM. The reading of these experiments was taken on an hourly basis till 5.00 pm and the cumulative distillate of the next 2 h was taken the next day at 9.00 am [8].

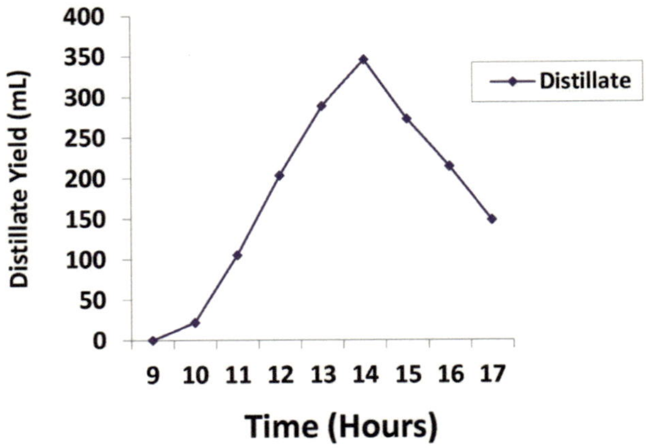

Figure 7.
Hourly variation of distillate yield at water depth-1.

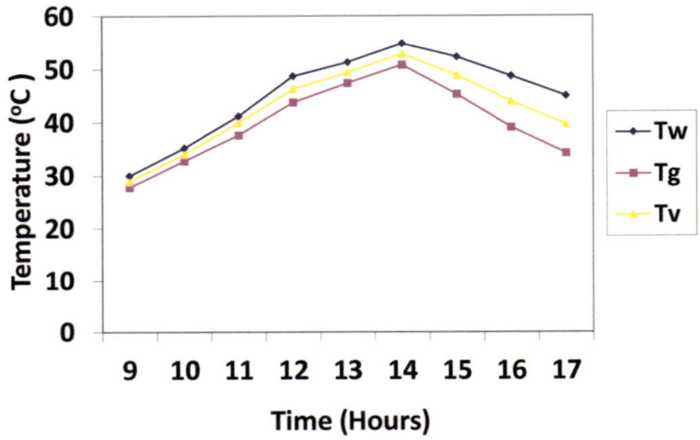

Figure 8.
Temperature variation at water depth-2.

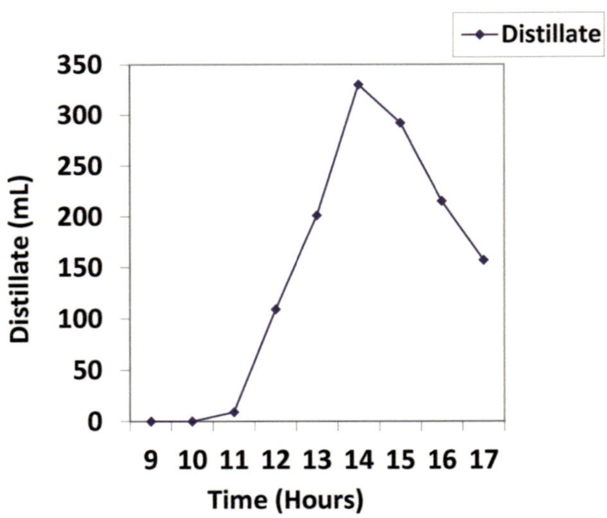

Figure 9.
Hourly variation of distillate yield at water depth-2.

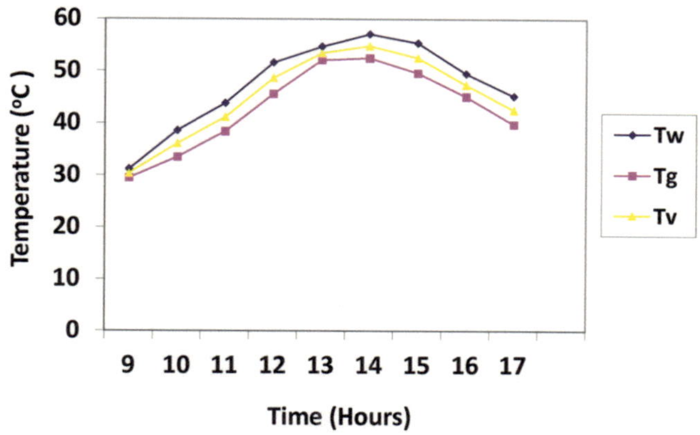

Figure 10.
Hourly variation of distillate with phenol as PCM-1.

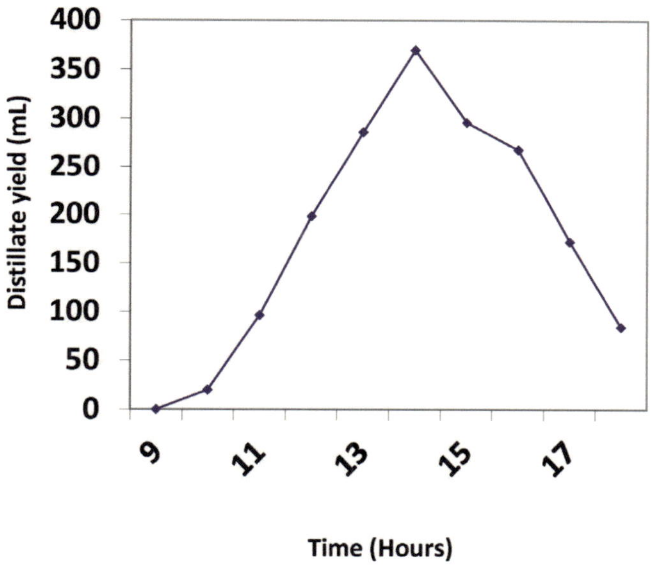

Figure 11.
Hourly variation of distillate yield with phenol as PCM-2.

3.2.1 Solar distillation with phenol as PCM

In **Figures 12** and **13**, it can be seen that the highest temperature attained by the water basin decreases with an increase in depth of water as in the case of PCM with and without PCM too (**Tables 1** and **2**). However, the standard deviation when Phenol as PCM is comparatively larger (**Tables 3** and **4**). In **Figures 10** and **11**, it can also be observed that for phenol, there has been a 4.1% drop in the maximum condensate surface temperature when the water depth has been increased from 5 cm to 7 cm. The decrease in water basin temperature with an increase in depth of water can be attributed to an increase in the volume of water. After sunset, due to a lack of solar radiation, the temperature of water in the basin decreases at a slower rate due to the use of stored heat energy from the PCM. The variation between water basin temperatures in two cases without phenol and with phenol at different water depths is subject to environmental conditions like fluctuation in solar radiation, wind speed, ambient temperature, and spatial wind barriers.

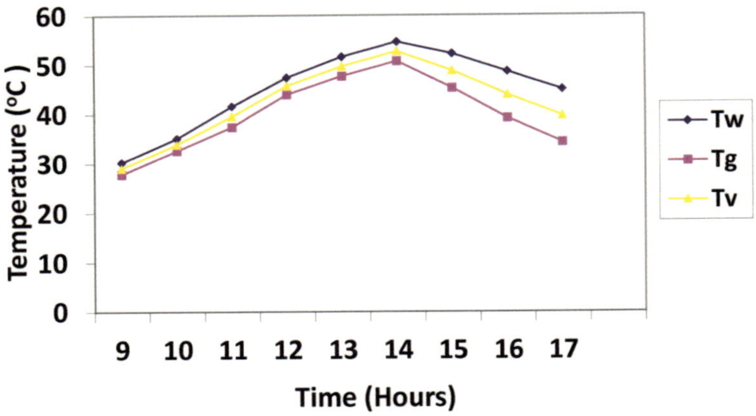

Figure 12.
Hourly variation of temperature with phenol-1.

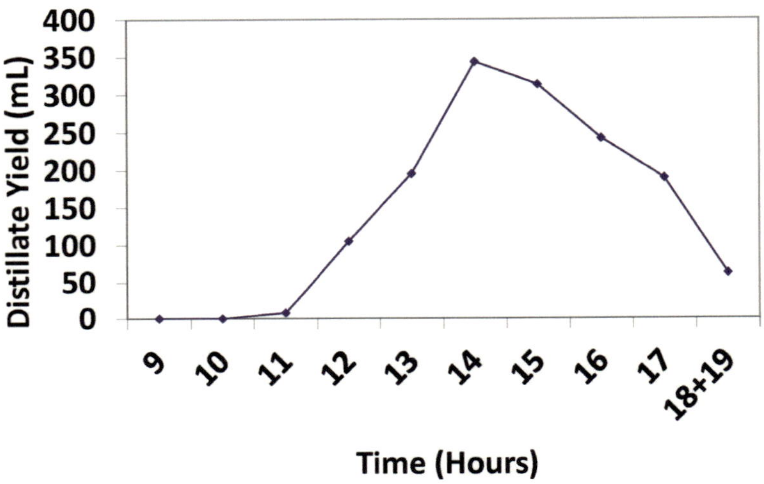

Figure 13.
Hourly variation of temperature with phenol-2.

S. No.	Time (h)	T_w (°C)	T_g (°C)	T_v (°C)	Distillate (mL)
1	9	31.8	29.5	30.35	0
2	10	38.6	33.5	36.05	28
3	11	43.8	38.4	41.1	95
4	12	51.6	45.6	48.6	193
5	13	54.7	52.1	53.4	286
6	14	57.1	52.5	54.8	370
7	15	55.4	49.6	52.5	297
8	16	49.5	45.1	47.3	268
9	17	45.2	39.8	42.5	1760
10	18 + 19	—	—	—	84

Water depth = 0.05 m.

Table 3.
Experimental results with phenol as PCM.

S. No.	Time (h)	T_w (°C)	T_g (°C)	T_v (°C)	Distillate (mL)
1	9	30.2	27.8	29	0
2	10	35.1	32.6	33.85	5
3	11	42.8	37.4	39.5	8
4	12	47.4	44	45.7	115
5	13	51.6	47.7	49.65	195
6	14	54.6	50.7	56.78	323
7	15	52.2	45.3	48.75	313
8	16	48.6	39.2	43.9	242
9	17	45	34.3	39.65	189

Water depth = 0.07 m.

Table 4.
Experimental results with phenol as PCM-2.

4. Discussion

The rationale behind conducting this study was to establish a clear relationship between incident solar radiation and the amount of fresh water produced. After developing the relationship and analyzing statistical information, we concur with the results from confirming data with two different depths [28, 29]. Then PCM was introduced, we chose phenol having attributions of economic availability and versatile properties. Phenol was also varied between two heights, and data was contrasted with that of water. Underlying factors were calculated as follows related to the heat and mass transfer [15].

We know from Dunkle [30], the hourly evaporation per m^2 from solar still is given by:

$$Q_{ew} = \mathbf{0{:}016} h_{cw}(P_w - Pg) \tag{6}$$

where,

$$\mathbf{Nu} = \mathbf{h_{cw} d/k} = \mathbf{C\,(Gr\,Pr)^n} \tag{7}$$

C and n are constants.
Also, the heat produced or

$$Q_{produced} = m_w{}^*L = UA\text{ʒ}T \ldots \text{ʒ}T = Ta - Tb, \tag{8}$$

which is approximately **(Ta − Tw)** *as basin and water temperature is almost the same, basin being assumed a black body.*

Above gives a model to find **Q** *and* **A** *for the required basin the fraction* **Q/A** *yields the power in kWh/m*^{2}.

4.1 Effect of the amount of water

For a fixed amount of water, the cumulative amount of freshwater produced had a steep rise as the sun goes higher until sunsets (**Table 5**). PCM, however, continue to heat the water even after the sunset giving the effect of evaporation a boost. We are assuming this as a unit operation under steady-state conditions because we are assuming that the feed water equals the sum of the rate of freshwater produced

S. No.	Height (m)	PCM	Distillate collected (mL)
1	0.05	No PCM	1595
2	0.05	Phenol	1788
3	0.07	No PCM	1333
4	0.07	Phenol	1478

Table 5.
Cumulative distillate collected results.

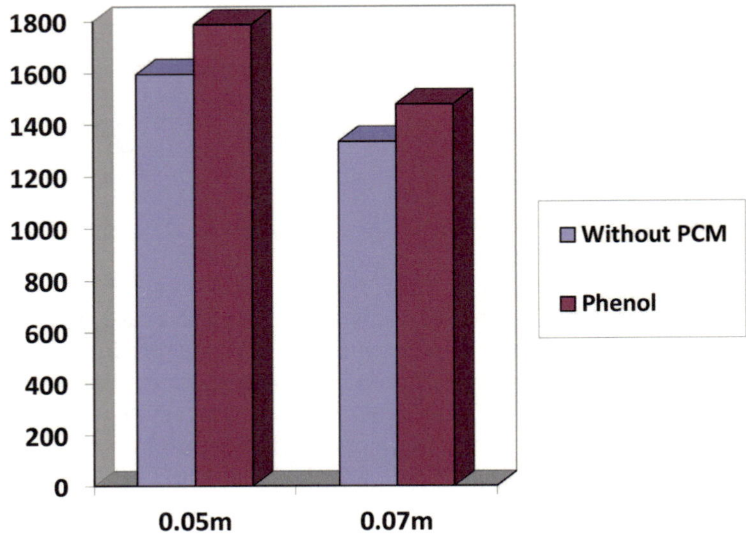

Figure 14.
Cumulative distillate collected at two heights.

and the rate of hot water leaving the unit. Accordingly, the productivity of the unit decreases since the vapor pressure decreases. One also can notice (**Figure 11**) that the rate of production is significant during the day time and gets lower after sunset. The outcome of variation of height is that the amount of distillate collected reduced with the increase in the height of water in the solar still (**Figure 14**) [31, 32].

4.2 Advantage of solar still with PCM over ordinary solar still

The potential advantages of solar still with PCM (phenol) are numerous, including flexibility, processability, low material cost, and independence on scarce resources. The flexibility as an advantage is shared with solar cells and solar energy storage panels and is a feature allowing the solar stills with PCM to be incorporated into applications where flexibility is an advantage. Such solar stills that can be rolled out onto a roof or other surfaces are one option. Processability is another major selling point of PCM infused solar stills. Both solar stills with and without PCM depend on distillation methods wherein sunrays are concentrated by glass requiring massive amounts of energy; with PCM based solar cells, on the other hand, energy is stored and for distillation, which yields distillate of desalinated water and complete setup are have a possibility of implementation on a larger platform. Flexibility and more energy storage capacity allows for up-scaling the production and thus reducing the cost per area of PCM solar stills. The promise of low material cost and minimal use of scarce materials can be realized with optimized PCM solar stills.

5. Conclusion and future scope of work

The rationale behind this research work is to apply and analyze the thermal energy engendered by the PCM using the incident solar radiation. The practical work conducted at two water depths concludes the inverse proportionality between the water level and heat released by water; we relate this to the volume occupied by the water in the still. It also follows that, as the water depth decreases, the distance between the top condensing cover and surface of the water also increases, affecting the distillate production.

- The cumulative distillate yield at 0.05 m in the double-slope solar still was 1595 mL, and 1788 mL when the experimented with no PCM, phenol as PCM.

- The cumulative distillate yield at 0.07 m in the double-slope solar still was 1333 mL, and 1478 mL, when the experimented with no PCM and phenol as PCM,

- The effectiveness of PCM to be used to enhance solar distillation intersected with the depth of water in solar still as the efficiency of still changed.

- Phenol gave an increase of distillate yield of nearly ~11.5%.

The data available could be used to prepare the theoretical model to predict the performance of solar stills for solar distillation under the climatic condition in the parts of the world (where the intensity of solar irradiation is around 5.44 kWh/m^2/day). Efficient and optimized stills and solar distillation systems are projected as replacing wood with carbon fiber or a more effective insulator. As we can observe in the world map, near the tropic of cancer where solar radiation is potent, and seashore is close, solar distillation is a viable option in case of water shortage. Solar stills are subject to further analysis to separate dirt particles and impurities. Stills can also be used in groundwater as well as tap water to improve the quality of water by removing dirt and unwanted particles. In essence, solar distillation would play a vital role in meeting world freshwater supply demands. The data obtained could be used to investigate the scope of solar distillation further. From this investigation, we discovered that for domestic application, double-basin single-slope cascade solar still is a suitable and economical design.

Nomenclature

C	constant
b	average spacing between water and glass surface (m)
d	the depth of the water (m)
Gr	Grashoff number (dimensionless)
A	area of the basin (m^2)
h_{cw}	convective heat transfer coefficient from water surface to glass (W/m^2 °C)
k	thermal conductivity (W/m °C)
L	latent heat of vaporization (J/kg)
m_w	yield of still per unit area per hour (kg/m^2/h)
Pg	partial vapor pressure at glass temperature (N/m^2)
E	energy of incident radiation
P_w	partial vapor pressure at water temperature (N/m^2)
Ta	ambient air temperature (°C)
Tb	basin temperature (°C)

T_g	average glass temperature (θ°C)
T_w	average water temperature (°C)
T_{w0}	temperature of basin water (°C)
U_L	overall heat transfer coefficient (W/m^2 °C)

Author details

Paritosh Kulkarni
Department of Chemical Engineering, Manipal Institute of Technology (MIT), Manipal Academy of Higher Education, Manipal, India

*Address all correspondence to: paritosh.chandrashekhar@learner.manipal.edu

References

[1] Wu S, Yan T, Kuai Z, Pan W. Preparation and thermal property analysis of a novel phase change heat storage material. Renewable Energy. 2020;**150**:1057-1065. DOI: 10.1016/j.renene.2019.11.002

[2] Hedayatizadeh M, Sarhaddi F, Pugsley A. A detailed thermal modeling of a passive single-slope solar still with improved accuracy. Groundwater for Sustainable Development. 2020;**11**:100384. DOI: 10.1016/j.gsd.2020.100384

[3] Abu-Arabi M, Al-harahsheh M, Ahmad M, Mousa H. Theoretical modeling of a glass-cooled solar still incorporating PCM and coupled to flat solar plate collector. Journal of Energy Storage. 2020;**29**:101372. DOI: 10.1016/j.est.2020.101372

[4] Sonker VK, Chakraborty JP, Sarkar A, Singh RK. Solar distillation using three different phase change materials stored in a copper cylinder. Energy Reports. 2019;**5**:1532-1542. DOI: 10.1016/j.egyr.2019.10.023

[5] Vigneswaran VS, Kumaresan G, Dinakar BV, Kamal KK, Velraj R. Augmenting the productivity of solar still using multiple PCMs as heat energy storage. Journal of Energy Storage. 2019;**26**:101019. DOI: 10.1016/j.est.2019.101019

[6] Zurigat YH, Abu-Arabithe MK. Modelling and performance analysis of a regenerative solar desalination unit. Applied Thermal Engineering. 2004;**24**:1061-1072

[7] Atul S, Tyagi VV, Chen CR, Buddhi D. Review on thermal energy storage with phase change materials and applications. Renewable and Sustainable Energy Reviews. 2009;**13**(2):318-345. DOI: 10.1016/j.rser.2007.10.005

[8] Gugulothu R, Somanchi NS, Vilasagarapu D, Banoth HB. Solar water distillation using three different phase change materials. Materials Today: Proceedings. 2015;**2**(4-5):1868-1875. DOI: 10.1016/j.matpr.2015.07.137

[9] Salem MR, Salem MR, Higazy MG, Abdrabbo MF. Performance enhancement of a solar still distillation unit: A field investigation. Solar Energy. 2020;**202**:326-341. DOI: 10.1016/j.solener.2020.03.098

[10] Bhardwaj R, ten Kortenaar MV, Mudde RF. Maximized production of water by increasing area of condensation surface for solar distillation. Applied Energy. 2015;**154**:480-490

[11] Karimi Estahbanati MR, Feilizadeh M, Jafarpur K, Feilizadeh M, Rahimpour MR. Experimental investigation of a multi-effect active solar still: The effect of the number of stages. Applied Energy. 2015;**137**:46-55

[12] Voropoulos K, Mathioulakis E, Belessiotis V. Solar stills coupled with solar collectors and storage tank—Analytical simulation and experimental validation of energy behavior. Solar Energy. 2003;**75**:199-205

[13] Maliani OD, Bekkaoui A, Baali EH, Guissi K, El Fellah Y, Errais R. Investigation on the novel design of solar still coupled with two axis solar tracking system. Applied Thermal Engineering. 2020;**172**:115144. DOI: 10.1016/j.applthermaleng.2020.115144

[14] Arunkumar T, Ao Y, Luo Z, Zhang L, Li J, Denkenberger D, et al. Energy efficient materials for solar water distillation - A review. Renewable and Sustainable Energy Reviews. 2019;**115**:109409. DOI: 10.1016/j.rser.2019.109409

[15] Tiwari GN, Singh HN, Tripathi R. Present status of solar distillation. Solar Energy. 2003;**75**(5):367-373. DOI: 10.1016/j.solener.2003.07.005

[16] Mani K, Salgaonkar BB, Das D, Bragança JM. Community solar salt production in Goa, India. Aquatic Biosystems. 2012;**8**(1):30. Published 2012 Dec 1. DOI: 10.1186/2046-9063-8-30

[17] Walker J, Resnick R, Halliday D. Halliday & Resnick Fundamentals of Physics. 10th ed. USA: John Wiley & Sons, Inc. 1960

[18] Al-Hayeka I, Badran OO. The effect of using different designs of solar stills on water distillation. Desalination. 2004;**169**:121-127

[19] Bouchekima B. A small solar desalination plant for the production of drinking water in remote arid areas of southern Algeria. Desalination. 2003;**159**:197-204

[20] Karimi Estahbanati MR, Ahsan A, Feilizadeh M, Jafarpur K, Ashrafmansouri S-S, Feilizadeh M. Theoretical and experimental investigation on internal reflectors in a single-slope solar still. Applied Energy. 2016;**165**:537-547

[21] El-Sebaii AA. Thermal performance of a triple-basin solar still. Desalination. 2005;**174**:23-37

[22] Minasian AN, Al-Karaghouli AA. An improved solar still: The wick-basin type. Energy Conversion and Management. 1995;**36**:213-217

[23] Badran AA, Al-Hallaq IA, Eyal Salman IA, Odat MZ. A solar still augmented with a flat-plate collector. Desalination. 2005;**172**:227-234

[24] Tiris C, Tiris M, Erdalli Y, Sohmen M. Experimental studies on a solar still coupled with a flat-plate collector and a single basin still. Energy Conversion and Management. 1998;**39**:853-856

[25] Mohamad MA, Soliman SH, Abdel-Salam MS, Hussein HMS. Experimental and financial investigation of asymmetrical solar stills with different insulation. Applied Energy. 1995;**52**(2-3):265-271

[26] Nafey AS, Abdelkader M, Abdelmotalip A, Mabrouk AA. Solar still productivity enhancement. Energy Conversion and Management. 2001;**42**:1401-1408

[27] Abu-Hijleh BAK, Rababa'h HM. Experimental study of a solar still with the sponge cubes in basin. Energy Conversion and Management. 2003;**44**:1411-1418

[28] Boukar M, Harmim A. Effect of climatic conditions on the performance of simple basin solar still: A comparative study. Desalination. 2001;**137**:15-22

[29] Al Hamadani AAF, Shulla SK. Water distillation using solar energy system with lauric acid as storage medium. International Journal of Energy Engineering. 2011;**1**(1):1-8

[30] Dunkle RV. Solar water distillation: The roof type still and a multiple effect diffusion still. In: Proceedings of International Heat Transfer Conference, University of Colorado, Boulder, Colorado, Part V. 1961. pp. 895-902

[31] Al-Karaghouli AA, Al Naser WE. Experimental comparative study of the double basin solar stills. Applied Energy. 2004;**77**:317-325

[32] Tiwari GN, Tripathi R. Performance evaluation of a solar still by using the concept of solar fractionation. Desalination. 2004;**169**:69-80

Chapter 10

Lithium Recovery from Brines Including Seawater, Salt Lake Brine, Underground Water and Geothermal Water

Samadiy Murodjon, Xiaoping Yu, Mingli Li, Ji Duo and Tianlong Deng

Abstract

Demand to lithium rising swiftly as increasing due to its diverse applications such as rechargeable batteries, light aircraft alloys, air purification, medicine and nuclear fusion. Lithium demand is expected to triple by 2025 through the use of batteries, particularly electric vehicles. The lithium market is expected to grow from 184,000 TPA of lithium carbonate to 534,000 TPA by 2025. To ensure the growing consumption of lithium, it is necessary to increase the production of lithium from different resources. Natural lithium resources mainly associate within granite pegmatite type deposit (spodumene and petalite ores), salt lake brines, seawater and geothermal water. Among them, the reserves of lithium resource in salt lake brine, seawater and geothermal water are in 70–80% of the total, which are excellent raw materials for lithium extraction. Compared with the minerals, the extraction of lithium from water resources is promising because this aqueous lithium recovery is more abundant, more environmentally friendly and cost-effective.

Keywords: thermodynamics, lithium energy, lithium recovery, adsorption, precipitation, membrane process

1. Introduction

Lithium and its compounds are widely used in manufactured glass, ceramics, greases, batteries, refrigerants, chemical reagents and other industries. World lithium reserves are about 14 million tons, mostly 70–80% is stored in salt lake brine, geothermal water and solid lithium contained in lithium ore. Currently, many researchers are turning their attention to 2600 billion tons of lithium-containing seawater, which is about 15,000 times more than solid lithium ores [1].

Figures for lithium resources and reserves differ considerably accordingly to the source, although there is a unanimous agreement that lithium resources in brine are much larger than those in hard rock [2–6]. The most recent figures from the US Geological Survey indicate total lithium resources (brine + hard rock) to be 54.1 million tons [5]. Approximate minimum and maximum hard rock lithium resources were reported at 12.8 and 30.7 million tons, respectively; while brine field data were reported as 21.3 and 65.3 million tons, respectively, for minimum and maximum estimates [3].

Lithium has various uses, but its abundance in nature is only 0.0018% [7]. The use of lithium on ceramics enriched with Li^6 is up to 15% for use in the production of tritium [8, 9]. In addition, enriched Li^6 is very expensive, what is commensurate with the value of gold. Consequently, it is necessary to extract and recycle lithium from the waste of solid breeding materials. Hence widespread use of lithium in various spheres, many studies have been conducted to extract lithium from various sources.

Lithium demand is expected to grow continuously and dramatically in the coming years as different types of lithium batteries are the most promising candidates for powering electric or hybrid vehicles [10, 11]. Lithium batteries include both current technologies such as lithium-ion and growing battery technologies such as lithium-sulfur or lithium-air [12–15].

Lithium demand is projected to increase by ~60% from 102,000 to 162,000 tonnes of lithium carbonate equivalent in the next 5 years, with battery applications taking a huge percentage of this growth [16, 17]. It was reported that the present lithium resource in continental and Salar brines is roughly 52.3 million tons of lithium equivalent, mainly in Argentina, Chile and Bolivia, from which 23.2 million tons can be extracted [18]. From the other side, lithium from mineral resources is 8.8 million tons, where there are huge deposits in the United States, Russia and China. Evans estimated lithium reserves and recoverable resources at 29.79 million tons [19].

Meanwhile, the general public mainly associates lithium batteries with portable electronics and electric and hybrid vehicles, large storage capacity lithium batteries are also a lead candidate for a possible energy storage solution for the electric grid, intelligent network, etc. Batteries with large capacity are needed to store green energy, wind, that is, sun and waves, all this by their nature intermittent sources of energy [20–30]. Nowadays battling to achieve a greater percentage of green energy, high-capacity batteries or energy banks are mandatory. Basically, if in the near future we want our energy matrix to be highly dependent on renewable energy, energy banks will be needed to provide continuous energy to the grid, during the time these intermittent energy sources are either off or not working completely (no wind, no waves, at night) [20–22]. After all, on its own of the energy source, high-capacity batteries are also an alternative for storing energy during periods of low demand, allowing this excess energy to be re-injected into the grid at high demand peaks [24].

Currently, lithium is relatively not expensive (a ton of Li_2CO_3 is about 15,000 USD), the market shows that, its price is rising with increasing demand [25].

In China, lithium prices have risen about 300% since 2016, and contract prices for existing manufacturers have risen to more than 16,000 USD per tonne.

Because of the exhaustion of lithium ores, recent studies have shown recovery of lithium from seawater, brine and geothermal water. Production of lithium from water resources has become more important due to its wide availability, ease of process and cost-effectiveness compared with its production from various resources [26].

Many methods for extracting lithium from seawater, brines and geothermal water have been reported [27]: solvent extraction, including precipitation, liquid-liquid extraction, selective membrane separation, electrodialysis, ion exchange adsorption, etc. [28–34]. Of these methods, the most attention was paid to ion exchange adsorption methods based on lithium-ion sieves because of their good lithium-ion selectivity and high adsorption properties [35–37]. From the point of view of cost and efficiency, extraction of lithium ions from solutions by ion exchange adsorption is an important method [38].

Various methods of removing lithium from water have been proposed in recent years. In their midst, adsorption has been proven to be a perfect way to extract lithium, offering significant benefits, such as availability, lower cost, profitability, efficiency and easy operation. For lithium removal, various Li adsorbent materials have previously been reported, including metal oxides, clay minerals, silicotitanates and zirconium phosphate. The main attention of the researchers was focused on the adsorbents of titanium-lithium ion sieves [39–43] manganese-lithium ion sieves [44–50] and aluminum salts [51, 52]. Adsorbents of aluminum salts [52] showed stable and high selectivity for Li^+ with lithium absorption of only 2–3 mg g^{-1}. By Chitrakar et al. [53] nanoscale H_2TiO_3 was synthesized by solid-phase reaction and its adsorption capacity with lithium reached 32 mg g^{-1} [41]. Tang et al. and Zhang et al. were synthesized H_2TiO_3 using different raw materials [41, 42]. Wang et al. synthesized lithium-enriched β-Li_2TiO_3 with a maximum lithium absorption of 76.7 mg g^{-1} in LiOH alkali solution [43]. Despite the fact the maximum absorption of Li^+ adsorbent H_2TiO_3 from the lithium-enriched solution has reached 76.7 mg g^{-1}, the high cost of synthesis and loss of dissolution of the titanium ion is still an obstacle. Chitrakar et al. by hydrothermal reaction synthesized $Li_{1.6}Mn_{1.6}O_4$ and with 52 mg g^{-1} Li^+ adsorption capacity [48]. By Xiao et al. synthesized spinel-structured hydrogen oxides of manganese, the saturated adsorption capacity of which was up to 42 mg g^{-1} [54]. 1-D MnO_2 was synthesized with a maximum adsorption capacity reaching 46.34 mg g^{-1} in LiOH solution (C_0 = 35 mg L^{-1}) [50]. In this case, the dismutation reaction during etching can lead to distortion of the lattice and dissolution of manganese, which will violate its cyclicality.

In addition to lithium and magnesium, the treated salt lake brines may contain significant concentrations of potassium, sodium and boron. Zhou et al. compared the competitive sequences for several cations using TBP/$FeCl_3$ in MIBK as the extractant [55–57]. However, quantitative correlations for competing for ion extractions, which are crucial in industrial design, were not reported.

The review is devoted to the extraction of lithium from brines, marine and geothermal waters, the collection of different methods of lithium extraction from water resources, which makes it possible to compare different methods that determine the optimal path for further research. Moreover, scientists around the world are challenged to find a way to extract lithium from water resources that are environmentally friendly, highly selective, economical, time-efficient and easy to process.

2. Lithium extraction

Lithium is comparatively abundant on the earth's crust, being the affluent 25th more element [58]. More than 150 minerals contain lithium, in solid sediments, in geothermal waters, in many continental brines and in seawater. The concentration of lithium in seawater is very low, with an average of 0.17 ppm [3, 59]. The change in concentration from 1 to 100 ppm shows geothermal waters around the world [2, 4]. Although lithium deposits in all of the above forms are widespread throughout the world, only a very few are large enough and/or concentrated to potentially allow their exploitation. Several high-grade lithium minerals and brines are the only ones currently manufacturing at lithium extraction [2–4].

Interest in the recycling of lithium batteries has grown in recent years. However, recycling is still not economically attractive if compared with the mining of the raw materials [60]. Facilities for recycling are now available in the USA, Canada, Belgium, Germany and Japan. However, lithium availability from recycling is insignificant as compared with mined raw materials [61].

Figures for lithium resources and reserves differ considerably accordingly to the source, although there is unanimously agreement that lithium resources in brine are much larger than those in hard rock [2–4, 6]. The latest data from the U.S. Geological Survey show that total lithium resources (brine + hard rock) are 54.1 million tons. It was reported that the minimum and maximum reserves of lithium in hard rocks were 12.8 and 30.7 million tons, respectively; while the brine field data were reported as 21.3 and 65.3 million tons, respectively, for the minimum and maximum evaluation [3].

3. Lithium resources

In contradistinction to the uses of lithium, it is necessary to discuss the question of responsibility for Li from a variety of sources. The economic efficiency of lithium is found in minerals, clays and brines. High-grade lithium ores and brines are the current sources for all commercial lithium manufacture. **Figure 1(a)** demonstrates the distribution of lithium over different resources. The figure shows that continental brine is the largest resource (59%) for lithium, followed by solid rock (25%). **Figure 1(b)** demonstrates the spread of lithium across countries. The largest of the studied lithium deposits are in Bolivia and Chile. **Figure 1(c)** demonstrates the distribution of lithium production across countries. The main producers and exporters of lithium ores are Chile and Australia. Chile and China have huge resources of lithium ore. Canada, Russia, Serbia and Congo (Kinshasa) have lithium ores of about 1 million tons each, and equal reserve for Brazil is total 180,000 tons [62].

It is estimated that the earth's crust contains an average of about 0.007% lithium. In nature, lithium does not occur freely, but it occurs in small quantities in almost all magmatic breeds and the ocean, in seawater, in the waters of many mineral springs. Of the approximately 20 known minerals containing lithium, only 4, that is, Lepidolite ($KLi_{1.5}Al_{1.5}[Si_3O_{10}][F,OH]_2$), Spodumene ($LiO_2{\cdot}Al_2O_3{\cdot}4SiO_2$), Petalite ($LiO_2{\cdot}Al_2O_3{\cdot}8SiO_2$) and Amblygonite ($LiAl[PO_4][OH,F]$) are known to occur in quantities sufficient for commercial interest as well industrial importance [63–66]. The spodumen ($LiAlSi_2O_6$) mineral is the most significant industrial lithium ore mineral. Minerals of lithium also exist as cookeite as ($LiAl_4(Si_3Al)O_{10}(OH)_8$) in fine hydrothermal veins of quartz. Taeniolite ($KLiMg_2\ Si_4O_{10}F_2$) is present in veins of smoky quartz in recrystallized novaculite, in manganese deposits the appearance of Lithiophorite ($(Al, Li)\ Mn^{4+}O_2(OH)$) is noted. Pegmatites, Taeniolite, Lithiophorite and Cookeite are considered to be economically inefficient sources of lithium [67–69]. A large part of the lithium is extracted from brine or seawater has a high concentration of lithium carbonate. In the earth's crust, there are brines called continental brines/subsurface brines are the main source for the production of lithium (lithium carbonate). The literature reports that lithium is also present in seawater at about 0.17 mg L^{-1}. Lithium is found in significant quantities in oil well brines and geothermal waters. These sources of brine and seawater are considered less expensive than mining from rocks such as spodumene, lepidolite, amblygonite and petalite containing lithium.

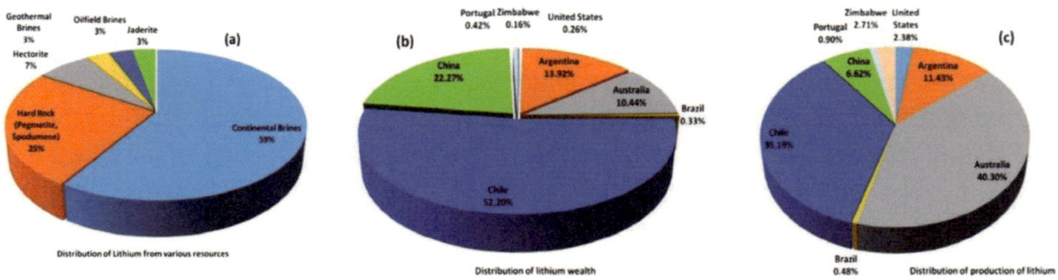

Figure 1.
The distribution of lithium (a) different natural resources, (b) worldwide distribution, (c) the number of producers around the world.

4. Lithium extraction from various resources

4.1 Lithium extraction from brine

Extracting lithium from brine is an important potential resource. When considered from an economic and scientific perspective, the following points are important to consider lithium recovery from brine: (1) suitability of pond soil and admissibility of the area for solar evaporation; (2) the concentration of lithium in brine; (3) the ratio of alkali metals and alkaline earth elements to lithium and (4) the complexity of the phase chemistry. The resources of brines containing lithium can be divided into three types: evaporative, geothermal and oilfield brines. In the process of evaporation of the brine about 50% of the original natural brine, lithium remains in the residual brine. This expression has been ascribed to the retention of lithium by precipitated salts. Residual brine is highly loaded with Mg^{2+} as compared with K^+ and Na^+, this makes it difficult to extract lithium from this residual brine [70].

The extraction of lithium from brine does not correspond to any general regularity since each process is specific depending on the composition of the brine field. Typical lithium production technology used for lithium extraction by Outotec®, where different methods such as precipitation, solvent extraction and flotation were used (**Figure 2**). Lithium extraction by Outotec® uses a lithium carbocation process to produce lithium [71]. Various lithium separation and purification methods have been reported in the literature, which is discussed below. By Chagnes and Swiatowska the general technological scheme of lithium production from brine and seawater is proposed [72]. In this method, liquid-liquid extraction, ion exchange, electrodialysis and adsorption are important hydrometallurgical processes necessary to concentrate lithium before production [72]. **Table 1** discusses and summarizes the extraction of lithium from both brine and synthetic brine in various ways.

4.2 Recovery of lithium from brine by precipitation

Pelly et al., Epstein et al. and Kalpan et al. it has been reported that lithium recovery as precipitation of lithium aluminate from Dead sea brine and final brine [73–75]. Pelly et al. have reported, it is necessary to control the pH of the brine through dilution to achieve 90% extraction efficiency end brine and Dead sea brine [73]. As indicated, the optimal pH should be in the range of 6.6–7.2 For Dead sea brine and 6.8–7.0 for end brine. The optimum reaction time should be 3 hours at room temperature. $AlCl_3{\cdot}6H_2O$ (30–40 g L^{-1}) was added to the brine. The negative effect was given by higher temperature, but better yields were obtained at room temperature and the yield decreased with increasing temperature [73]. The importance of extracting lithium from the Dead sea by precipitation as lithium aluminate followed by liquid-liquid extraction to separate lithium from aluminum with economic evaluation was reported [74]. Kaplan et al. reported on the process of extracting lithium by lithium aluminate from Dead sea brine by precipitation [75]. A small amount of lithium, which is mainly present as LiCl, was precipitated as a lithium aluminate precipitate using ammonia and aluminum salt at room temperature. Although subsequent reduction processes both by dissolving lithium in sulfuric acid and followed by precipitation with calcium chloride lithium were reduced as alum [75]. An et al. reported on the process of extracting lithium from brine collected in Salar de Uyuni, Bolivia. Mg and Ca were extracted from the brine as $Mg(OH)_2$ and gypsum ($CaSO_4{\cdot}2H_2O$) using sulfate and lime. Both CAO and MgO were extracted using oxalic acid followed by firing using residual Ca and Mg. In the end, by heating at 80–90°C lithium was recovered in the form of Li_2CO_3.

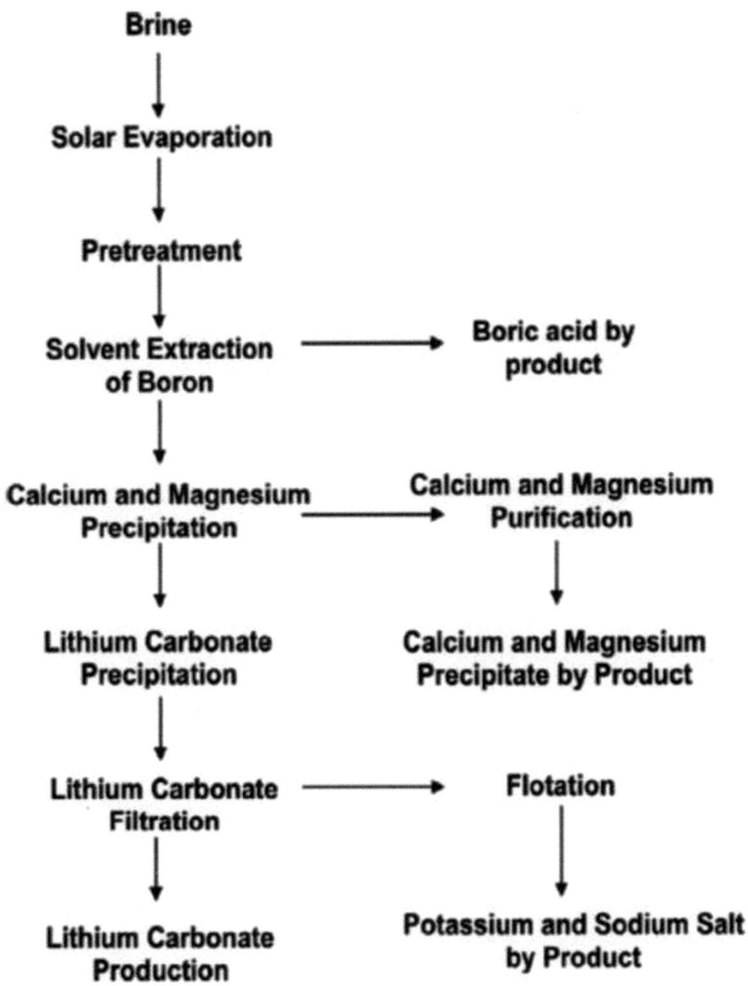

Figure 2.
Outotec lithium production technology from brine. Reproduced with permission from Outotec.

As a result of precipitation processes with a high content of pure (99.55%) and crystalline Li_2CO_3 were reduced [93].

4.3 Ion-exchange method of lithium recovery from brine

Through the use of a specially made resin/aluminates composite/inorganic ion exchanger, from brine lithium can be recovered productively. Bukowski et al. reported through a process of carbocation and ion exchange an extract of pure LiCl from brines containing higher levels of $CaCl_2$ and $MgCl_2$ [77]. Three different ion exchange resins Y80-N Chemie AG (Chemie AG Bitterfeld-Wolfen)), TP207 resin (Bayer AG), (MC50 resin, (Chemie AG Bitterfeld-Wolfen) for lithium extraction from synthetic brine were investigated. As a result of the conducted researches, it is established that it is possible to clean LiCl solutions with Y 80 resin at room temperature and with TP 207 resin at 50°C [77]. Hui et al. synthesis of H_2TiO_3 ion exchanger and extraction of lithium from the brine of natural gas wells have been reported [76]. Ion exchanger H_2TiO_3 was synthesized from Li_2CO_3 and TiO_2 or precipitation of LiOH and TiO_2 followed by calcination at 400–800°C. Ion exchanger H_2TiO_3 provided high selectivity for Li^+ at an exchange capacity of Li^+ 25.34 mg g^{-1} in mixtures

Resources	Process	Reagents	Mechanism	Reference
End brine and Dead Sea brine	Precipitation	Lithium aluminate	Precipitaion	[73]
Dead Sea brine	Precipitation	Lithium aluminate	Precipitaion	[74]
Dead Sea brine	Precipitation	Lithium aluminate	Precipitaion	[75]
Salar de Uyini, Bolivia	Precipitation		Precipitaion	[76]
Dead Sea brine	Gel permeation chromatography	Polyacrymidegel, Bio-Gel P-2 and Blue Dextran 2000	Column chromatography	[77]
From other alkali metal	Reversed phase chromatography	Polytetrafluroethylene tribytyle phosphase (TBP), dibenzoylmethane (DBM) and trioctylephosphine oxide (TOPO)	Column chromatography	[76]
Seawater, hydrothermal water	Column chromatography	Titanium (IV) antimonate cation exchanger (TiSbA)	Column chromatography	[78]
Synthetic brine	Chelating resins	MC50 resin, TP207 resin, Y80-N Chemie AG	Ion exchange	[79]
Brine of natural gas wells	Inorganic ion exchanger	H_2TiO_3 ion exchanger	Ion exchange	[80]
Salt Lake brine	Inorganic ion exchanger	H_2TiO_3 ion exchanger	Ion exchange	[81]
Synthetic brine	Liquid-liquid extraction	n-Butanol		[82]
Synthetic brine	Liquid-liquid extraction	2-Ethyle-1,3-hexanediol, isoamyle alcohol, di-isopropyl ether, diethyl ether		[83]
Brine	Liquid-liquid extraction	With tributyl phosphate (TBP)		[84]
Brine	Liquid-liquid extraction	Heptofluorodimethyloctunedione, peniafluorodimeihylhepiunedione, trifluorodimethylhexanedione, dibenzoylmethane and tetrameihylheptonedione		[85]

Resources	Process	Reagents	Mechanism	Reference
Salt Lake brine	Ionic liquid liquid-liquid extraction	-Butyl-3-methylimidazolium bis[(trifluoromethyl)-sulfonyl]-imide, 1-ethyl-3-methyl-imidazolium bis[(trifluoromethyl)-sulfonyl]-imide and 1-butyl-3-methylimidazolium hexafluorophosphate		[86]
Brine	Ionic liquid liquid-liquid extraction	1-Alkyl-3-methylimidazolium-based ionic liquids (ILs), in which the alkyl chain lengths were 4-butyl (C4), 5-pentyl (C5), 6-hexyl (C6), 7-heptyl (C7), 8-octyl (C8) or 9-nonyl (C9)		[87]
Salt Lake brine	Ionic liquid liquid-liquid extraction	1-Octyl-3-methyl-imidazolium hexafluorophosphate and tributyl phosphate (TBP)		[88]
Salt Lake brine	Ionic liquid liquid-liquid extraction	Bis(trifluoromethylsulfonyl) imide in TBP		[88]
Brine	Electro-electrodialysis	Bipolar membranes		[89]
Salt Lake brine	Membrane electrolysis	Bipolar membranes		[93]
Dead Sea brine	Solvent impregnated membrane	Solvent-polymeric membranes (2-ethylhexyl)-diphenyl phosphate		[66]
Brine	Desalination	Nanofiltration membrane		[91]
Salt Lake brine	Desalination	Nanofiltration (NF90 membrane XLE membrane)		[92]

Table 1.
Recovery of lithium from brines by various processes.

of alkaline earth metal and an alkali metal. From brine ion exchanger H_2TiO_3 showed 97% exchange rate and 98% elution rate for Li^+ [76]. Chitrakar et al. reported with the same H_2TiO_3 ion exchanger extract lithium from Salt lake brine. Adsorption of lithium ions by H_2TiO_3 ion exchanger according to Langmuir model having exchange capacity for Li^+ 25.34, 32.6 mg g^{-1} at pH 6.5 from brine was reported [53].

4.4 Liquid-liquid method of lithium recovery from brine

Many studies have provided my traditional liquid-liquid extraction and liquid-liquid extraction by ionic liquids (ILs) have been reported for lithium extraction from brine. Gabra et al. using synthetic solutions of nbutanol containing different amounts of lithium, potassium, calcium and sodium chloride, a laboratory-scale of LiCl extraction process were developed. A method for lithium reduction for separation and LiCl reduction is proposed, derived from distribution coefficients, separation coefficients and the presentation of McCabe-Thiel results. According to this method, 99.6% purity of LiCl can be restored [78]. Liquid-liquid extraction of lithium from brines by alcohol such as isoamyl alcohol and n-butanol, combined with precipitation of the lithium-aluminum complex reported by Bukowski et al. The amount of LiCl extraction from brine at pH 5.4 with different alcohol follow the order: 2-ethyl-1,3-hexanediol > isoamyl alcohol > di-isopropyl ether > diethyl ether and can extract 32.8, 25.2, 11.4, 9.1% lithium, respectively, along with Na, Mg and Ca. Lithium extraction was also studied using a binary mixture of the above compounds in a 1:1 ratio at a pH of 5.4. 2-ethyl-1,3-hexanediol mixed with isoamyl alcohol is suitable for 90% LiCl reduction as well as suppression of metal co-extraction [79]. Zhou et al. reported the extraction of lithium from brine sources using tributyl phosphate (TBP) in three different diluents [55]. Three salt solutions ($ZnCl_2$, $FeCl_3$ and $CrCl_3$) were selected as co-extractors to investigate the possibility of extracting lithium metal from brine sources. The method of liquid-liquid extraction equilibrium of lithium with tributyl phosphate (TBP) in methylisobutyl ketone (MIBC), TBP in kerosene and TBP in 2-octanol was analyzed. In liquid–liquid extraction, lithium equilibrium is investigated by $FeCl_3$ solution as a co-extractor. The results showed that the extraction efficiency followed the sequence: TBP/2-octanol < TBP/kerosene < TBP/MIBK. It was significantly larger than the TBP/2-octanol system than the TBP/MIBK and TBP/kerosene systems for lithium recovery [55]. A method for extracting lithium from neutral brines using beta-diketone and trioctyl phosphine oxide in benzene was patented by Baldwin and Seeley [80]. The mechanism of extraction was discussed in more detail with scientists [81].

4.5 Liquid-liquid extraction using ionic liquid method to extract lithium from brine

Unlike traditional liquid-liquid extraction, ionic liquid extraction is considered not only as a solvent but also as a co-extraction reagent. Gao et al. reported the extraction of lithium from salt lake brine using tri-isobutyl phosphate in ionic liquid and kerosene [82]. Three ionic liquids (ILs) have been reported, that is, 1-ethyl-3-methyl-imidazolium-bis[(trifluoromethyl)-sulfonyl]-imide, 1-butyl-3-methylimidazolium-bis[(trifluoromethyl)-sulfonyl]-imide and 1-butyl-3-methylimidazolium-hexafluorophosphate with triisobutyl phosphate (TIBP) and kerosene for ion recovery lithium from salt lake brine. The results show that the best selective lithium extraction was obtained using IL 1-ethyl-3-methylimidazole bis[(trifluoromethyl)-sulfonyl] imide. Under optimal extraction conditions, the one-stage efficiency of lithium ion extraction was 83.71% and the one-stage distillation efficiency was 85.61% at 1.0 mol L^{-1} HCl in 1.0 mol L^{-1} NaCl as a stripping agent at (O/A) = 2 [82]. Also, lithium extraction from brine is performed using imidazolium-containing ionic liquids with varying alkyl chain lengths in a series of ionic liquids based on 1-alkyl-3-methylimidazolium (ILs), in which the alkyl chain lengths are 4-butyl (C_4), 5-pentyl (C_5), 6-hexyl (C_6), 7-heptyl (C_7), 8-octyl (C_8) or 9-nonyl (C_9), in the presence of tri-isobutyl phosphate (TIBP) and kerosene systems presented by Gao et al. [83]. Studies have shown that the shorter the alkyl chain length of imidazolibased ILs, the higher the lithium recovery efficiency. Optimal lithium extraction can be achieved using ionic liquids based on n-butyl (C_4) based on 1-alkyl-3-methylimidazoline (ILs). With a single contact of extraction and distillation, the efficiency of lithium extraction under optimal conditions was 74.14 and 86.37%, respectively. And the optimal condition was ionic liquids based on N-butyl-3-methylimidazole: TIBP: kerosene = 1:8:1 (vol/vol), pH = 5.0, O/A = 2.0 at the extraction stage using 1 mol L^{-1} HCl at O/A = 3 at the distillation stage [83]. Separation of lithium and magnesium from Salt lake brine by liquid-liquid extraction using ILs containing tributyl phosphate, reported Chenglong et al. [84, 85]. Tributyl phosphate (TBP), ILs and 1-octyl-3-methylimidazolium hexafluorophosphate, respectively, were used as the extraction medium and extractant for lithium extraction from Salt lake brine. The most suitable conditions for the extraction of this system were the ratio of TBP/ILs at 9/1(vol), O/A at 2:1. The pH of the brines of salt lakes is maintained constant. The obtained data show that the efficiency of single-stage extraction of lithium and magnesium was 80.64 and 5.30%, respectively. The total extraction efficiency of 99.42% was achieved by three-stage countercurrent extraction. With a one-stage method of removing lithium and magnesium, the efficiency was at A/O phase ratio of 298.78 and 99.15%, respectively, at 80°C. Provisional result showed that ILs has the potential to replace volatile organic solvents in liquid-liquid recovery of lithium cations [84]. At room temperature, ionic liquid solvent extraction of lithium cations using TBP was reported by Chenglong et al. The authors used TBP against the widely used ILs bis(trifluoromethylsulfonyl) imide and quantitative reduction of lithium [85].

4.6 Membrane process of extraction of lithium from brine

The extraction of lithium from brine by membrane method is a relatively modern and novel technology reported by various authors, which are discussed below. Through electroelectrodialysis with bipolar membranes, the production of lithium hydroxide from brines has been reported by Jiang et al. [86]. In a laboratory-scale process, a sequentially configured electro-electrodialysis with a bipolar membrane was installed with a permutation of the conventional electrodialysis stack. Standard electrodialysis stacks were reconfigured using five cation exchange membranes and four anion exchange membranes. With conventional electrodialysis and Na_2CO_3, through preconcentrating and precipitating brine, respectively, 98% pure Li_2CO_3 powder can be recovered. The authors investigated the influence of current density and raw material concentration on the production of lithium hydroxide (LiOH). Cost-effective was electro-electrodialysis with bipolar membranes at a current density of 30 mA/cm^2 and a feed concentration of 0.18 MPa. Jiang et al. argued that the process is environmentally friendly and cost-effective [86]. The extraction of lithium from salt lake brine by membrane electrolysis was reported by Liu et al. [87]. Different technological parameters are optimized: the distance between the anode and the cathode, the initial concentration of lithium in the analyte, the electrolyte temperature, the electrolysis time and the surface density of the active substrate. The electrode demonstrates a remarkable Li^+ 38.9 mg g^{-1} exchange capacity and an analyte pH value below 8.00 at optimal conditions [87]. Extraction of lithium from Dead sea brine by membrane separation using an ion-exchange hybrid process reported by Jagur-Grodzinski and Schori [88]. Lithium cations can be selectively permeated by solvent-polymer membranes. Better selectivity of Li^+ transport by Mg^{2+} and Ca^{2+} gave membranes with (2-ethylhexyl)-diphenyl phosphate. No significant changes in membrane permeability and selectivity were observed during the 6 months of operation. Preliminary concentration of lithium and subsequent selective separation of lithium by membrane and ion exchange fusion were described by Jagur-Grodzinski and Schori. The expediency of lithium separation by combination of ion exchange process and membrane is substantiated [88]. The processes of concentration and separation of lithium from brine by reverse osmosis, nanofiltration was used. Sun et al. reported the isolation of lithium and magnesium from brine using a desalination nanofiltration membrane [89]. Magnesium lithium rejection rate was estimated by optimizing various operational parameters such as pressure, supply water temperature, pH and Mg^{2+} to Li^+ ratio. Lithium extractions from salt lake brine using RO and NF processes have also been investigated. Studies show that the separation of magnesium and lithium was strongly dependent on the operating pressure, Mg^{2+}/Li^+ ratio and pH [89]. Lithium recovery from salt lake brine has been reported using NF and a low-pressure RO membrane by Somrani et al. [90]. Lithium selective membrane NF90 compared with XLE with low-pressure reverse osmosis membrane. For Li^+ extraction, the NF90 membrane is more efficient than the XLE on the low pressure RO membrane due to its higher permeability to clean water and 0.1 m NaCl solution. A lower critical pressure (Pc = 0) and higher selectivity were obtained at a low operating transmembrane pressure (<15 bar) between monovalent cations (40%). Nf90 membrane showed 100% magnesium rejection in the initial step separation from dilute brine (15% for Li^+, 10 times dilution). An 85% separation between Mg^{2+}/Li^+ was achieved in the final. Lithium can easily be separated by dialysis from the solution [90].

5. Lithium extraction from seawater

In the near future, to meet the needs of the world community in lithium, the ocean is considered the most important and promising resource for lithium [66]. It is reported that the total amount of lithium reserves in the oceans is approximately 2.6×10^{11} t [91]. Lithium extraction from hydromineral sources is carried out on a semi-industrial and industrial scale in the USA from salt lakes [66, 92, 94, 95], in Japan from thermal waters [96, 97], in Israel from the Dead sea [66, 73]. The extraction of lithium metal from geothermal and brine has also been studied in Russia, Germany, Bulgaria and Korea [98]. Typically, lithium is extracted from seawater by these two processes: (1) co-precipitation and extraction process and (2) ion exchange and sorption process.

Various methods have emerged with the development of technology, such as liquid-liquid extraction, a membrane process is used to extract lithium from seawater **Table 2**. The process of lithium extraction from both brine and synthetic brine has been considered and generalized through various processes such as liquid–liquid extraction, ion exchange and sorption, co-deposition and membrane processes.

5.1 Co-precipitation method for extracting lithium from seawater

Like other methods, it has not received wide application the extraction process of lithium recovery and extraction by co-precipitation. For lithium recovery, an important problem is the presence of higher concentrations of alkali and alkali metals in seawater. The alkali metal group has a very similar parameter, which creates problems for lithium recovery. The problems associated with lithium recovery from seawater and terrestrial hydromineral resources are very similar [66]. To extract lithium from seawater, various reagents such as potassium, iron sulfates and aluminum hydroxides, are successfully used to co-precipitate lithium [66, 96]. To obtain lithium concentrate, the dissolution of the co-precipitate after an ion exchange process is used. A hydrometallurgical process for extracting lithium from seawater using an adsorption process with a manganese oxide adsorbent followed by a deposition process reported by Um and Hirato [99]. By this method, at a temperature of (25–90°C), $MgCl_2$ and $CaCl_2$ from seawater were precipitated as $Mg(OH)_2$ and $Ca(OH)_2$. Using the NaOH, pH was managed between 7 and 14 with an initial concentration of $CaCl_2$, $MgCl_2$ and $MnCl_2$ (10 and 100 mmol/dm^3). Followed by the second stage Li_2CO_3 was recovered through carbonation using Na_2CO_3 by neutralization using HCl [99].

Resources	Process	Reagents	Mechanism	Reference
Seawater	Precipitation	Na_2CO_3 + HCl	Precipitation	[99]
Seawater	Adsorption	k-MnO_2 adsorbent	Sorption	[100]
Seawater	Adsorption	$Al(OH)_3$ layer	Sorption	[101]
Seawater	Adsorption	(HMnO) ion-sieve (microporous)	Sorption	[102]
Seawater	Adsorption	k-MnO_2	Sorption	[103]
Seawater	Adsorption	MnO_2	Sorption	[104]
Seawater	Adsorption	HMnO	Sorption	[105]
Seawater	Adsorption	Nanostructure MnO_2 ion-sieve	Sorption	[18]
Seawater	Adsorption	MnO_2 adsorbent	Sorption	[106]
Seawater	Adsorption	$H_{1.6}Mn_{1.6}O_4$	Sorption	[48]
Seawater	Liquid-liquid extraction	Cyclohexane and tri-octyloxyphosphine		[101, 107]
Seawater	Liquid-liquid extraction	Thenoyltrifluoroacetone (TTA) and TOPO		[108]
Seawater	Membrane process	Mixed matrix nanofiber as a flow-through membrane	Adsorption	[102]
Seawater	Membrane process	Inorganic adsorbent containing polymeric membrane	Adsorption	[103]
Seawater	Membrane process	Inorganic adsorbent containing polymeric membrane	Adsorption	[109]
Seawater	Membrane process	Recyclable composite nanofiber adsorbent	Adsorption	[110]
Seawater	Membrane process	Li ionic superconductor functioning as a Li separation membrane	Dialysis	[111]
Seawater	Membrane process	Ionic liquid membrane	Electrodialysis	[112, 113]
Seawater	Membrane process	Membrane distillation and crystallization	Osmotic and vacuum configuration	[114]
Seawater	Membrane process	Mixed matrix nanofiber as a flow-through membrane	Adsorption	[115]

Table 2.
Recovery of lithium from seawater by various processes.

5.2 Ion exchange and sorption method for extracting lithium from seawater

Although various mega-industries are interested in extracting lithium from seawater in the present decade, extracting lithium from seawater has become increasingly attractive to researchers over several years through ion exchange and sorption. Several alternative methods of lithium extraction from seawater using ion-exchange after solar evaporation and fractional crystallization of NaCl, KCl and $CaSO_4$ are also proposed. According to this method, organic and inorganic sorbents are similar to the compounds used to extract lithium. Reports explaining this method are discussed below. Obtained by treating a Dowex-1 type microporous anion exchanger with a lithium-selective aluminum-containing resin with a saturated solution of ammonia, $AlCl_3$, and finally a solution of lithium halide before heating to produce a composite matrix of the microcrystalline resin $LiX{\cdot}2Al(OH)_3$ is an example of such products that have been patented in the United States [66, 116, 117]. High selectivity for lithium extraction was synthesized with sorbents based on antimony, tin, dioxides based on titanium and zirconium [118], mixed oxides of titanium and iron, titanium and chromium, titanium arsenate and magnesium and thorium [66]. To extract lithium from seawater, only manganese oxide-based cation exchange yields effective results in a wide range of lithium-selective ion exchange materials. Russian scientists use manganese oxide and mixed oxides of manganese and aluminum, known as ISM-1 and ISMA-1, respectively, to reduce lithium [66, 119]. For Li^+ in mixtures of alkali metal and alkali metal ions, the H_2TiO_3 ion exchanger resulted in high selectivity. Achieving the exchange capacity of Li^+ was 25–34 mg g^{-1}. High selectivity for lithium cations by synthetic inorganic materials of titanium (IV) antimonate cation exchanger (TiSbA) ion exchange has been reported by Abe et al. Recovery of lithium cations from hydrothermal water as well as seawater can be successfully applied. Using the periodic method, the effect of K^+, Mg^{2+} and Ca^{2+} cations on the adsorption of lithium cations on TiSbA has been reported by Abe et al. They showed that lithium adsorption decreases significantly with increasing concentrations of K^+, Mg^{2+} and Ca^{2+} cations. Lithium from the sea and hydrothermal water is enriched through TiSbA columns. To separate lithium cations from seawater and hydrothermal water TiSbA exchanger potentially be reused. With HNO_3 solution as the eluent, the adsorbed lithium can be eluted [120].

Selective extraction of lithium from seawater using two sequential ion exchange processes has been reported by Nishihama et al. [100]. By bench chromatographic operation with adsorbent k-MnO_2, lithium was concentrated from seawater, which has a 33% lithium recovery efficiency. A combination of ion exchange using resin and solvent impregnated resin is carried out lithium purification from the concentrated liquor of the reference unit. The cleaning process consists of the removal of divalent metal ions with a strong acid cation exchange resin accompanied by the removal of Na^+ and K^+ with b-diketone/TOPO impregnated resin; finally, the reduction of Li^+ as Li_2CO_3 precipitates using a saturated solution $(NH_4)^{2-}CO_3$. According to the method, the concede was 56%, and the cleanness was 99.9% [100]. Takeuchi reported on a new method of extracting lithium from seawater, also supported [101]. At a temperature of 50°C, almost 70% for lithium ion recovery is achieved in a periodic mode with a high selectivity of the $Al(OH)_3$ layer [101].

Several authors have reported that the extraction of lithium from seawater by sorption/desorption is a fairly common process, which is discussed below [108]. Many studies based on manganese oxide sorbate are used for the sorption/desorption of lithium from seawater. Japanese scientists have developed a sorbet based on hydrated c-oxides of manganese and mixed oxide of manganese and magnesium [102, 103]. By Ooi et al. lithium extraction from seawater using manganese oxide ion sieve (HMnO) was investigated. Maximum (7.8 mg g^{-1}) absorption of lithium

HMnO from seawater was achieved [102]. A study using ISMA-1 sorbents to extract lithium from seawater shows the following information: (1) The Li^+ cation distribution ratio is 4×10^4. (2) Sorbents are easily regenerated by nitric acid. (3) They exhibit a high capacity for lithium cations of about 20 mg/m. (4) Lithium concentrates containing up to 1 g L^{-1} of lithium can be achieved under optimal conditions. A two stage scheme for obtaining Li_2CO_3 from seawater using this information of a pilot plant with a capacity of 3 m^3 of seawater per hour has been developed and presented [66]. ISMA-1 sorbents provide higher chemical stability, but manganese oxide degradation associated with ion exchange remains the most serious drawback for their large-scale application in the lithium reduction process. A Japanese researcher developed a composite material by introducing a fine powder k-MnO_2 with spinel structure into polyvinyl chloride to improve the kinetic properties of manganese oxide sorbents [121]. Sorbents ISM and ISM-1, synthesized in Russia, are also a composite material obtained using a polymer binder [66, 119]. In Korea, it has also been reported to recover lithium from seawater using an ion exchange type of manganese oxide adsorbent. To recover dissolved lithium in seawater a highly efficient ion exchange adsorbent was prepared according to their method. A highly efficient ion exchange type adsorbate was synthesized as a result of the solid state reaction of Li_2CO_3 and $MgCO_3$. The ion sieve is formed after treatment of seawater with adsorbate, which is reduced by acid treatment. The lithium-ion sieve was produced by 3 cycles of 0.5 m HCl treatment with 24 h/cycle stringing, which shows 25.7 mg L^{-1} lithium absorption from artificial seawater [98]. Extraction of lithium from seawater by manganese oxide ion-sieve reported by Liu et al. The most promising method of industrial application was considered to be the extraction of lithium from seawater by adsorption using manganese oxide-ion sieves [104]. The sorption properties of HMnO in seawater and wastewater have been studied by Park et al. [105]. Lithium recovery from lake Urmia by the MnO_2 ion sieve, where more than 90% lithium recovery can be achieved, was reported by Zandevakili et al. [122]. Wajima et al. studied the adsorption behavior of lithium from seawater using the adsorbent manganese oxide [106]. In studies using a pseudo-second-order kinetic model, a higher adsorption Kinetics of lithium cations in seawater was observed [106]. Reduction of lithium from seawater using manganese oxide adsorbent synthesized from $Li_{1.6}Mn_{1.6}O_4$ precursor studied by Chitrakar et al. Manganese oxide adsorbent $LiMnO_2$ was synthesized from $H_{1.6}Mn_{1.6}O_4$ at 400°C by hydrothermal and reflux method. $H_{1.6}Mn_{1.6}O_4$ was synthesized from precursor $Li_{1.6}Mn_{1.6}O_4$. The sufficiently effective adsorbent can absorb lithium up to 40 mg g^{-1} from seawater [48].

5.3 Liquid-liquid extraction method of recovery of lithium from seawater

Scientists reported that liquid-liquid lithium extraction from seawater and liquid-liquid extraction are considered a potential process for extracting lithium from seawater. The use of liquid-liquid extraction to extract lithium from seawater is very limited, but the separation, purification and extraction of lithium by liquid-liquid extraction have been considered by several authors [81]. On the basis of the reference liquid-liquid extraction of lithium from seawater discussed below.

Scientists have used several extractants, such as primary alcohol C_3-C_5 and aliphatic alcohol C_6-C_8, to extract lithium from seawater, but the most effective and promising is Isobutanol. Japanese scientists have developed the most interesting method of extraction and in fact the most modern technology [107, 115]. In these methods, lithium is first extracted with cyclohexane and trioctyloxyphosphine, then the lithium reacts with hydrochloric acid and potassium phosphate, followed by lithium precipitation. The product obtained by this method has a purity of more than 95%. Synergistic extraction of lithium from seawater using a TTA-TOPO mixture has been reported by Harvianto et al. [123]. About 93% of the lithium can be recovered by TTA-TOPO. With acidic solutions, the lithium ion can be easily removed, but the removal efficiency decreases with increasing pH of acidic solutions. The type of acid does not affect the stripping efficiency. Similarly, by liquid-liquid extraction, 65% of lithium can be extracted from seawater, in the liquid-liquid extraction process, a magnesium ion is precipitated in advance. The recovery efficiency of lithium ions is negated by other metal ions in seawater [123].

5.4 Membrane process recovery of lithium from seawater

In recent years a number of authors have studied lithium recovery using different types of membranes. The membrane process of lithium reduction is a fairly advanced process that is gaining the attention of various researchers around the world.

Polysulfone (PSf)-based mixed matrix nanofiber dispersed with particulate lithium ion sieves as a flow-through membrane Li^+ absorber has developed by Park et al. A mixed matrix of electro-spinning nanofibers was prepared by thermal annealing, where lithium-ion sieves were activated by acid pickling as $Li_{0.67}H_{0.96}Mn_{1.58}O_4$ or MO. PSF based mixed matrix nanofiber effectively improved Li^+ selectivity. At minimal trans-membrane pressure, the mixed matrix nanofiber membranes were very permeable to water. By supporting the dynamic adsorption capacity of Li^+ mixed matrix nanofibers, the shorter adsorption–desorption cycle time (24 h) was successfully controlled by continuous streaming operations. In a small volume of acid solution, Li^+ enrichment was successfully achieved by repeated desorption of Li^+ [115]. Recovery of lithium from seawater using an inorganic adsorbent containing a polymer membrane reservoir system, reported Chung et al. To extract lithium from seawater, Chung et al. used three different membranes: a PSf nonwoven membrane, a PSF nonwoven composite membrane and a Kimtex® composite membrane.

The proposed system has the advantage of direct application in the seawater eliminates the use of a pressurized flow system [124]. The proposed system can have a direct application in seawater using a pressure flow system. In addition, lithium extraction from seawater using an inorganic adsorbent containing a polymer membrane has been reported by Umeno et al. Lithium recovery from seawater desalination retentate using composite poly(acrylonitrile) nanofibers with $H_{1.6}Mn_{1.6}O_4$ (HMO) lithium ion sieves was reported by Park et al. [110]. To obtain nanofibers, HMO/PAN dope solutions in N,N-dimethylformamide (DMF) with different HMO loads were used, and nanofibers were obtained by electrospinning. For efficient lithium extraction from seawater desalination retentate the material may be a potential membrane (Park et al.) [110]. The use of poly(vinyl chloride) (PVC) membrane adsorbent spinel-type manganese oxide by solvent exchange reported by Umeno et al. Poly(vinyl chloride) was dissolved in DMF solution, then lithium manganese oxide (spinel type) was mixed with DMF to obtain a suspension. The cured PVC film was prepared by applying a suspension to a thin film and immersed in water. To extract lithium obtained as a membrane type adsorbent the membrane was treated with HCl solution. Lithium extraction has been reported to be highly dependent on the method of preparation [109]. Hoshino reported on the recovery of lithium by dialysis and electrodialysis of seawater [111–113]. Selective extraction of lithium from seawater under laboratory conditions was investigated by electrodialysis using an ionic liquid (PP13-TFSI) impregnated with a membrane. The lithium recovery process was developed using a membrane process [112, 113]. Lithium extraction from seawater was selectively achieved by dialysis using a lithium ion superconducting membrane. For appropriate industrial lithium mass production applications, the dialysis process can be energy efficient and easily scalable [111]. Recovery of lithium by membrane desalination followed by crystallization was reported by Quist-Jensen et al. [114]. Extraction of lithium chloride and comparison of membrane crystallization in direct contact, vacuum and osmotic configuration were carried out. In their environment, the necessary supersaturation for crystallization was achieved for the simultaneous production of pure water and lithium by vacuum-membrane distillation [114].

6. Lithium ion-sieve effect

In 1971, ion-sieve oxides were first prepared by Volkhin et al. [125] since ion-sieve oxides have received increasing attention in the last few decades due to the special properties and performance as metal ions [126–130]. To extract specific metal ions with effective ion-sieve characteristics, ion-sieve oxides are fine adsorbents. Ion-sieve oxide adsorbents are obtained from corresponding precursors containing ions of the target metal. Characteristically, precursors are stable molecular structures, even if target ions are removed from their crystal sites, free crystal sites can still be retained. Thus, the resulting free crystal regions can only contain ions whose ionic radii are less than or equal to the radii of the target ions. In fact, only lithium ions can re-enter the free spaces of lithium ion sieves because lithium has the smallest ionic radius among all metal ions.

The study shows that only lithium ions can be adsorbed when LISs are placed in aqueous solutions containing different kinds of metal ions. **Figure 3** shows how LIS works. The main stage is the formation of LIS with hydrogen filled state [LIS (H)] by removing lithium ions from the lithium filled state [LIS (Li)], principally through Li-H ion exchange, then the adsorption isolation of lithium ions LIS from Li^+-containing solutions based on the steric effect. The spent LIS (H) is then regenerated to form LIS (Li) by adsorption of lithium ions. In General, the process can be called "LIS effect" [131–135].

7. Lithium ion-sieve

In fact, two types of chemical elements can be used, such as LISs, lithium manganese oxide type (LMO type) and lithium titanium oxide type (LTO type). LMO-type LISs are the most popular selective lithium adsorbents at present because of superior lithium absorption abilities, magnificent regeneration performance and high lithium selectivity. In addition, the extraction of lithium from aqueous solutions has recently improved significantly through the use of electrochemical methods. However, the LISs type suffers from the dissolution of manganese in aqueous solutions, which in industrial conditions can lead to serious water contamination. In this regard, LISs type LTO can overcome this problem, can be easily removed from the aqueous solution, and titanium compounds are not harmful to the aquatic environment [136–138]. In addition, LTO-type LISs has much more stable molecular structures due to the high energy of the titanium-oxygen bond compared with LMO-type LISs. But when an electrical potential is applied LISs of type LTO have limited use in extracting lithium from an aqueous solution. This restriction may prevent future industrial use of LISs type LTO. Thus, LMO-type and LTO-type LISs have their own unique benefits and

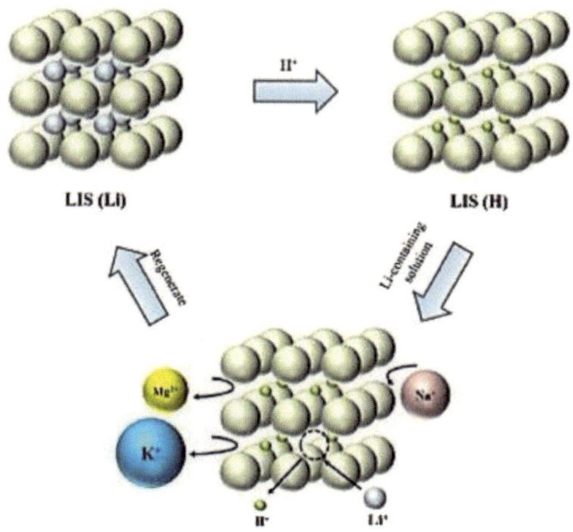

Figure 3.
Schematic representation of LIS process.

problems. Therefore future research to meet large scale industrial applications may focus on minimizing their respective disadvantages.

7.1 Lithium recovery by LMO type lithium-ion sieves

7.1.1 Study of ternary phase diagram of Li-Mn-O

With research [44], several LMO-type LISs have been well developed by many scientists [48]. As a rule, their precursors demonstrate a spinel structure. Because of multiple valence states of manganese, several lithium manganese oxides with different crystal structures can be formed. At 25°C, the phase diagram Li-Mn-O shows the isothermal cross-section **Figure 4** [139–143].

In the blue area in **Figure 4a**, the stoichiometric spinel phase defect is defined by the triangle $Mn_3O_4 \cdot Li_4Mn_5O_{12} \cdot \lambda\text{-}MnO_2$. Using the general formula $Li_xMn_{3-x}O_4$ $(0 \leq x \leq 1.33)$, one can imagine the stoichiometric spinel phases lying on the bond between Mn_3O_4 and $Li_4Mn_5O_{12}$. According to the general formula $Mn_{3-x}O_4$ $(0 \leq x \leq 1)$ defective spinels of manganese oxides located between Mn_3O_4 and $\lambda\text{-}MnO_2$ are presented. In accordance with the general formula $Li_2O \cdot yMnO_2$ $(y > 2.5)$, the defect of lithium-manganese-oxide spinel is expressed and the communication line lies between $Li_4Mn_5O_{12}$ and $\lambda\text{-}MnO_2$. At this point, in $LiMn_2O_4 \cdot Li_2Mn_4O_9 \cdot Li_4Mn_5O1_2$ the blue triangle in **Figure 4b** is the active area for preparing the precursors of LMO-type LISs. Therefore, it is possible to obtain high Li-Mn precursors such as $Li_5Mn_4O_9$ and $Li_7Mn_5O_{12}$ in principle, implying that high Li^+ capacity LISs may be obtained in the future.

Currently, only a few LMO-type LIS precursors with high Li^+ adsorption capacities such as $\lambda\text{-}MnO_2$, $MnO_2 \cdot 0.31H_2O$ and $MnO_2 \cdot 0.5H_2O$, which are derived from $LiMn_2O_4$, $Li_4Mn_5O_{12}$ and $Li_{1.6}Mn_{1.6}O_4$, respectively, were prepared. As shown in **Figure 5**, a phase diagram consisting of additional proton-type manganese oxides depending on the valence state of manganese, molar Li/Mn and H/Mn ratios constructed by Chitrakar et al. [47].

As shown in the figure, LIS precursors of the LMO-type can be classified into two types of reactions and are represented in two perpendicular planes: the vertical plane represents the redox reaction region, and the horizontal plane represents the ion exchange region. **Table 3** mainly summarizes their main properties for the absorption of lithium from aqueous solutions.

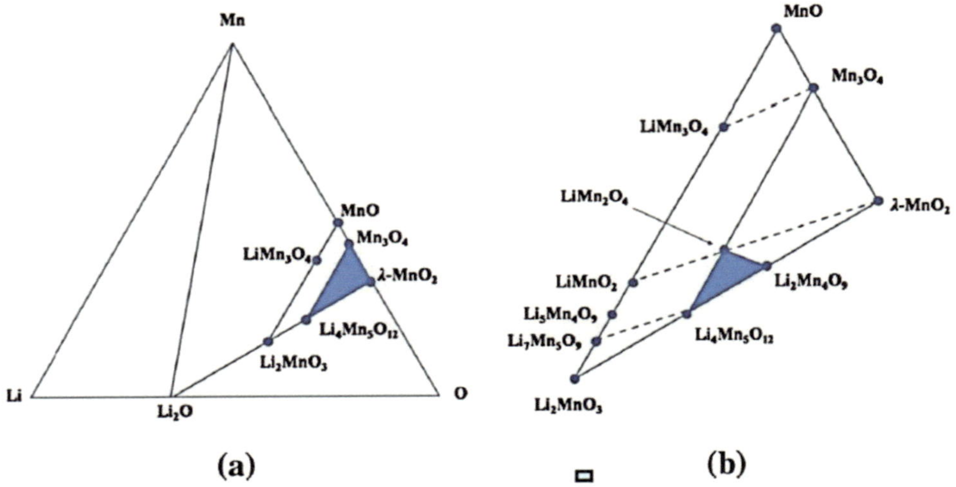

Figure 4.
(a) An isothermal cross section of the Li-Mn-O phase diagram at 25°C and (b) an expanded region of the Li-Mn-O phase diagram.

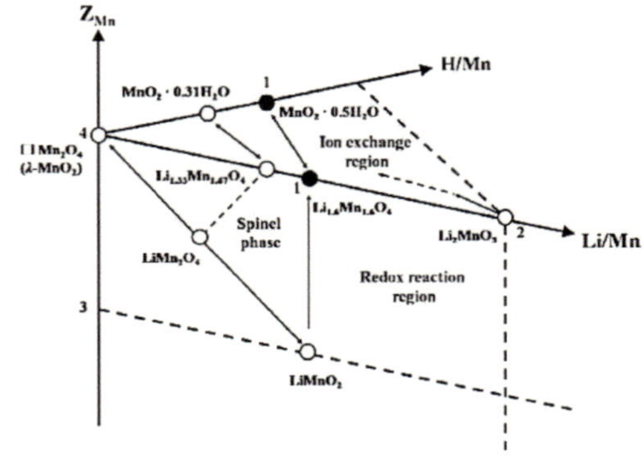

Figure 5.
Phase diagram of LMO and their delithiated products [47]. Reproduced from Ref. [47].

7.1.2 The spinel structure of the precursors of LMO

Inevitably, the chemical properties depend on the chemical structures to be determined, so the extraction of lithium by LMO precursors is explained by their peculiar chemical structure. Actually, all synthesized precursors of LMOs have spinel structures [144–152]. Among these, the $LiMn_2O_4$ structure is the most representative one, as shown in **Figure 6**.

Spinel $LiMn_2O_4$ has a cubic crystal structure that belongs to the spatial group *Fd3m*. The structure shows that the tetrahedron's 8a sites occupy lithium ions. At a molar ratio of 1:1, Mn^{3+} and Mn^{4+} ions are randomly distributed over 16d sites of octahedra, and oxygen anions occupy 32e sites of the face-centered cubes. Accordingly, the formula $(Li)_{8a}[Mn(III)Mn(IV)]_{16d}O_4$ can be represented by spinels $LiMn_2O_4$, which can be described by the general spinel formula (AB_2O_4). From other side, the $LiMn_2O_4$ unit cell can be viewed as a complex cubic structure: oxygen atoms are 32 and 16 manganese atoms occupy half of the octahedral pore (16d), while the other half of the sections (16c) are free. Here are 8 of the lithium atoms occupy 1/8 of tetrahedral interstices plot (8a). Li^+ can intercalate/deintercalate in three-dimensional networks of free octahedral and octahedral gaps along the

Materials		Li^+ uptake capacity	Regenerations	Li^+ selectivity	Ref.
Precursors	Ion sieves				
$LiMn_2O_4$	λ-MnO_2	16.9 mg g^{-1} (LiCl solution, pH = 9.19)	—	Equilibrium distribution (Kd): $Li^+ \gg Ca^{2+} > Mg^{2+} > K^+ > Na^+$	[140]
1D nano $LiMn_2O_4$	1D nano λ-MnO_2	ca. 20 mg g^{-1} (10 mmol L^{-1} LiCl solution, pH = 10.1)	—	K_d: $Li^+ \gg Ca^{2+} > Mg^{2+} > Na^+ > K^+$	[141]
$Li_4Mn_5O_{12}$	MnO_2–$0.4H_2O$	39.6 mg g^{-1} (10 mmol L^{-1} LiCl solution, pH = 10.1)	The Li^+ adsorption capacity of the spherical ion sieve after 55 cycle's adsorption-desorption was 0.4 mmol g^{-1}	The adsorption capacities of other ions were almost zero except for Mg^{2+}; the ratio of Mg/Li was reduced to less than 1 from 746	[54]
$Li_4Mn_5O_{12}$	$H_4Mn_5O_{12}$	49.6 mg g^{-1} (0.1 mol L^{-1} LiCl + LiOH solution)	—	K_d: $Li^+ \gg Mg^{2+} > Ca^{2+} > K^+ > Na^+$	[142]
$Li_1.6Mn_{1.6}O_4$	MnO_2-$0.5H_2O$	42.1 mg g^{-1} (initial Li^+ concentration 10 mmol L^{-1}, pH = 10.1)	The Li^+ adsorption capacity reduces from 4.08 mmol g^{-1} to 3.62 mmol g^{-1} after six times	K_d: $Li^+ \gg Mg^{2+} > Na^{2+} > K^+ > Ca^+$	[45]
$Li_{1.6}Mn_{1.6}O_4$	MnO_2-$0.5H_2O$	40.9 mg g^{-1} (seawater)	The Li^+ adsorption capacity is 35.4 mg g^{-1}, and Li^+ desorption rate still reaches 96% at the second cycle operation	K_d: $Li^+ \gg Cg^{2+} > K^+ > Mg^{2+} > Na^+$	[48]
$Li_{1.5}Mn_2O_4$	$H_{1.5}Mn_2O_4$	ca. 15.3 mg g^{-1} (pH = 8.1)	—	400 times higher conc. of Li^+ could be achieved while most of Na^+ remains in artificial seawater that contains 10 ppm of Li^+ and 10,000 ppm of Na^+ by chromatographic separation	[135]
$Li_{1.51}Mn_{1.63}O_4$	$H_{1.36}Li_{0.07}Mn_{1.65}O_4$	34.07 mg g^{-1} (pH = 12.01)	—	—	[140]
$Li_{1.57}Mn_{1.65}O_4$	$H_{1.39}Li_{0.01}Mn_{1.65}O_4$	ca. 37 mg g^{-1} (pH = 12)	—	—	[143]

Table 3.
Classification for some LMO type LISs.

8a-16c-8a-16c channel, what is the structural basis of Li^+ intercalation/deintercalation in $LiMn_2O_4$ spinel [145].

The 1:2 ratio shows a spinel $LiMn_2O_4$ of the two metal cations Li and Mn; although the stoichiometric proportion may be somewhat weakened in some circumstances. For example, in **Figure 7** it is shown that manganese ions in 16d sites can be replaced by lithium ions without changing the entire crystal framework.

Since more lithium ions can be extracted or inserted, the corresponding LIS of the substituted precursor $Li_{1.33}Mn_{1.67}O_4$ (or $Li_4Mn_5O_{12}$) is theoretically a higher lithium capacity than λ-MnO_2. Ammundsen et al. [148] the results of neutron diffraction studies of the lithium reinsertion process are given only for tetrahedral sites and not for octahedral sites, which indicates that the lithium extraction/insertion reaction can be expressed by the equation below:

$$(Li)[Li_{0:33}Mn_{1:67}]O_4 + H^+ \leftrightarrow (H)[Li_{0:33}Mn_{1:67}]O_4 + Li^+ \quad (1)$$

Another typical lithium-rich precursor to LMO is $Li_{1.6}Mn_{1.6}O_4$ (or $Li_2Mn_2O_5$), which are relevant LIS is $MnO_2{\cdot}0.5H_2O$. Among all available manganese, LISs $MnO_2{\cdot}0.5H_2O$ has the highest theoretical lithium capacity (*ca.* 72.3 mg g^{-1}). With this composition, the ratio of cations and anions (4:5) differs from that of typical spinel compounds (3:4), meaning that additional lithium ions are likely to be found in interstitial regions of the spinel structure with a single-digit arrangement [143]. Chitrakar et al. [47] proposed three hypothetical models through a preliminary Rietveld analysis, since there is still no published structural model for $Li_{1.6}Mn_{1.6}O_4$: (1) $(Li)_{8a}[Li_{0.2}]_{16c}[Li_{0.4}Mn_{1.6}]_{16d}O_4$ site at the of 16c model with excess Li; (2) a $(Li)_{8a}[Li_{0.5}Mn_{1.5}]_{16d}O_{3.75}$ model with oxygen deficiency and (3) a hexagonal lattice model with cation deficiency $(Li_{0.8}\square_{0.2})_{3b}(Mn_{0.8}\square_{0.2})_{3a}O_2$ (the "□" are the free areas in the spinels). The modulation results showed that all models traced the X-ray peaks of the heat-treated sample, but the third model (a hexagonal lattice with a deficit of cations) accurately traced the relative intensity of the X-ray peaks. By Ariza et al. [147] showed that X-ray absorption spectroscopy of $Li_{1.6}Mn_{1.6}O_4$ samples does not result in the complete displacement of the manganese absorption edge after lithium extraction/reintroduction. In addition, the structural arrangement and oxidation state of manganese remained unchanged during lithium extraction and re-administration, confirming the ion exchange mechanism for lithium extraction and re-administration. Thus, there is still some disagreement on the crystal structure of $Li_{1.6}Mn_{1.6}O_4$. Possible future research by scientists should focus on this issue to link the development of LIS to the excellent absorption properties of lithium.

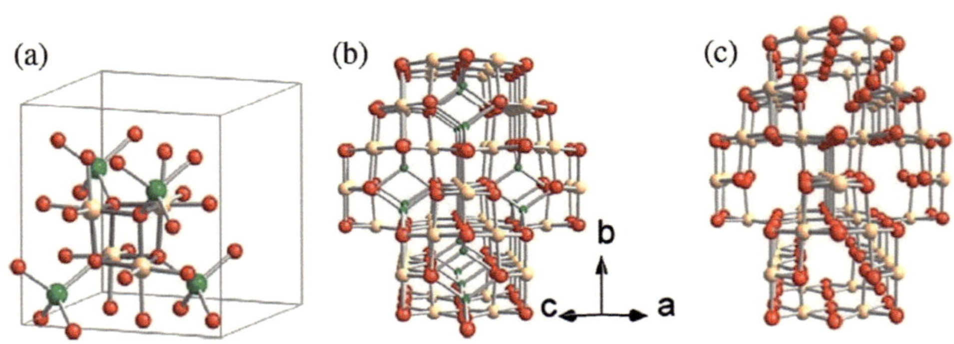

Figure 6.
Promising type (a) cubic core in spinel unit cell $LiMn_2O_4$, (b) $LiMn_2O_4$ of extended three-dimensional frame structure and (c) λ-MnO_2 with voids after Li ions removal. Green, pink and red represent Li, Mn and O atoms, respectively [146].

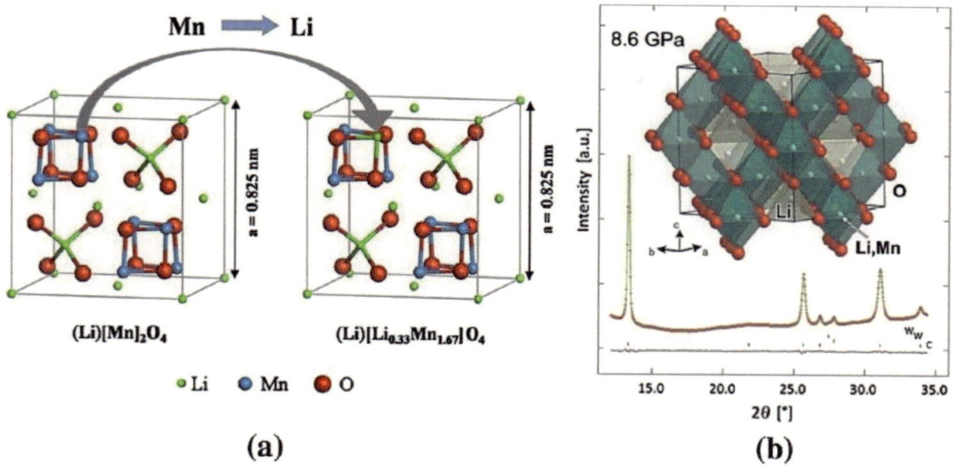

Figure 7.
(a) Cubic spinel lithium manganese oxide quadrants were comparison and (b) recorded under the 8.6 GPa. C, cubic spinel phase (spatial group Fd3m); W, tungsten strip model of polyhedral structure and structure refinement by Rietveld X-ray diffraction powder sample for $Li_{1.33}Mn_{1.67}O_4$ (or $Li_4Mn_5O_{12}$).

7.1.3 The doping modification

Because of the specific configuration of the 3d electron orbit, t_{2g}^3-e_g^1, Mn^{3+} can cause the Jahn-teller effect, which can cause severe distortions in the octahedral structure of MnO_6. This distortion will be accompanied by a decrease in LMO stability and a decrease in the efficiency of the intercalation/deintercalation process of Li^+ [153–157]. Much more seriously in industrial operations dissolving large amounts of manganese in water can lead to water contamination. Consequently, some alloying modifications have been proposed to replace Mn^{3+} with other metal ions, which is more efficient.

In the field of lithium-ion batteries, a wide variety of cation substitution (including Co^{2+}, Ni^{2+}, Cr^{3+}, Mg^{2+}, Al^{3+}, Fe^{3+} and ions of rare earth element) has been applied to inhibit capacity fading and improve electrochemical performance [158–183]. Analogously, modifications of LIS by doping with metal ions to improve the absorption properties of lithium in aqueous solutions are proposed. The effect of $Li_mMg_xMn(III)_yMn(IV)_zO_4$ ($0 \leq x \leq 0.5$) on the dissolution of manganese within acid treatment, the results showed that the adsorption capacity of lithium and the chemical stability of protonated samples increased with the mg/MN ratio studied by Chitrakar et al. [181]. Mild chemical method of Mg^{2+} doped lithium-manganese spinel synthesized by Tian et al. [36]. During the periodic experiment, it was found that the sorption of Li^+ showed a high pH and a dependence profile of the initial concentration. In addition, kinetic experiments have shown that the adsorption process followed by a pseudo-second-order model. Li^+ extraction process in both compound $LiMg_{0.5}Mn_{1.5}O_4$ spinel and $LiZn_{0.5}Mn_{1.5}O_4$ spinel studied by Feng et al. [182, 183]. Discovered that the extraction and insertion of Li^+ are topotaxically through ion exchange mechanisms. In addition, with $LiAlMnO_4$ and $LiFeMnO_4$ spinel Li^+ extraction/insertion reactions in the aqueous phase, also follow the ion exchange mechanisms tested by Liu et al. [184]. $LiM_xMn_{2-x}O_4$ spinel series (M = Ni, Al, Ti; $0 \leq x \leq 1$) and comparison of their lithium reduction properties in aqueous solutions prepared by Ma et al. [185]. Studies have shown that $LiAl_{0.5}Mn_{1.5}O_4$ spinels exhibit relatively high Li extraction coefficient and relatively low Mn and Al extraction coefficients when treated with acid, and

$LiNi_{0.5}Mn_{1.5}O_4$ and $LiTi0_{.5}Mn_{1.5}O_4$ spinels do not exhibit satisfactory Li^+ extraction and adsorption properties because of substantial cell contraction or expansion. By Chitrakar et al. Sb-doped LMO spinel was synthesized for the first time [186]. Samples received $Li_{1.16}Sb(V)_{0.29}Mn(III)_{0.77}Mn(IV)_{0.77}O_4$ was a well-crystallized spinel-type structure, in the following order of affinity K < Na ≪ Li and exchange capacity reaching 5.6 mmol g^{-1} for Li^+. In a subsequent study, a series of Li-Sb-Mn composite oxides with various Sb/Mn molar ratios by solid-state reactions obtained by Ma et al. [187]. Studies have shown that the molar ratio Sb/Mn of composite oxides Li-Sb-Mn is a decisive factor in the identification of their structure and extraction-adsorption properties Li^+. Hereinafter, the acid-treated composite spinel oxide Li-Sb-Mn with a molar ratio Sb/Mn of 0.05 showed in lithium solution a high adsorption capacity of Li^+ 33.23 mg g^{-1}. By Chitrakar et al. the ion-exchange property of iron-doped lithium manganese oxides $Li_{1.33}Fe_xMn_{1.67-x}O_4$ (x = 0.15, 0.30 and 0.40) in Bolivian brine was studied [38]. Studies have shown that the adsorbent with a Fe/Mn ratio of 0.1, obtained by calcining the precursor at 450°C, has the highest extractability of lithium with HCl solution. Finally, from crude brine at a final pH of 2.0, the adsorbent showed lithium absorption of 18.1 mg g^{-1} with an increase in absorption to 28 mg g^{-1} at a final pH of 7.2 after adding 1 mol L^{-1} NaOH l-1 solution to the crude brine.

Study of the description of the LMO-doped spinels, it is obvious that doping modifications can effectively improve the adsorption properties of lithium. Nevertheless, little attention has been paid to refining LIS compared with the great progress of ion-doped manganese oxide spinels in the field of electrochemistry. At present, just several studies of LISs doped with a single metal have been studied. Lithium adsorption property of multicharged ions doped LISs, including several cation-doped, several anion-doped and cation-anion-doped LISs in aqueous solution, still an untouched area for research. Early research of numerous ion-doped $LiMn_2O_4$ showed high capacity retention, high discharge capacity, and lithium ion batteries good cycling performance. This is due to the fact that multiple ions doped $LiMn_2O_4$, have increased structural stability [188–197]. Besides, as cathodes, co-doping has a synergistic effect on increasing the cyclic durability of materials, which can for single ion-doped $LiMn_2O_4$ discourage all factors responsible for capacity loss [198–201]. Similarly, it has been convincingly shown that multiple ion doping has a beneficial effect on improving the regeneration efficiency and absorption capacity of lithium LISs in aqueous solutions. Prospective studies should focus on the synergistic effects of different ions on the reductive properties of lithium.

7.2 About LTO-type LISs

There are currently two categories of LTO-type LISs: layered structure H_2TiO_3 and spinel structure $H_4Ti_5O_{12}$. Albeit the amount of LTO-type LISs is confined, there is great potential to develop these green lithium adsorbents for application in the industry, avoiding water pollution.

7.2.1 Study of layered H_2TiO_3 LISs

The chemical structure of layered H_2TiO_3 is shown in **Figure 8**. From the layered precursor Li_2TiO_3 a layered H_2TiO_3 is obtained. One can better describe as $Li[Li_{1/3}Ti_{2/3}]O_2$ the crystal structure of this precursor; precisely, when metal atoms are placed in octahedral voids the structure can be represented as cubic close packed oxygen atoms. In the structure of Li_2TiO_3 two types of layers form Li and Ti. The first layer (Li) is inhabited only by lithium atoms, while the other layer ($LiTi_2$) occupies Li 1/3 and

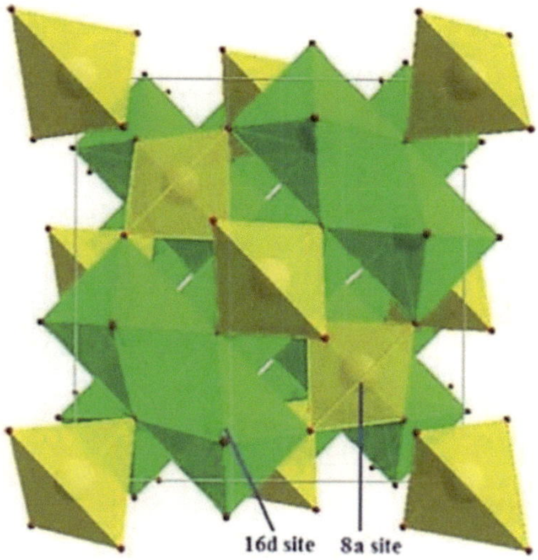

Figure 8.
Crystal structure of $Li_4Ti_5O_{12}$ (yellow tetrahedra represent lithium, and green octahedra represent disordered lithium and titanium) [39]. Reproduced from Ref. [39].

Ti 2/3. In fact, in the structure of Li_2TiO_3, lithium ions in the layers make up 75% of the total amount of lithium, while the surviving 25% are in layers ($LiTi_2$) [53].

Accordingly, whole lithium ions are changed by protons in the layered structure of H_2TiO_3. Accordingly, in early studies, some researchers believed that the structure of H_2TiO_3 was converted from layered Li_2TiO_3 by topotactic substitution of lithium ions by protons. The authors explore the composition of H_2TiO_3 by reviewing the variation among Li_2TiO_3 and H_2TiO_3 and modeling the XRD patterns of $H_xLi_{2-x}TiO_3$ ($0 \leq x \leq 2$), they pointed out that a structure with a layered double hydroxide type with a sequence of $3R_1$ oxygen layers is more acceptable for H_2TiO_3, and H_2TiO_3 can be described as laying charge-neutral metal oxyhydroxide plates $[(OH)_2OTi_2O(OH)_2]$ [202]. In advanced research, requires additional experimental testing to confirm the well-honed structure.

In 1988, Onodera et al. first obtained Li_2TiO_3 [203], many kinds of research have been conducted on its electrochemical application [204–208] and in the degradation of pollutants the photocatalytic applications [209–211]. Chitrakar et al. investigated the behavior of ion exchange in salt lake brines [53]. While the rate of adsorption of lithium was relatively slow (it took 1day to reach equilibrium at room temperature), at pH 6.5 the capacity of the Li^+ can reach up to 32.6 mg g^{-1}, that is among the adsorbents of lithium the greatest value is studied in an acidic solution. Besides, H_2TiO_3 has been found to be able to efficiently absorb lithium ions from Na^+, K^+, Mg^{2+} and Ca^{2+} containing competitive cations in brine. With ionic radii exceeding Li^+ (0.074 nm), it is not possible to introduce sites into the LTO adsorbent, since exchange sites have radii sizes Na^+ (0.102 nm), K^+ (0.138 nm) and Ca^{2+} (0.100 nm), which do not allow adsorption due to the large size of the ionic radii. Although the ionic radius of Mg^{2+} (0.072 nm) is close to the ionic radii of Li^+, dehydration of magnesium ions requires high energy to enter the exchange nodes, since the free hydration energy for Mg ($\Delta G^0{}_h = -1980$ kJ mol^{-1}) is four times greater than for Li ($\Delta G^0{}_h = -475$ kJ mol^{-1}) [212]. In addition, the Li-Mg separation ratio reached 102.4 on the 8th adsorption cycle, that in salt lake brines represents the excellent separation of Li^+ and Mg^{2+} found by Shi et al. [40]. In designing the orthogonal test, the maximum absorption of lithium by H_2TiO_3 reached 57.8 mg g^{-1} at the optimal state studied by He et al. [213].

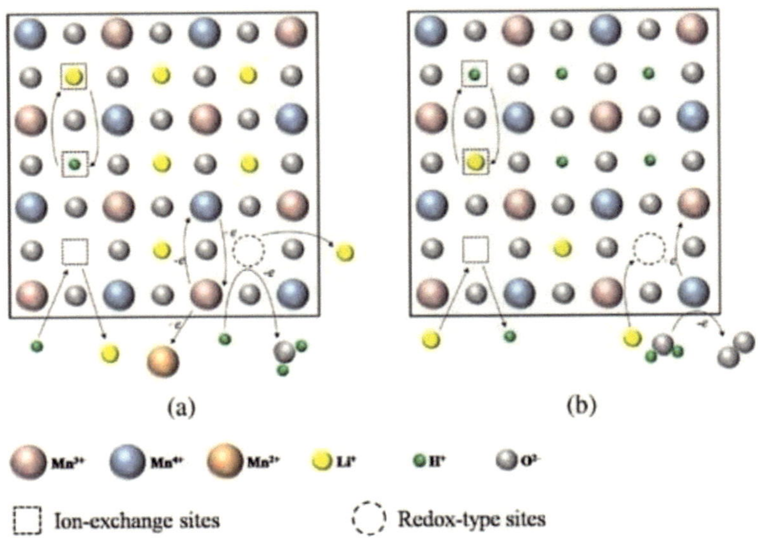

Figure 9.
Schematic representation in spinel manganese oxides by the composite mechanism (a) Li^+ extraction reactions and (b) Li^+ insertion reactions.

7.2.2 Study of spinel titanium oxides

The LTO-type LISs represent the different types of spinel titanium oxides that are derived from spinel precursors $Li_4Ti_5O_{12}$. In the field of lithium-ion batteries, spinel $Li_4Ti_5O_{12}$ is seen as one of the most promising future anode candidates for large-scale lithium-ion batteries used for hybrid electric vehicles or power electric vehicles. Through high efficient due to high potential during charge and discharge of about 1.55 V (vs. Li/Li^+), good cycle property and good heat resistance and security [214–216]. There is great potential for the development of spinel $Li_4Ti_5O_{12}$ in the extraction of lithium from aqueous solutions. High capacity lithium has on LIS ($H_4Ti_5O_{12}$) and due to stronger Ti–O bond cycling performance is better than that of manganese-type LISs. Withal, $Li_4Ti_5O_{12}$ has an identical chemical structure like $Li_4Mn_5O_{12}$ (**Figure 9**).

Nevertheless, as far as we know, there are currently very limited reports on the property of extracting lithium from $H_4Ti_5O_{12}$. A three-dimensionally ordered precursor to nano $Li_4Ti_5O_{12}$ using colloidal PMMA crystal matrices developed by Dong et al. [217]. High selectivity for Li^+, 56.81 mg g^{-1} showed corresponding ion sieve and good stability to acid. LISs $H_4Ti_5O_{12}$ with nanotube morphology synthesized by an ordinary two-stage hydrothermal process presented a lithium capacity of 39.43 mg g^{-1} in a 120 mg L^{-1} in lithium solution reported by Moazeni et al. [39].

8. Conclusions

Lithium is one of the rarest metals with various applications and the demand for lithium will increase with the ever-increasing use of electric and electronic devices and hybrid electric vehicles.

Therefore, the search for ways to obtain lithium from water sources suitable for the production of lithium compounds is a serious and very important problem.

Various methods have been given in the literature for lithium recovery from brines, seawater and geothermal water: including precipitation, solvent extraction, selective membrane separation, liquid-liquid extraction, ion exchange adsorption, electro dialysis and so on.

The recovery of lithium by the absorption method shows promising results for future production. Because of the adsorption method, evaporation, crystallization process can be avoided. That is why it is necessary to develop and recommend a technically and economically feasible, environmentally friendly and sustainable process.

Scientists and manufacturers are faced with the task to solve several problems: the ion sieve has a relatively low ion exchange capacity and weak stability; lithium absorption reaches from 16 to 26–28 mg g^{-1}, the theoretical adsorption capacity is 54 mg g^{-1}; dissolution of sorbents. Weight loss was observed in almost all compositions; low stability during cycling; the appearance of secondary waste in the regeneration of acids; the process takes a long time.

To solve this problem, scientists of the world have carried out many scientific works to improve the stability of sorbents, increase the absorption capacity, selectivity, acceleration of sorption time, for this purpose, many methods were used, including organic chemicals, synergies, binders, various composites. But none of them makes it possible to industrialize the method of lithium adsorption. That is why there is still a goal to find ways to improve the method of lithium adsorption. Lithium adsorption extraction may be an alternative option to meet future demand, energy sustainability, environment and circular economy.

Acknowledgements

The authors gratefully acknowledge partial financial supports from the National Natural Science Foundation of China (U1607123 and 21773170), the Key Projects of Natural Science Foundation of Tianjin (18JCZDJC10040), the Major Special Projects of Tibet Autonomous Region (XZ201801-GB-01) and the Yangtze Scholars and Innovative Research Team of the Chinese University (IRT_17R81).

Author details

Samadiy Murodjon[1,2], Xiaoping Yu[1,2], Mingli Li[3], Ji Duo[3] and Tianlong Deng[1,2]*

1 College of Marine and Environmental Science, Tianjin University of Science and Technology, TEDA, Tianjin, P.R. China

2 College of Chemical Engineering and Materials Science, Tianjin University of Science and Technology, TEDA, Tianjin, P.R. China

3 Central Laboratory of Geological Mineral Exploration and Development Bureau of Tibet Autonomous Region, Tibet, P.R. China

*Address all correspondence to: tldeng@tust.edu.cn

References

[1] Ji Z-Y, Yang F-J, Zhao Y-Y, Liu J, Wang N, Yuan J-S. Preparation of titanium-base lithium ionic sieve with sodium persulfate as eluent and its performance. Chemical Engineering Journal. 2017;**328**:768-775

[2] Kesler SE, Gruber PW, Medina PA, Keoleian GA, Everson MP, Wallington TJ. Global lithium resources: Relative importance of pegmatite, brine and other deposits. Oregon Geology. 2012;**48**:55-69

[3] Vikström H, Davidsson S, Höök M. Lithium availability and future production outlooks. Applied Energy. 2013;**110**:252-266

[4] Kunasz I. Lithium resources. In: Kogel JE et al., editors. Industrial Minerals and Rocks—Commodities, Markets, and Uses, 7th Edition Society for Mining, Metallurgy, and Exploration (SME). 2006

[5] U.S. Geological Survey. Mineral Commodity Summaries. 2018. Avaiable from: https://minerals.usgs.gov/minerals/pubs/commodity/lithium/mcs-2018-lithi.pdf [Accessed: 16 May 2018]

[6] Munk L, Hynek S, Bradley DC, Boutt D, Labay KA, Jochens H. LithiumBrines: A Global Perspective (Chapter 14). 2016

[7] Glasstone S, Senonske A. Nuclear Reactor Engineering. 3rd ed. Delhi: CBS Publisher & Distributor; 1986

[8] Tsuchiya K, Kawamura H. Fabrication and Characterization of 6Li-enriched Li_2TiO_3 Pebbles for a High Li-burnup Irradiation Tests. Japan Atomic Energy Agency; 2006

[9] Alvani C, Carconi PL, Casadio S, Conitini V, Dibartolomeo A, Pierdominici F, et al. Journal of Nuclear Materials. 2001;**289**:303-307

[10] Opitz A, Badami P, Shen L, Vignarooban K, Kannan AM. Can Li-ion batteries be the panacea for automotive applications? Renewable & Sustainable Energy Reviews. 2017;**68**:685-692

[11] Evans RK. Lithium's future supply, demand. The Northern Miner. 2010;**96**(35):11-12

[12] Winter M, Brodd RJ. What are batteries, fuel cells, and supercapacitors? Chemical Reviews. 2004;**104**(10):4245-4269

[13] Bruce PG, Freunberger SA, Hardwick LJ, Tarascon JM. $Li-O_2$ and Li-S batteries with high energy storage. Nature Materials. 2012;**11**(1):19-29

[14] Van Noorden R. The rechargeable revolution: A better battery. Nature. 2014;**507**:26-28

[15] Arias AN, Tesio AY, Flexer V. Review—Non-carbonaceous materials as cathodes for lithium-sulfur batteries. Journal of the Electrochemical Society. 2018;**165**(1):A6119-A6135

[16] Hykawy J. Looking at lithium: Discussing market demand for lithium in electronics. Materials World. 2010;**18**(5):34-35

[17] Siame E, Pascoe D. Extraction of lithium from micaceous waste from China clay production. Minerals Engineering. 2011;**24**(14):1595-1602. DOI: 10.1016/j.min.eng.2011.8.013

[18] Yaksic A, Tilton JE. Using the cumulative availability curve to assess the threat of mineral depletion: The case of lithium. Resources Policy. 2009;**34**:185-194

[19] Evans JK. An Abundance of Lithium. 2008. Available from: http://www.worldlithium.com/An_Abundance_of_Lithium_1_files/

An_Abundance _of_Lithium.pdf; http://www.worldlithium.com/AN_ABUNDANCE_OF_LITHIUM_-_Part_2.html [Accessed 27 November 2011]

[20] Brouwer AS, van den Broek M, Zappa W, Turkenburg WC, Faaij A. Least-cost options for integrating intermittent renewables in low-carbon power systems. Applied Energy. 2016;**161**:48-74

[21] Pellow MA, Emmott CJM, Barnhart CJ, Benson SM. Hydrogen or batteries for grid storage? A net energy analysis. Energy & Environmental Science. 2015;**8**(7):1938-1952

[22] Sternberg A, Bardow A. Power-to-what?—Environmental assessment of energy storage systems. Energy & Environmental Science. 2015;**8**(2):389-400

[23] Bazán J, Rieradevall J, Gabarrell X, Vázquez-Rowe I. Low-carbon electricity production through the implementation of photovoltaic panels in rooftops in urban environments: A case study for three cities in Peru. Science of the Total Environment. 2018;**622-623**:1448-1462

[24] Rahimi-Eichi H, Ojha U, Baronti F, Chow MY. Battery management system: An overview of its application in the smart grid and electric vehicles. IEEE Industrial Electronics Magazine. 2013;7(2):4-16

[25] An Increasingly Precious Metal. The Economist. 2018. Available from: https://www.economist.com/news/business/21688386-amid-surge-demand-rechargeable-batteries-companies-are-scrambling-supplies [Accessed 30 July 2018]

[26] Swain B. Recovery and recycling of lithium: A review [J]. Separation and Purification Technology. 2016;**172**:388-403

[27] Luo QP, Guo PC, Li CZ, Chen L. Distribution of lithium resources and research status on lithium extraction technology. Hydrometallurgy of China. 2012;**31**:67-70

[28] Grosjean C, Miranda PH, Perrin M, Poggi P. Assessment of world lithium resources and consequences of their geographic distribution on the expected development of the electric vehicle industry. Renewable & Sustainable Energy Reviews. 2012;**16**(3):1735-1744

[29] Hamzaoui HA, Jamoussi B, M'Nif A. Lithium recovery from highly concentrated solutions: Response surface methodology (RSM) process parameters optimization. Hydrometallurgy. 2008;**90**(1):1-7

[30] Liu Q, Ai H-M. Sodium benzoate as a green, efficient, and recyclable catalyst for knoevenagel condensation. Synthetic Communications. 2012;**43**(51):3004-3010

[31] Liu Q, Ai H, Li Z. Potassium sorbate as an efficient and green catalyst for knoevenagel condensation. Ultrasonics Sonochemistry. 2011;**18**(2):477-479

[32] Nie X-Y, Sun S-Y, Song X, Yu J-G. Further investigation into lithium recovery from salt lake brines with different feed characteristics by electrodialysis. Journal of Membrane Science. 2017;**530**:185-191

[33] Nie X-Y, Sun S-Y, Sun Z, Song X, Yu J-G. Ion-fractionation of lithiumions from magnesium ions by electrodialysis using monovalent selective ion-exchange membranes. Desalination. 2017;**403**:128-135

[34] Stamp A, Lang DJ, Wäger PA. Environmental impacts of a transitiontoward e-mobility: The present and future role of lithium carbonate production. Journal of Cleaner Production. 2012;**23**(1):104-112

[35] Ji Z-Y, Yang F-J, Zhao Y-Y, Liu J, Wang N, Yuan J-S. Preparation

of titanium-base lithium ionic sieve with sodium persulfate as eluent and itsperformance. Chemical Engineering Journal. 2017;**328**:768-775

[36] Tian L, Ma W, Han M. Adsorption behavior of Li^+ onto nano-lithium ionsieve from hybrid magnesium/ lithium manganese oxide. Chemical Engineering Journal. 2010;**156**(1):134-140

[37] Zhu G, Wang P, Qi P, Gao C. Adsorption and desorption properties of Li^+ on PVC-$H_{1.6}Mn_{1.6}O_4$ lithium ion-sieve membrane. Chemical Engineering Journal. 2014;**235**(1):340-348

[38] Chitrakar R, Makita Y, Ooi K, Sonoda A. Synthesis of iron-doped manganese oxides with an ion-sieve property: Lithium adsorption from Bolivianbrine. Industrial & Engineering Chemistry Research. 2014;**53**(9):3682-3688

[39] Moazeni M, Hajipour H, Askari M, Nusheh M. Hydrothermal synthesis and characterization of titanium dioxide nanotubes as novel lithium adsorbents. Materials Research Bulletin. 2015;**61**(61):70-75

[40] Shi XC, Zhang ZB, Zhou DF, Zhang LF, Chen BZ, Yu LL. Synthesis of Li^+ adsorbent (H_2TiO_3) and its adsorption properties. Transactions of Nonferrous Metals Society of China. 2013;**23**(1):253-259

[41] Tang DH, Zhou DL, Zhou JB, Zhang P, Zhang LY, Xia Y. Preparation of H_2TiO_3-lithium adsorbent using low-grade titanium slag. Hydrometallurgy. 2015;**157**:90-96

[42] Zhang LY, Zhou DL, Yao QQ, Zhou JB. Preparation of H_2TiO_3-lithium adsorbent by the sol-gel process and its adsorption performance. Applied Surface Science. 2016;**368**:82-87

[43] Wang SL, Li P, Cui WW, Zhang HL, Wang HY, Zheng SL, et al. Hydrothermal synthesis of lithium-enriched β-Li_2TiO_3 with an ion-sieve application: Excellent lithium adsorption. RSC Advances. 2016;**6**(104):102608-102616

[44] Hunter JC. Preparation of a new crystal form of manganese dioxide: λ-MnO_2. Journal of Solid State Chemistry. 1981;**39**(2):142-147

[45] Xiao JL, Sun SY, Wang J, Li P, Yu JG. Synthesis and adsorption properties of $Li_{1.6}Mn_{1.6}O_4$ spinel. Industrial & Engineering Chemistry Research. 2013;**52**(34):11967-11973

[46] Yang XJ, Kanoh H, Tang WP, Ooi K. Synthesis of $Li_{1.33}Mn_{1.67}O_4$ spinels with different morphologies and their ion adsorptivities after delithiation. Journal of Materials Chemistry. 2000;**10**(8):1903-1909

[47] Chitrakar R, Kanoh H, Miyai Y, Ooi K. A new type of manganese oxide ($MnO_2 \cdot 0.5H_2O$) derived from $Li_{1.6}Mn_{1.6}O_4$ and its lithium ion-sieve properties. Chemistry of Materials. 2000;**12**(10):3151-3157

[48] Chitrakar R, Kanoh H, Miyai Y, Ooi K. Recovery of lithium from seawater using manganese oxide adsorbent ($H_{1.6}Mn_{1.6}O_4$) derived from $Li_{1.6}Mn_{1.6}O_4$. Industrial & Engineering Chemistry Research. 2001;**40**(9):2054-2058

[49] Shi XC, Zhou DF, Zhang ZB, Yu LL, Xu H, Chen BZ, et al. Synthesis and properties of $Li_{1.6}Mn_{1.6}O_4$ and its adsorption application. Hydrometallurgy. 2011;**110**(1):99-106

[50] Zhang QH, Li SP, Sun SY, Yin XS, Yu JG. Lithium selective adsorption on 1-D MnO_2 nanostructure ion-sieve. Advanced Powder Technology. 2009;**20**(5):432-437

[51] Rjabtsev AD, Kotsupalo NP, Kishkan LN. Method of Producing Lithium Hydroxide from Brines and Plant for Method Embodiment. 2003. RU2193008C2

[52] Bauman WC, Burba JL, Burba III. Recovery of Lithium Values from Brines. 2013. US Patent 5,599,516, 1997

[53] Chitrakar R, Makita Y, Ooi K, Sonoda A. Lithium recovery from salt lake brine by H_2TiO_3. Dalton Transactions. 2014;**43**(23):8933-8939

[54] Xiao JL, Nie XY, Sun SY, Song XF, Li P, Yu JG. Lithium ion adsorption desorption roperties on spinel $Li_4Mn_5O_{12}$ and pH-dependent ion-exchange model. Advanced Powder Technology. 2015;**26**(2):589-594

[55] Zhou ZY, Qin W, Fei WY. Extraction equilibria of lithium with tributyl phosphate in three diluents. Journal of Chemical & Engineering Data. 2011;**56**(9):3518-3522

[56] Zhou ZY, Qin W, Liu Y, Fei WY. Extraction equilibria of lithium with tributyl phosphate in kerosene and $FeCl_3$. Journal of Chemical & Engineering Data. 2011;**57**:82-86

[57] Zhou ZY, Qin W, Liang SK, Tan YZ, Fei WY. Recovery of lithium using tributyl phosphate in methyl isobutyl ketone and $FeCl_3$. Industrial and Engineering Chemistry Research. 2012;**51**:12926-12932

[58] Taylor SR, McLennan SM. The Continental Crust: Its Composition and Evolution. Palo Alto, CA: Blackwell Scientific Pub; 1985

[59] Talens PL, Villalba MG, Ayres RU. Lithium: Sources, production, uses, and recovery outlook. JOM. 2013;**65**(8):986-996

[60] Chen H, Shen J. A degradation-based sorting method for lithium-ion battery reuse. PLoS One. 2017;**12**(10):e0185922

[61] Jaskula BW. USGS: 2015 Minerals Yearbook: Lithium [advanced released]. Available from: https://minerals.usgs.gov/minerals/pubs/commodity/lithium/myb1-2015-lithi.pdf [Accessed: 30 July 2018]

[62] Jaskula BWUS. Geological Survey. In: Jewell S, Kimball SM, editors. Mineral Commodity Summaries. 2015

[63] Sadoway D. Toward new technologies for the production of lithium. JOM. 1998;**50**:24-26

[64] Mahi P, Smeets AAJ, Fray DJ, Charles JA. Lithium-metal of the future. JOM. 1986;**38**:20-26

[65] Bohner HO. A. Metallurgical Society of, C. Light Metals, Light metals. In: 1985 Proceedings of the Technical Sessions Sponsored by the TMS Light Metals Committee at the 114th Annual Meeting, New York, NY, February 24-28, 1985, Metallurgical Society of AIME, Warrendale, PA

[66] Marinsky JA, Marcus Y. Ion Exchange and Solvent Extraction: A Series of Advances. Taylor & Francis; 1995

[67] Kogel JEM. Society for Mining, Exploration, Industrial Minerals & Rocks: Commodities, Markets, and Uses, Society for Mining, Metallurgy, and Exploration. 2006

[68] Survey G. United States Geological Survey Professional Paper. U.S. Government Printing Office; 1961

[69] Miser HD, Stevens RE. Taeniolite from Magnet Cove, Arkansas. American Mineralogist. 1938;**23**:104-110

[70] Hamzaoui AH, M'Nif A, Hammi H, Rokbani R. Contribution to the lithium

recovery from brine. Desalination. 2003;**158**:221-224

[71] Outotec® Lithium Production Technologies, Outotec Oyj. 2016

[72] Chagnes A, Swiatowska J. Lithium Process Chemistry: Resources, Extraction, Batteries, and Recycling. Elsevier Science; 2015

[73] Pelly I. Recovery of lithium from Dead Sea brines. Journal of Applied Chemistry & Biotechnology. 1978;**28**:469-474

[74] Epstein JA, Feist EM, Zmora J, Marcus Y. Extraction of lithium from the Dead Sea. Hydrometallurgy. 1981;**6**:269-275

[75] Kaplan D. Process for the extraction of lithium from Dead Sea solutions. Israel Journal of Chemistry. 1963;**1**:115-120

[76] Hui Z. Property of H_2TiO_3 type ion-exchangers and extraction of lithium from brine of natural gas wells. Chinese Journal of Applied Chemistry. 2000;**17**:307-309

[77] Bukowsky H, Uhlemann E, Steinborn D. The recovery of pure lithium chloride from "brines" containing higher contents of calcium chloride and magnesium chloride. Hydrometallurgy. 1991;**27**:317-325

[78] Gabra GG, Torma AE. Lithium chloride extraction by n-butanol. Hydrometallurgy. 1978;**3**:23-33

[79] Bukowsky H, Uhlemann E. Selective extraction of lithium chloride from brines. Separation Science and Technology. 1993;**28**:1357-1360

[80] Baldwin W, Seeley F. Extraction of Lithium from Neutral Brines Using a Beta Diketone and Trioctylphosphine Oxide. Atomic Energy Commission; 1974

[81] Swain B. Separation and purification of lithium by solvent extraction and supported liquid membrane, analysis of their mechanism: A review. Journal of Chemical Technology and Biotechnology. 2016;**91**:25-49

[82] Gao D, Yu X, Guo Y, Wang S, Liu M, Deng T, et al. Extraction of lithium from salt lake brine with triisobutyl phosphate in ionic liquid and kerosene. Chemical Research in Chinese Universities. 2015;**31**:621-626

[83] Gao D, Guo Y, Yu X, Wang S, Deng T. Extracting lithium from the high concentration ratio of magnesium and lithium brine using imidazoliumbased ionic liquids with varying alkyl chain lengths. Journal of Chemical Engineering of Japan. 2016;**49**:104-110

[84] Shi Chenglong JY, Yan JING. Lithium and magnesium separation from salt lake brine by ionic liquids containing tributyl phosphate. CIESC Journal. 2015;**66**:253-259

[85] Shi C, Jing Y, Jia Y. Solvent extraction of lithium ions by tri-n-butyl phosphate using a room temperature ionic liquid. Journal of Molecular Liquids. 2016;**215**:640-646

[86] Jiang C, Wang Y, Wang Q, Feng H, Xu T. Production of lithium hydroxide from lake brines through electro–electrodialysis with bipolar membranes (EEDBM). Industrial and Engineering Chemistry Research. 2014;**53**:6103-6112

[87] Liu X, Chen X, He L, Zhao Z. Study on extraction of lithium from salt lake brine by membrane electrolysis. Desalination. 2015;**376**:35-40

[88] Jagur-Grodzinski J, Schori E. Solvent-polymeric membranes for separation of Li^+ from other alkali metal and alkaline earth ions. Israel Journal of Chemistry. 1985;**26**:65-70

[89] Sun S-Y, Cai L-J, Nie X-Y, Song X, Yu J-G. Separation of magnesium and lithium from brine using a Desal nanofiltration membrane. Journal of Water Process Engineering. 2015;7:210-217

[90] Somrani A, Hamzaoui AH, Pontie M. Study on lithium separation from salt lake brines by nanofiltration (NF) and low pressure reverse osmosis (LPRO). Desalination. 2013;**317**:184-192

[91] Mero JL. The Mineral Resources of the Sea. Elsevier Science; 1965

[92] James GM. Recovery of Lithium from Brines, Google Patents. 1966

[93] An JW, Kang DJ, Tran KT, Kim MJ, Lim T, Tran T. Recovery of lithium from Uyuni Salar brine. Hydrometallurgy. 2012;**117-118**:64-70

[94] James GM. Process for the Recovery of Lithium and Potassium from Great Salt Lake Brine, Google Patents. 1967

[95] Repsher WJ, Rapstein KT. Recovery of Lithium from Brine. Dow Chemical Co.; 1981

[96] Yanagase K, Yoshinaga T, Kawano K, Matsuoka T. The recovery of lithium from geothermal water in the Hatchobaru area of Kyushu, Japan. Bulletin of the Chemical Society of Japan. 1983;**56**:2490-2498

[97] Yoshinaga T, Kawano K, Imoto H. Basic study on lithium recovery from lithium containing solution. Bulletin of the Chemical Society of Japan. 1986;**59**:1207-1213

[98] Kang-Sup Chung J-CL, Jeong J-k, Kim E-J, Kim Y-S. Journal of Korean Geoscience Engineering. 2003;**40**:402-408

[99] Um N, Hirato T. Precipitation behavior of $Ca(OH)_2$, $Mg(OH)_2$, and $Mn(OH)_2$ from $CaCl_2$, $MgCl_2$, and $MnCl_2$ in $NaOH$-H_2O solutions and study of lithium recovery from seawater via two-stage precipitation process. Hydrometallurgy. 2014;**146**:142-148

[100] Nishihama S, Onishi K, Yoshizuka K. Selective recovery process of lithium from seawater using integrated ion exchange methods. Solvent Extraction and Ion Exchange. 2011;**29**:421-431

[101] Takeuchi T. Extraction of lithium from sea water with metallic aluminum. Journal of Nuclear Science and Technology. 1980;**17**:922-928

[102] Ooi K, Miyai Y, Katoh S. Recovery of lithium from seawater by manganese oxide adsorbent. Separation Science and Technology. 1986;**21**:755-766

[103] Miyai Y, Ooi K, Katoh S. Recovery of lithium from seawater using a new type of ion-sieve adsorbent based on $MgMn_2O_4$. Separation Science and Technology. 1988;**23**:179-191

[104] Liu L, Zhang H, Zhang Y, Cao D, Zhao X. Lithium extraction from seawater by manganese oxide ion sieve $MnO_2 \times 0.5H_2O$. Colloids and Surfaces A: Physicochemical and Engineering Aspects. 2015;**468**:280-284

[105] Park H, Singhal N, Jho EH. Lithium sorption properties of HMnO in seawater and wastewater. Water Research. 2015;**87**:320-327

[106] Wajima T, Munakata K, Uda T. Adsorption behavior of lithium from seawater using manganese oxide adsorbent. Plasma and Fusion Research. 2012;7:2405021

[107] Kitamura T, Wada H. Properties of adsorbents composed of hydrous aluminum oxide, and its selective adsorption of lithium from sea water. Bulletin of the Society of Sea Water Science, Japan. 1978;**32**:78-81

[108] Lemaire J, Svecova L, Lagallarde F, Laucournet R, Thivel P-X. Lithium recovery from aqueous solution by sorption/desorption. Hydrometallurgy. 2014;**143**:1-11

[109] Umeno A, Miyai Y, Takagi N, Chitrakar R, Sakane K, Ooi K. Preparation and adsorptive properties of membrane-type adsorbents for lithium recovery from seawater. Industrial and Engineering Chemistry Research. 2002;**41**:4281-4287

[110] Park MJ, Nisola GM, Beltran AB, Torrejos REC, Seo JG, Lee S-P, et al. Recyclable composite nanofiber adsorbent for Li^+ recovery from seawater desalination retentate. Chemical Engineering Journal. 2014;**254**:73-81

[111] Hoshino T. Innovative lithium recovery technique from seawater by using world-first dialysis with a lithium ionic superconductor. Desalination. 2015;**359**:59-63

[112] Hoshino T. Development of technology for recovering lithium from seawater by electrodialysis using ionic liquid membrane. Fusion Engineering and Design. 2013;**88**:2956-2959

[113] Hoshino T. Preliminary studies of lithium recovery technology from seawater by electrodialysis using ionic liquid membrane. Desalination. 2013;**317**:11-16

[114] Quist-Jensen CA, Ali A, Mondal S, Macedonio F, Drioli E. A study of membrane distillation and crystallization for lithium recovery from highconcentrated aqueous solutions. Journal of Membrane Science. 2016;**505**:167-173

[115] Park MJ, Nisola GM, Vivas EL, Limjuco LA, Lawagon CP, Seo JG, et al. Mixed matrix nanofiber as a flow-through membrane adsorber for continuous Li^+ recovery from seawater. Journal of Membrane Science. 2016;**510**:141-154

[116] Lee JM, Bauman WC. Recovery of Lithium from Brines, Google Patents. 1978

[117] Repsher WJ, Rapstein KT. Recovery of Lithium from Brine. The Dow Chemical Company; 1981

[118] Alberti G, Massucci MA. Crystalline insoluble acid salts of tetravalent metals–IX. Journal of Inorganic and Nuclear Chemistry. 1970;**32**:1719-1727

[119] Volkhin GVLVV, Onorin SA, Khodyashev NB, Kudryavtsev PG, Shvetsova TI. Chemistry and Technology of Inorganic Sorbents. Perm: P.P.I/publish; 1980 (Russian)

[120] Abe M, Chitrakar R. Synthetic inorganic ion-exchange materials. XLV. Recovery of lithium from seawater and hydrothermal water by titanium(IV) antimonate cation exchanger. Hydrometallurgy. 1987;**19**:117-128

[121] Miyai Y, Ooi K, Sakakibara J, Katoh S. Physical and lithium-adsorptive properties of manganese oxide adsorbent granulated with polyvinyl chloride (PVC), recovery of lithium from sea water by manganese oxide adsorbent (part 14). Bulletin of the Society of Sea Water Science, Japan. 1991;**45**:193-198

[122] Zandevakili S, Ranjbar M, Ehteshamzadeh M. Recovery of lithium from Urmia Lake by a nanostructure MnO_2 ion sieve. Hydrometallurgy. 2014;**149**:148-152

[123] Harvianto G, Kim S-H, Ju C-S. Solvent extraction and stripping of lithium ion from aqueous solution and its application to seawater. Rare Metals. 2015;**35**:948-953

[124] Chung K-S, Lee J-C, Kim W-K, Kim SB, Cho KY. Inorganic adsorbent containing polymeric membrane reservoir for the recovery of lithium from seawater. Journal of Membrane Science. 2008;**325**:503-508

[125] Volkhin V, Shulga E, Zilberman M. Ion-exchange properties of mixed ferrocyanides of some transition metals. Newsletters of USSR Academy of Sciences, Inorganic Materials. 1971;7:77-81

[126] Komarneni S, Roy R. A cesium-selective ion sieve made by topotactic leaching of phlogopite mica. Science. 1988;**239**:1286-1288

[127] Diamond D, Svehla G, Seward EM, Mckervey MA. A sodium ion-selective electrode based on methyl p-t-butylcalix[4]aryl acetate as the ionophore. Analytica Chimica Acta. 1988;**204**:223-231

[128] Burr GW, Kurdi BN, Scott JC, Lam CH, Gopalakrishnan K, Shenoy RS. Overview of candidate device technologies for storage-class memory. IBM Journal of Research and Development. 2008;**52**:449-464

[129] Anthony RG, Philip CV, Dosch RG. Selective adsorption and ion exchange of metal cations and anions with silico-titanates and layered titanates. Waste Management. 1993;**13**:503-512

[130] Feng Q, Kanoh H, Miyai Y, Ooi K. Hydrothermal synthesis of lithium and sodium manganese oxides and their metal ion extraction/insertion reactions. ChemInform. 1995;**26**:1226-1232. [ChemInform Abstract]

[131] Eliad L, Salitra G, Soffer A, Aurbach D. Ion sieving effects in the electrical double layer of porous carbon electrodes: Estimating effective ion size in electrolytic solutions. The Journal of Physical Chemistry. B. 2001;**105**:6880-6887

[132] Miyake M, Sakata S, Endo K, Suzuki T. Cesium ion sieve effect of sodium-substituted hectorite. Chemistry of Materials. 1994;**6**:1599-1600

[133] Marcus Y, Sengupta A, Marinsky JA. Ion exchange and solvent extraction. A series of advances. Brain Dev. 2002;**25**:315-321. [Volume 15. No to Hattatsu]

[134] Liu Y, Feng Q, Kenta O. The synthesis and ion-exchange property of Li^+ memorized spinel $LiAIMnO_4$. Ion Exchange and Adsorption. 1995;**11**:216-222

[135] Kitajou A, Suzuki T, Nishihama S, Yoshizuka K. Selective recovery of lithium from seawater using a novel MnO_2 type adsorbent. II—Enhancement of lithium ion selectivity of the adsorbent. Ars Separatoria Acta. 2003;**2**:97-106

[136] Přibil R, Veselý V. Separation of titanium from iron and aluminium. Talanta. 1963;**10**:233-235

[137] Das J, Pobi M. Separation of titanium, iron and aluminium on a chelating resin with benzoylphenylhydroxylamine group and application to bauxite and clay. Fresenius' Journal of Analytical Chemistry. 1990;**336**:578-581

[138] Chen L, Wen S, Xu G, Xie H. A novel process for titanium sand by magnetic separation and gravity concentration. Mineral Processing and Extractive Metallurgy Review. 2013;**34**:139-150

[139] Thackeray MM, Kock AD, Rossouw MH, Liles D, Bittihn R, Hoge D. Spinel electrodes from the Li-Mn-O system for rechargeable

lithium battery applications. Journal of the Electrochemical Society. 1992;**139**:363-366

[140] Zhang QH, Sun S, Li S, Jiang H, Yu JG. Adsorption of lithium ions on novel nanocrystal MnO_2. Chemical Engineering Science. 2007;**62**:4869-4874

[141] Zhang QH, Li SP, Sun SY, Yu XS, Yin JG. $LiMn_2O_4$ spinel direct synthesis and lithium ion selective adsorption. Chemical Engineering Science. 2010;**65**:169-173

[142] Sun S-Y, Song X, Zhang Q-H, Wang J, Yu J-G. Lithium extraction/insertion process on cubic Li-Mn-O precursors with different Li/Mn ratio and morphology. Adsorption. 2011;**17**:881-887

[143] Wang L, Meng CG, Ma W. Study on Li^+ uptake by lithium ion-sieve via the pH technique. Colloids and Surfaces, A: Physicochemical and Engineering Aspects. 2009;**334**:34-39

[144] Wang L, Meng C, Ma W. Preparation of lithium ion-sieve and utilizing in recovery of lithium from seawater. Frontiers of Chemical Engineering in China. 2009;**3**:65-67

[145] Hirayama M, Sonoyama N, Ito M, Minoura M, Mori D, Yamada A, et al. Characterization of electrode/electrolyte interface with X-ray reflectometry and epitaxial-film $LiMn_2O_4$ electrode. Journal of the Electrochemical Society. 2007;**154**:A1065-A1072

[146] Greedan JE, Raju NP, Wills AS, Morin C, Shaw SM, Reimers JN. Structure and magnetism in k-MnO_2. Geometric frustration in a defect spinel. Chemistry of Materials. 1998;**10**:3058-3067

[147] Ariza MJ, Jones DJ, Rozière J, Chitrakar R, Ooi K. Probing the local structure and the role of protons in lithium sorption processes of a new lithium-rich manganese oxide. Chemistry of Materials. 2006;**18**:1885-1890

[148] Ammundsen B, Jones DJ, Rozière J, Berg H, Tellgren R, Thomas JO. Ion exchange in manganese dioxide spinel: Proton, deuteron, and lithium sites determined from neutron powder diffraction data. Chemistry of Materials. 1998;**10**:1680-1687

[149] Ozawa K. Lithium ion Rechargeable Batteries: Materials, Technology, and New Applications. Wiley-VCH. 2010

[150] Robinson DM, Go YB, Greenblatt M, Dismukes GC. Water oxidation by k-MnO_2: Catalysis by the cubical Mn_4O_4 subcluster obtained by delithiation of spinel $LiMn_2O_4$. Journal of the American Chemical Society. 2010;**132**:11467-11469

[151] Darul J, Nowicki W, Piszora P. Unusual compressional behavior of lithium-manganese oxides: A case study of $Li_4Mn_5O_{12}$. Journal of Physical Chemistry C. 2012;**116**:17872-17879

[152] Sharma N, Yu D, Zhu Y, Wu Y, Peterson VK. Non-equilibrium structural evolution of the lithium-rich $Li_{1+y}Mn_2O_4$ cathode within a battery. Chemistry of Materials. 2013;**25**:754-760

[153] Tan T-Y, Kennedy BJ, Zhou Q, Ling CD, Miiller W, Howard CJ, et al. Impact of Jahn-teller active Mn^{3+} on strain effects and phase transitions in $Sr_{0.65}Pr_{0.35}MnO_3$. Physical Review B. 2012;**85**:104107

[154] Alonso JA, Martínez-Lope MJ, Casais MT, Fernández-Díaz MT. Evolution of the Jahn_Teller distortion of MnO_6 octahedra in $RMnO_3$ perovskites (R = Pr, Nd, Dy, Tb, Ho, Er, Y): A neutron diffraction study. Inorganic Chemistry. 2000;**39**:917-923

[155] Ghose S, Kersten M, Langer K, Rossi G, Ungaretti L. Crystal field spectra and Jahn Teller effect of Mn^{3+} in clinopyroxene and clinoamphiboles from India. Physics and Chemistry of Minerals. 1986;**13**:291-305

[156] Yamada A, Tanaka M. Jahn-Teller structural phase transition around 280 K in $LiMn_2O_4$. Materials Research Bulletin. 1995;**30**:715-721

[157] Lufaso MW, Woodward PM. Jahn-Teller distortions, cation ordering and octahedral tilting in perovskites. Acta Crystallographica B. 2004;**60**:10-20

[158] Li C, Fan Y, Li S, Xie B, Bi L, Yang S. Synthesis and electrochemical properties of Th-doped $LiMn_2O_4$ powders for lithium-ion batteries. Rare Metals. 2006;**25**:58-61

[159] Helan M, Berchmans LJ, Syamala Kumari VS, RaviSankar R, Shanmugam VM. Molten salt synthesis of $LiGd_{0.01}Mn_{1.99}O_4$ using chloride–carbonate melt. Materials Research Innovations. 2011;**15**:130-134

[160] BalajiSRK,MutharasuD,ShanmuganS, SubramanianNS,RamanathanK.Influence of Sm^{3+} ion in structural, morphological, and electrochemical properties of $LiMn_2O_4$ synthesized by microwave calcination. Ionics. 2010;**16**:351-360

[161] Suryakala K, Marikkannu K, Kalaignan GP, Vasudevan T. Synthesis and electrochemical characterization of $LiMn_2O_4$ and $LiNd_{0.3}Mn_{1.7}O_4$ as cathode for lithium ion battery. International Journal of Electrochemical Science. 2008;**3**:136-144

[162] Iqbal MJ, Ahmad Z. Electrical and dielectric properties of lithium manganate nanomaterials doped with rare-earth elements. Journal of Power Sources. 2008;**179**:763-769

[163] Ha H-W, Yun NJ, Kim K. Improvement of electrochemical stability of $LiMn_2O_4$ by CeO_2 coating for lithium-ion batteries. Electrochimica Acta. 2007;**52**:3236-3241

[164] Sun H, Chen Y, Xu C, Zhu D, Huang L. Electrochemical performance of rare-earth doped $LiMn_2O_4$ spinel cathode materials for Li-ion rechargeable battery. Journal of Solid State Electrochemistry. 2012;**16**:1247-1254

[165] Tu J, Zhao XB, Cao GS, Tu JP, Zhu TJ. Improved performance of $LiMn_2O_4$ cathode materials for lithium ion batteries by gold coating. Materials Letters. 2006;**60**:3251-3254

[166] Wang L, Zhao J, Guo S, He X, Jiang C, Wan C. Investigation of SnO_2-modified $LiMn_2O_4$ composite as cathode material for lithium-ion batteries. International Journal of Electrochemical Science. 2010;**5**:1113-1126

[167] Son JT, Park KS, Kim HG, Chung HT. Surface-modification of $LiMn_2O_4$ with a silver-metal coating. Journal of Power Sources. 2004;**126**:182-185

[168] Wang H, Tan TA, Yang P, Lai MO, Lu L. High-rate performances of the Ru-doped spinel $LiNi_{0.5}Mn_{1.5}O_4$: Effects of doping and particle size. Journal of Physical Chemistry C. 2011;**115**:6102-6110

[169] Wu HM, Belharouak I, Abouimrane A, Sun YK, Amine K. Surface modification of $LiNi_{0.5}Mn_{1.5}O_4$ by ZrP_2O_7 and ZrO_2 for lithium-ion batteries. Journal of Power Sources. 2010;**195**:2909-2913

[170] Liu D-Q, He Z-Z, Liu X-Q. Synthesis and characterization of $LiGaxMn_{2-x}O_4$ (06x60.05) by triethanolamine-assisted sol–gel method. Journal of Alloys and Compounds. 2007;**440**:69-73

[171] Liu H, Cheng C, Zongqiuhu D, Zhang K. The effect of ZnO coating on $LiMn_2O_4$ cycle life in high temperature for lithium secondary batteries. Materials Chemistry and Physics. 2007;**101**:276-279

[172] Ein-Eli Y, Urian RC, Wen W, Mukerjee S. Low temperature performance of copper/nickel modified $LiMn_2O_4$ spinels. Electrochimica Acta. 2005;**50**:1931-1937

[173] Wu HM, Tu JP, Chen XT, Li Y, Zhao XB, Cao GS. Effects of Ni-ion doping on electrochemical characteristics of spinel $LiMn_2O_4$ powders prepared by a spray-drying method. Journal of Solid State Electrochemistry. 2007;**11**:173-176

[174] Sakunthala A, Reddy MV, Selvasekarapandian S, Chowdari BVR, Selvin PC. Synthesis of compounds, $Li(MMn_{11/6})O_4$ (M = $Mn_{1/6}$, $Co_{1/6}$, $(Co_{1/12}Cr_{1/12})$, $(Co_{1/12}Al_{1/12})$, $(Cr_{1/12}Al_{1/12})$) by polymer precursor method and its electrochemical performance for lithium-ion batteries. Electrochimica Acta. 2010;**55**:4441-4450

[175] Malyovanyi SM, Andriiko AA, Monko AP. Synthesis and electrochemical behavior of Fe-doped overstoichiometric $LiMn_2O_4$-based spinels. Journal of Solid State Electrochemistry. 2003;**8**:7-10

[176] Xu W, Yuan A, Tian L, Wang Y. Improved high-rate cyclability of sol–gel derived Cr-doped spinel $LiCryMn_{2-y}O_4$ in an aqueous electrolyte. Journal of Applied Electrochemistry. 2011;**41**:453-460

[177] Liu D-Q, Liu X-Q, He Z-Z. The elevated temperature performance of $LiMn_2O_4$ coated with $Li_4Ti_5O_{12}$ for lithium ion battery. Materials Chemistry and Physics. 2007;**105**:362-366

[178] Amaral FA, Bocchi N, Brocenschi RF, Biaggio SR, Rocha-Filho RC. Structural and electrochemical properties of the doped spinels $Li_{1.05}M_{0.02}Mn_{1.98}O_{3.98}N_{0.02}$ (M = Ga^{3+}, Al^{3+}, or Co^{3+}; N = S^{2-} or F^-) for use as cathode material in lithium batteries. Journal of Power Sources. 2010;**195**:3293-3299

[179] Singh P, Sil A, Nath M, Ray S. Synthesis and characterization of $Li[Mn_{2-x}Mg_x]O_4$ (x = 0.0-0.3) prepared by sol-gel synthesis. Ceramics-Silikáty. 2010;**54**:38-46

[180] Liu Q, Wang S, Tan H, Yang Z, Zeng J. Preparation and doping mode of doped $LiMn_2O_4$ for Li-ion batteries. Energies. 2013;**6**:1718-1730

[181] Chitrakar R, Yoji M, Kenta O, Akinari S. Magnesium-doped manganese oxide with lithium ion-sieve property: Lithium adsorption from salt lake brine. Bulletin of the Chemical Society of Japan. 2013;**86**:850-855

[182] Feng Q, Kanoh H, Miyai Y, Ooi K. Li^+ extraction/insertion reactions with $LiZn_{0.5}Mn_{1.5}O_4$ spinel in the aqueous phase. Chemistry of Materials. 1995;7:379-384

[183] Feng Q, Miyai Y, Kanoh H, Ooi K. $Lithium^{(1+)}$ and $magnesium^{(2+)}$ extraction and $lithium^{(1+)}$ insertion reactions with lithium magnesium manganese oxide ($LiMg_{0.5}Mn_{1.5}O_4$) spinel in the aqueous phase. Chemistry of Materials. 1993;**5**:311-316

[184] Liu Y-F, Feng Q, Ooi K. Li^+ extraction/insertion reactions with $LiAlMnO_4$ and $LiFeMnO_4$ spinels in the aqueous phase. Journal of Colloid and Interface Science. 1994;**163**:130-136

[185] Ma L-W, Chen B-Z, Shi X-C, Zhang W, Zhang K. Stability and Li^+ extraction/adsorption properties of

$LiMxMn_{2-x}O_4$ (M = Ni, Al, Ti; 0 6 x 6 1) in aqueous solution. Colloids and Surfaces, A: Physicochemical and Engineering Aspects. 2010;**369**:88-94

[186] Chitrakar R, Kanoh H, Makita Y, Miyai Y, Ooi K. Synthesis of spinel-type lithium antimony manganese oxides and their Li extraction/ion insertion reactions. Journal of Materials Chemistry. 2000;**10**:2325-2329

[187] Ma L-W, Chen B-Z, Shi X-C, Zhang K. Li^+ extraction/adsorption properties of Li-Sb-Mn composite oxides in aqueous medium. Transactions of Nonferrous Metals Society of China. 2011;**21**:1660-1664

[188] Shiu J-J, Pang WK, Wu S-H. Preparation and characterization of spinel LiNi0.5-xMgxMn1.5O4 cathode materials via spray pyrolysis method. Journal of Power Sources. 2013;**244**:35-42

[189] Yang MC, Xu B, Cheng JH, Pan CJ, Hwang BJ, Meng YS. Electronic, structural, and electrochemical properties of $LiNi_xCu_yMn_{2-x-y}O_4$ ($0 < x < 0.5$, $0 < y < 0.5$) high-voltage spinel materials. Chemistry of Materials. 2011;**23**:2832-2841

[190] Verrelli R, Scrosati B, Sun YK, Hassoun J. Stable, high voltage $Li_{0.85}Ni_{0.46}Cu_{0.1}Mn_{1.49}O_4$ spinel cathode in a lithium-ion battery using a conversion-type CuO anode. ACS Applied Materials & Interfaces. 2014;**6**:5206-5211

[191] Kim W-K, Han D-W, Ryu W-H, Lim S-J, Eom J-Y, Kwon H-S. Effects of Cl doping on the structural and electrochemical properties of high voltage $LiMn_{1.5}Ni_{0.5}O_4$ cathode materials for Li-ion batteries. Journal of Alloys and Compounds. 2014;**592**:48-52

[192] Ebin B, Gürmen S, Lindbergh G. Preparation and electrochemical properties of spinel $LiFe_xCu_yMn_{1.2}O_4$ by ultrasonic spray pyrolysis. Ceramics International. 2014;**40**:1019-1027

[193] Li R, Gong F, Wang W. Co-precipitation synthesis and performance of multi-doped $LiCr_xNi_xMn_{2-2x}O_{4-2}F_2$ cathode materials for lithium ion batteries. Ionics. 2006;**12**:353-363

[194] Sang HP, Park KS, Yang KS, Nahm KS. Synthesis and characterization of a new spinel, $Li_{1.02}Al_{0.25}Mn_{1.75}O_{3.97}S_{0.03}$, operating at potentials between 4.3 and 2.4 V. Journal of the Electrochemical Society. 2000;**147**:2116-2121

[195] Sun YK, Park GS, Lee YS, Yoashio M, Nahm KS. Structural changes (degradation) of oxysulfide $LiAl_{0.24}Mn_{1.76}O_{3.98}S_{0.02}$ spinel on high-temperature cycling. Journal of the Electrochemical Society. 2001;**148**:A994-A998

[196] Xiao J, Zhu H-L, Chen Z-Y, Peng Z-D, Hu G-R. Preparation and property of spinel $LiMn_2O_4$ material by co-doping anti-electricity ions. Transactions of Nonferrous Metals Society of China. 2006;**16**:467-472

[197] Xiong L, Xu Y, Tao T, Goodenough JB. Synthesis and electrochemical characterization of multi-cations doped spinel $LiMn_2O_4$ used for lithium ion batteries. Journal of Power Sources. 2012;**199**:214-219

[198] Zhang HL, Xiang N. Structural and electrochemical investigation of $Li_{1.02}Mn_{1.92}Al_{0.02}Fe_{0.02}Cr_{0.024-x}F_x$ (x = 0, 0.08) synthesized by solid-state method. Advances in Materials Science and Engineering. 2013;**1**:1-7

[199] Sahan H, Göktepe H, Patat S. Synthesis and cycling performance of double metal doped $LiMn_2O_4$ cathode materials for rechargeable lithium ion batteries. Inorganic Materials. 2008;**44**:420-425

[200] Bonino F, Panero S, Satolli D, Scrosati B. Synthesis and characterization of $Li_2M_xMn_{4-x}O_8$ (M = Co, Fe) as positive active materials for lithium-ion cells. Journal of Power Sources. 2001;**97-98**:389-392

[201] Göktepe H, Sahan H, Patat S, Ülgen A. Enhanced cyclability of triple-metal-doped $LiMn_2O_4$ spinel as the cathode material for rechargeable lithium batteries. Ionics. 2008;**15**:233-239

[202] Yu C-L, Wang F, Cao S-Y, Gao D-P, Hui H-B, Guo Y-Y, et al. The structure of H_2TiO_{3-a} short discussion on "lithium recovery from salt lake brine by H_2TiO_3". Dalton Transactions. 2015;**44**:15721-15724

[203] Onodera Y, Iwasaki T, Hayashi H, Torii K. A new inorganic material with high selective adsorbability for lithium ions. Chemistry and Industry (London). 1988(24):786

[204] Aceves JM, West AR. Electrochemical decomposition of Li_4SiO_4 and Li_2TiO_3 in solid-state thermal cells. Journal of the Chemical Society, Faraday Transactions 1: Physical Chemistry in Condensed Phases. 1982;**78**:2599-2608

[205] Johnson CS, Kim J-S, Kropf AJ, Kahaian AJ, Vaughey JT, Thackeray MM. Structural and electrochemical evaluation of (1-x) $Li_2TiO_3\cdot(x)LiMn_{0.5}Ni_{0.5}O_2$ electrodes for lithium batteries. Journal of Power Sources. 2003;**119-121**:139-144

[206] Zhao E, Liu X, Hu Z, Sun L, Xiao X. Facile synthesis and enhanced electrochemical performances of Li_2TiO_3-coated lithium-rich layered $Li_{1.13}Ni_{0.30}Mn_{0.57}O_2$ cathode materials for lithium-ion batteries. Journal of Power Sources. 2015;**294**:141-149

[207] You NK, Yun CK, Park SB. Preparation and electrochemical properties of nanometer-sized Li_2TiO_3-$LiCrO_2$ nanocomposite cathode powders by spray pyrolysis. International Journal of Electrochemical Science. 2013;**8**:2504-2514

[208] Chauvaut V, Cassir M. Behaviour of titanium species in molten Li_2CO_3 + Na_2CO_3 and Li_2CO_3 + K_2CO_3 in the anodic conditions used in molten carbonate fuel cells: II. Electrochemical intercalation of Li^+ in Li_2TiO_3 at 600 and 650 °C. Journal of Electroanalytical Chemistry. 1999;**474**:9-15

[209] Grzechulska J, Hamerski M, Morawski AW. Incorporation of lithium into TiO_2 host and its application in photocatalysis. Molecular Crystals and Liquid Crystals. 2000;**341**:243-248

[210] Song H, Jiang H, Liu T, Liu X, Meng G. Preparation and photocatalytic activity of alkali titanate nano materials $A_2Ti_nO_{2n+1}$ (A = Li, Na and K). Materials Research Bulletin. 2007;**42**:334-344

[211] Yao H, Fan M, Wang Y, Luo G, Fei W. Magnetic titanium dioxide based nanomaterials: Synthesis, characteristics, and photocatalytic application in pollutant degradation. Journal of Materials Chemistry A. 2015;**3**:17511-17524

[212] Marcus Y. Thermodynamics of solvation of ions. Part 5. Gibbs free energy of hydration at 298.15 K. Journal of the Chemical Society, Faraday Transactions. 1991;**87**:2995-2999

[213] He G, Zhang L, Zhou D, Zou Y, Wang F. The optimal condition for H_2TiO_3–lithium adsorbent preparation and Li^+ adsorption confirmed by an orthogonal test design. Ionics. 2015;**21**:2219-2226

[214] Park JH, Lee S, Kim SS, Kim JH, Park JH. Effect of conductive additives on the structural and electrochemical properties of $Li_4Ti_5O_{12}$ spinel. Bulletin

of the Korean Chemical Society. 2012;**33**:188-192

[215] Yi TF, Jiang LJ, Shu J, Yue CB, Zhu RS, Qiao HB. Recent development and application of $Li_4Ti_5O_{12}$ as anode material of lithium ion battery. Journal of Physics and Chemistry of Solids. 2010;**71**:1236-1242

[216] Yi TF, Yang SY, Xie Y. Recent advances of $Li_4Ti_5O_{12}$ as a promising next generation anode material for high power lithium-ion batteries. Journal of Materials Chemistry A. 2015;**3**:5750-5777

[217] Dong DQ, Wang WX, Wang ML. Preparation of 3DON $Li_4Ti_5O_{12}$ and behavior of Li^+ ion exchange. Applied Mechanics and Materials. 2014;**618**:175-179

Chapter 11

Fuel Cells as a Source of Green Energy

Rabea Q. Nafil and Munaf S. Majeed

Abstract

A fuel cell is an effective tool for extracting chemical energy from a special type of gaseous fuel other than fossil fuels. It is expected to be a replacement for thermal engines and rechargeable batteries within the next few years as they are emission-free and not subjected to Carnot restrictions. The fuel cell can be manufactured in different sizes depending on the amount of energy required, where it can be too small to be used in precision equipment or large enough to work as electrical stations. This proposal shows a demonstration of the principle of work involved in the fuel cells, structure components, and practice ideas to enhance the output power.

Keywords: energy extraction, fuel cell, green energy, zero emission, hydrogen energy

1. Introduction

Most of the power generators use conventional fuel and they obey the same operation principle that is transforming chemical energy to mechanical energy [1]. The heat engine (which is a very common conventional generator) burns or oxidizes the fuel in the air [2]. The chemical energy of fossil fuel is converted into thermal energy and this is the first step of its operation. The rising temperature and pressure lead to expanding engine gases. These excited gases rotate a crankshaft (conversion energy from thermal to mechanical energy step). Finally, the electricity is generated by the mechanical work done by the crankshaft [2]. Now it becomes clear that the production of electricity by heat engine passes through three stages. The overall conversion efficiency declines as the number of stages increases and this is one of the main disadvantages of heat engines [3].

Thermal engines have not been at the top of the list of energy sources despite their ability to provide high energy. This is due to several reasons, including that thermal engines cause increased global warming and environmental pollution by emitting carbon dioxide. Thermal engines often use fossil fuels which are originally threatened with depletion. Also, the sustainability of damaged mechanical parts adds an economic cost to the manufacture and operation of conventional engines. All these reasons prompted the search for new and sustainable energy sources that can compensate for the deficiencies in traditional sources of energy generation and storage. The solution embodies the use of fuel cells, which are not only a source of green energy but also fill the lack of advantages of both the thermal engines and rechargeable batteries. Their simple construction and non-use of fossil fuels gave them the ability to work for a long time without the need to replace their damaged parts. Fuel cells have unlimited operating time, higher power density and therefore greater storage capacity to outperform the advantages of a simple battery and rechargeable battery [4].

Fuel cells are zero pollution' power generators. They work as an electrochemical converter to chemically produce electrical energy from gas fuels. The conversion occurs directly by the chemical reactions and without burning of the fuel. Even better, they are not following the thermodynamic laws that limit most of the power plants. This chapter presents the operation principle of fuel cells, their main components, and finally the newer trends in developing these cells for commercial uses.

2. The fuel cells' principle of work

The main components of a fuel cell are two electrodes isolated from each other with a membrane (either solid or liquid electrolyte). The added parts are usually to improve the functioning of the cell. According to Eq. (1) [5], the gas fuel (hydrogen) oxidized at the anode to produces electrons and cations during entering of the oxidant from the cathode. These released electrons depart the anode through an external circuit and create an electric current before reaching the cathode. The electrons current can be utilized in working any load [5].

$$2H_2 \rightarrow 4H^+ + 4e^- \tag{1}$$

The positive hydrogen ions diffuse through the membrane and recombine with the exhausted electrons that reached at the cathode to reproduce H_2 gas. Finally, the produced hydrogen atoms combine with the oxygen atoms to generate water as the only waste of the complete reaction. Eqs. (2) and (3) show the reduction of oxygen and the fuel cell reaction, respectively [6, 7].

$$O_2 + 4H^+ + 4e^- \rightarrow 2H_2O \tag{2}$$

$$2H_2 + O_2 \rightarrow 2H_2O + heat \tag{3}$$

Fuel cell' efficiency is greatly influenced by the quality of the electrolyte. It increases with increasing the membrane ability to block electrons generated from the oxidation of the anode from reaching the cathode. Therefore, the electrolyte must be chosen with caution as it acts as a filter that allows only positive ions to pass

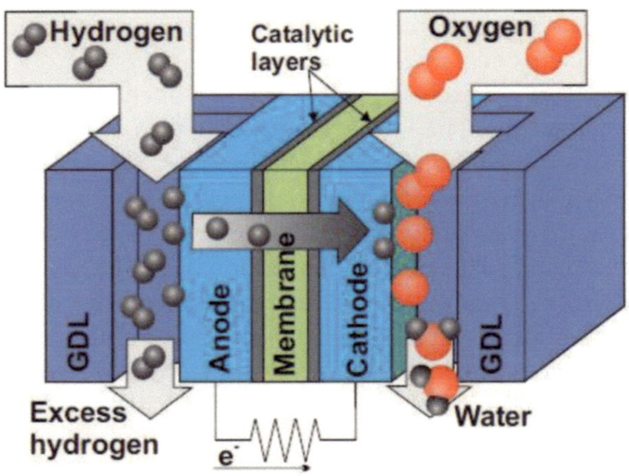

Figure 1.
The components of fuel cell [11].

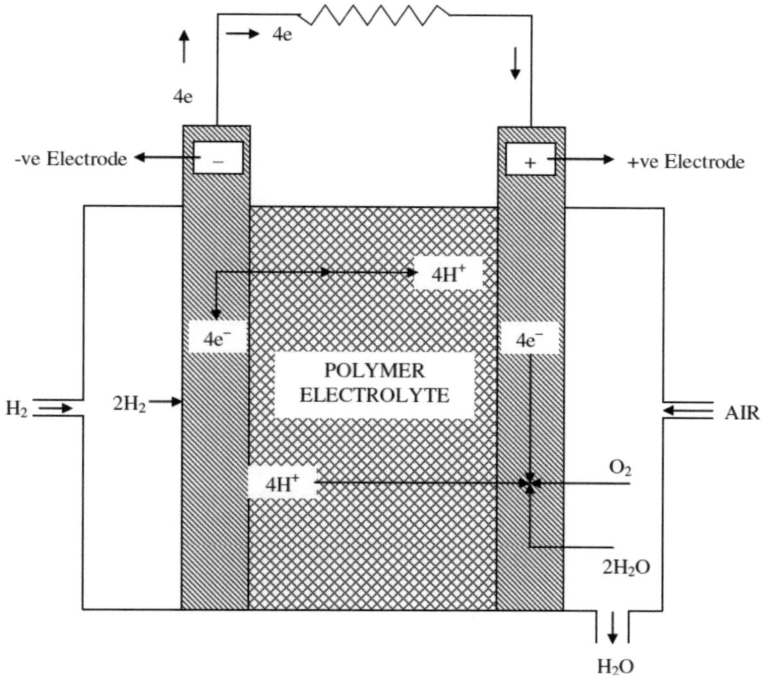

Figure 2.
A scheme of oxidization processes at the electrodes [13].

through it in one direction only which is the direction of the cathode [8]. To reduce the reaction energy on the two electrodes, a catalyst is added [9, 10]. The structure of the fuel cell is displayed in **Figure 1**.

Theoretically, the anodic substance could be any material that gives up electrons such as hydrogen, methane, kerosene, or even carbon. But among all, hydrogen is preferred due to its availability, has a high energy density, and good reactivity with catalysts [12]. Also, the best chose of oxidant agent is oxygen gas because it is easily subjected to a reduction reaction and is much available in the air. **Figure 2** illustrates the oxidization reaction schematically.

3. The main components of fuel cells

In addition to the two electrodes and their separator electrolyte, other additional parts are required for specific tasks such as gases flow and electrical insulation or connection. These parts are shown in **Figure 3**.

3.1 The end plates

The outer sides of the anode and the cathode are supported mechanically by plastic plates which are called the endplates. Their main jobs are to save the fuel cell component and to prevent the reacting gasses from leaking [15].

3.2 The bipolar plates

A fuel cell has two bipolar plates each one consists of a current collector and separator plate. They are manufactured from a metal or a carbon-filled composite (conductive polymer) [16]. Researchers suggested several designs of

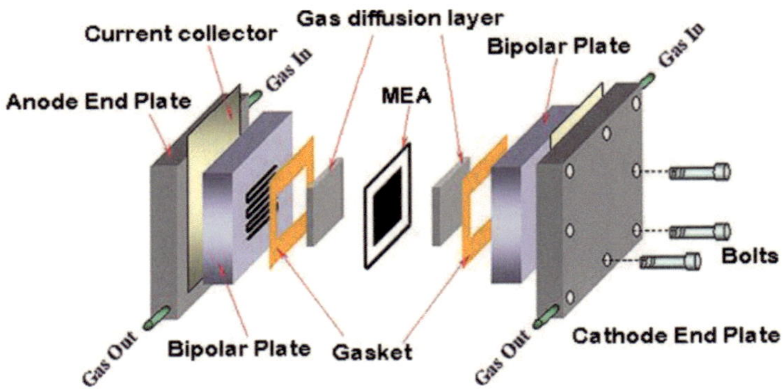

Figure 3.
The component of fuel cell [14].

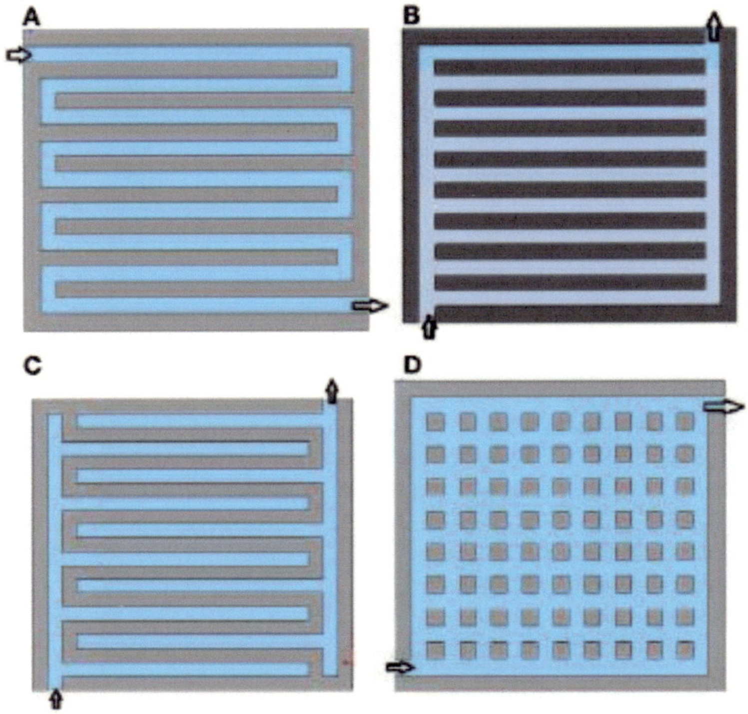

Figure 4.
Various designs of bipolar plates [17].

the bipolar plate to feed the gaseous fuel to the electrodes uniformly as shown in **Figure 4**. Their designs differ by the shape and the distributions of the gas flow channels.

When many cells are connected in one stack, the bipolar plates conduct current between the adjacent cells. Another function of the bipolar plate is to dissipate the generated heat during the chemical reactions [18, 19].

3.3 Gaskets

Gaskets are sheets of rubbery polymer used to prevent leaking of reactant gases and coolant. They are carefully selected to be able to withstand the propellant of gaseous fuels, resist operating temperature up to 100°C, not affected by ambient air humidity or moisture of reaction gases, non-corrosive by liquid acids, and also withstand mechanical pressures resulting from prolonged operation [20].

3.4 The gas diffusion layers

Two pieces of a conductive material and they are representing important parts of the fuel cell. They are manufactured in a porous form using carbon fibers weaved into carbon papers or clothes [21]. Their major works are to transfer the reactants between the bipolar plates and the catalyst layers, assist the dissipation of the produced water and heat from the catalysts and protect them from corrosion [22].

3.5 Membrane electrodes assemble (MEA)

A compact unit which is responsible for producing the output power of the fuel cells. Its thickness approaches a few hundred microns and consists of; the electrolyte, the two electrodes with their catalysts, with/without GDL [23]. These components are:

3.5.1 The electrodes

As in any electric cell, the fuel cell has two electrodes that are cathode and anode. At these parts of fuel cell the chemical reactions which produce electricity occur. Therefore, the catalyst is required for each electrode. When hydrogen atoms enter a fuel cell at the anode, the catalyst strips them of their electrons in the oxidation process. The oxygen reduction occurs at the cathode. The electrodes that have a large active surface area of electrode to catalyst are regarded as a good electrode. The overall performance of the electrode should be stabile during operation time, and this can be satisfied by choosing a suitable catalyst for the specific electrode material, distribution the catalyst over the electrode uniformly, remove the produced excess heat and water, and most importantly for the best connection with the external electrical circuit.

3.5.2 The electrolyte (membrane)

The electrolytes are thin plastic-like or solid polymer membranes. Their main function is to pass the hydrogen positive ions from the anode to the cathode while completely block the free electrons from moving in this direction [18]. If the free electrons could pass through the electrolyte, they would hold up the chemical reaction. The filtration process is based on the semi-permeability of the membrane. There are several mechanisms of ions permeability through the membrane including absorption and adsorption, distillation, extraction, physical filtration, and stripping.

The compositions of the electrolytes are perfluorosulfonic acids, which are Teflon-like fluorocarbon polymers that have side chains ending in sulfonic acid groups ($–SO_{32}–$). Such acidic polymer electrolytes require water to conduct hydrogen ions. Therefore, the reaction gas (H_2) in contact with the electrolyte must be saturated with water. The commercial name of Teflon based polymer electrolyte is Nafion [20]. It has a high ionic conductivity at 80°C, good thermal and chemical stability [24–27], and high chemical resistance. Its chemical chain is shown in **Figure 5** [28].

Fuel cells performances improve largely with increasing temperature to 90o C due to the decline in ohmic resistance of their electrolytes [29]. But at higher temperatures, the fuel cell will dry. Since Nafion ionic conductivity is strongly dependent on water content, the Nafion membranes will have an osmotic swelling problem and maybe potentially dissolve in methanol solution when increasing methanol concentration and temperature.

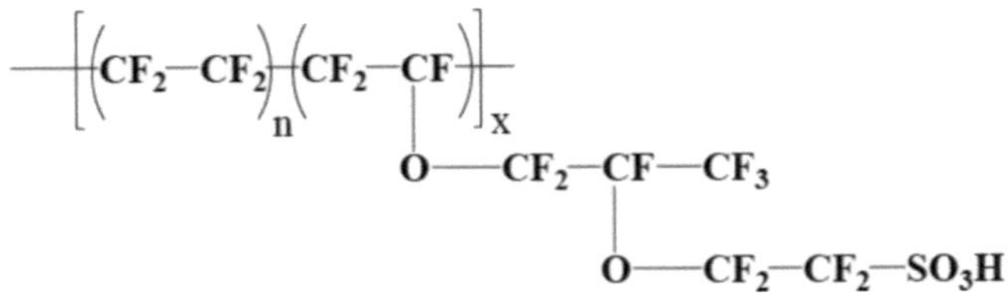

Figure 5.
Nafion chemical chain [28].

3.5.3 The catalysts

The catalysts are special chemical materials used to increase the rate of reactions without being consumed. The main condition of the material to be used as a catalyst is to stay unchanged after reactions. Therefore, they could be recovered from the reaction mixture chemically. The existing of catalysts reduces the reaction energy barrier and thus speeds up the reaction at low temperatures [30]. Most types of fuel cells used platinum (Pt) and its group as a catalyst at the two electrodes. At the anode, the Pt catalyst helps in splitting the hydrogen molecules into free electrons and positive cations. While at the cathode, it enables oxygen reduction [31]. A perfect catalyst for a fuel cell is the one with high electric conductivity, stable during contact with the reactant gases and the electrolyte.

4. The new trend in enhancement fuel cell performance

All types of fuel cells are a good source of green energy, but none is yet suitable enough to replace traditional sources. That leads the designers to spend their efforts on improving efficiency, size, weight, and cost. Some new trends to enhance fuel cell performance are explained as the following:

A single fuel cell generates a little electrical current, so a group of cells is connected in a geometrical form called a stack. The bipolar plates in any fuel cell stack cause most of the weight and participate in the cost. Hence, the dimensions and manufacturing cost of the stick can be significantly reduced if the bipolar plate is removed. Some designers used laser ablation to sculpture microchannels on the cell electrodes to increase the gas flow rate. They found that by this simple idea the bipolar plates could be excluded [32].

Oxidation and reduction process occurs on the surface of the catalyst, which necessitates an increase in the reaction space between the catalyst and the reaction gases. But this contradicts the continuous tendencies to lower the cost of fuel cell since Platinum is very expensive. Although platinum is preferred as a catalyst in most fuel cell types because it is a stable and active substance, a less expensive alternative should be found. So it was replaced by palladium. One way to increase the active surface areas of the catalysts is to reduce particles size. For this, nanotechnology had been used in many research works to manufacture porous nanocatalyst to exclude bulk catalyst and gas diffusion layer [32].

5. Conclusions

According to the explanations given in the previous syllabuses of the chapter, the following conclusions can be reached.

1. Energy extraction by the burning of fossil fuels goes through three stages and in each stage is spent energy to complete the process. Therefore, the efficiency

of internal combustion generators is low. Whereas in fuel cells, production is direct and not subject to Carnot limitations, which increases fuel cell efficiency.

2. Fuel cells use hydrogen gas, which is available in air and can be produced in secondary ways such as the electrolysis of water. Therefore, fuel cells are economical and there are no difficulties in manufacturing them in countries which have not petroleum.

3. It does not cause harmful gas emissions and thus it is considered environmentally friendly.

4. Using laser and nanotechnology, fuel cells can be manufactured in very small sizes that can be included in micro devices such as mobile phones and medical devices.

Acknowledgements

The authors are pleased to thank their academic institutions, which are the University of Technology/Department of Applied Sciences and the AL-Nahrain Nanorenewable Energy Research Center/AL-Nahrain University for their continued sponsorship of academics.

Author details

Rabea Q. Nafil[1*] and Munaf S. Majeed[2]

1 Applied Physics Branch, Applied Sciences Department, University of Technology, Baghdad, Iraq

2 AL-Nahrain Nanorenewable Energy Research Center, AL-Nahrain University, Baghdad, Iraq

*Address all correspondence to: 100068@uotechnology.edu.iq

References

[1] Jia Q, Zhang C, Deng B, et al. Performance improvement for proton exchange membrane fuel cell using hydrogen pressure pulsation approach. Journal of Fuel Cell Science and Technology. 2015;**12**:041008-041001

[2] Heywood J. Internal Combustion Engine Fundamentals. 1st ed. USA: McGraw-Hill; 1988

[3] Niroumand A. PEM Fuel Cell Low Flow Rate Diagnostic [Thesis]. University of Simon Fraser; 2009

[4] Weydah H, Gilljam M, Lian T, et al. Fuel cell systems for long-endurance autonomous underwater vehicles e challenges and benefits. International Journal of Hydrogen Energy. 2019;**10**(1016):05-035

[5] Li C, Liu Y, Xu B, et al. Finite time thermodynamic optimization of an irreversible proton exchange membrane fuel cell for vehicle use. PRO. 2019;**7**:419. DOI: 10.3390/pr7070419

[6] Manoharan Y, Hosseini S, Butler B, et al. Hydrogen fuel cell vehicles; current status and future Prospect. Applied Sciences. 2019;**9**:296. DOI: 10.3390/app9112296

[7] Joshi N. Development in direct methanol–oxygen fuel cell (DMFC). IOSR-JAC. 2014;7(9):24-26

[8] Lu Y, Zhu B, Cai Y, et al. Progress in electrolyte-free fuel cells. Frontiers in Energy Research. 2017;**4**:17

[9] Chong L, Wen J, Kubal J, et al. Ultralow-loading platinum-cobalt fuel cell catalysts derived from imidazolate frameworks. Science. 2018;**362**(6420):1276-1281

[10] Bele M, Gatalo M, Jovanovi P, et al. Insight on single cell proton exchange membrane fuel cell performance of Pt-Cu/C cathode. Catalysts. 2019;**9**:544

[11] Zakrisson E. The Effect of Start/Stop Strategy on PEM Fuel Cell Degradation Characteristics [Thesis]. Sweden: Chalmers Technology University; 2011

[12] Flores G, Varaldo H, Feria O, et al. Improvement of microbial fuel cell performance by selection of anodic materials and enrichment of inoculum. Journal of New Materials for Electrochemical Systems. 2015;**18**:121-129

[13] Kumar K. Fabrication and electrical study of PEMFuel cell Basedon Nano–crystalline PEO based conducting polymer electrolyte system. Journal of Petroleum Technology and Alternative Fuels. 2011;**2**(1):11-20

[14] Hogarth M, Ralph T. Catalysis for low temperature fuel cells. Platinum Metals Review. 2002;**46**:146-164

[15] Habibnia M, Tamami P, Davini H. Design and investigation of honeycomb end plates for PEM fuel cells. Iranian Journal of Hydrogen and Fuel Cell. 2017;**3**:189-199

[16] Dey T, Deshpande J, Singdeo D, et al. Study of PEM fuel cell end plate design by structural analysis based on contact pressure. Hindawi Journal of Energy. 2019;**2019**:3821082

[17] Liu H, Li P, Robles D, et al. Experimental study and comparison of various designs of gas flow fields to PEM fuel cells and cell stack performance. Frontiers in Energy Research. 2014;**2**:2

[18] Mohapatra A, Tripathy S. A critical review of the use of fuel cells towards sustainable management of resources. IOP Conference Series:

Materials Science and Engineering. 2018;**377**:012135

[19] Elyasi M, Ghadikolaee H, Hosseinzadeh M. Fabrication of metallic bipolar plates in PEM fuel cell using semi-stamp rubber forming process. The International Journal of Advanced Manufacturing Technology. 2017;**92**:765-776

[20] Fili M, Habibnia M, Tamami P. Modeling and experimental study on the sealing gasket of proton exchange membrane fuel cells. Iranian Journal of Hydrogen and Fuel Cell. 2016;**3**:213-220

[21] Davis S. Polymer Electrolyte Membrane (PEM) Fuel Cell Seals Durabilty [Thesis]. University of Loughborough; 2015

[22] Omrani R, Shabani B. Gas diffusion layers in fuel cells and lectrolysers: A novel semi-empirical model to predict electrical conductivity of sintered metal Fibres. Energies. 2019;**12**:855

[23] Liang H, Su H, Pollet B, et al. Development of membrane electrode assembly for high temperature proton exchange membrane fuel cell by catalyst coating membrane method. Journal of Power Sources. 2015;**288**:121e127

[24] Rolfi A, Oldani C, Merlo L, et al. New perfluorinated ionomer with improved oxygen permeability for application in cathode polymeric electrolyte membrane fuel cell. Journal of Power Sources. 2018;**396**:95-101

[25] Chen Y, Zhong Q, Li G, et al. Electrochemical study of temperature and Nafion effects on interface property for oxygen reduction reaction. Ionics. 2018;**24**:3905-3914

[26] Su H, Sita C, Pasupathi S. The effect of gas diffusion layer PTFE content on the performance of high temperature proton exchange membrane fuel cell. International Journal of Electrochemical Science. 2016;**11**:2919-2926

[27] Toupin M, Malek K, Mokrini A. Techno-economics of a new high throughput process for proton exchange membranes manufacturing. World Electric Vehicle Journal. 2016;**8**:431-442

[28] Feng M, Qu R, Wei Z. Characterization of the thermolysis products of Nafion membrane: A potential source of perfluorinated compounds in the environment. Scientific Reports. 2015;**5**:9859

[29] Shi Y, Janßen H, Lehnert W. A transient behavior study of polymer electrolyte fuel cells with cyclic current profiles. Energies. 2019;**12**:2370

[30] Chen J, Yan L, Song W, et al. Catalytic oxidation of synthesis gas on platinum at low temperatures for power generation applications. Energies. 2018;**11**:1575

[31] Gebauer C, Jusys Z, Behm R. On the role of the support in Pt anode catalyst degradation under simulated H2 fuel starvation conditions. Journal of the Electrochemical Society. 2018;**165**(15):J3342-J3349

[32] Majeed M, Nafil R, Dhahir M. Nanosecond laser ablation for PEM fuel cell scaled down. Optik: International Journal for Light and Electron Optics. 2018;**172**:855-860

Chapter 12

Fuel Cells: Alternative Energy Sources for Stationary, Mobile and Automotive Applications

Irina Petreanu, Mirela Dragan and Silviu Laurentiu Badea

Abstract

This paper presents a classification and also an overview of fuel cells, including the working principles, the equations of the governing reactions, and the main applications. A brief exposure of thermodynamics and electrochemical theory describe the functioning of the fuel cells. Further, the proton exchange membrane fuel cells assembly, starting with the schematic presentation of the main components, the role of each component in fuel cell, the specific materials and their requested properties, and the way of assembling the components into device will be detailed. In conclusion, the challenges related to reliability and the cost and the targets for future development of the proton exchange membrane fuel cells for mobile and stationary applications will be presented.

Keywords: fuel cells, proton exchange membrane, catalyst layer, gas diffusion layer, membrane electrode assembly, bipolar plate

1. Introduction

The stringent request for green energy sources and technology turned the attention of the scientific world to the chemical energy conversion systems known as fuel cells (FC). The possibility to convert chemical energy to electricity was discovered in the first half of nineteenth century and the invention of the fuel cells was attributed to Sir William Grove [1, 2].

Fuel cells are electrochemical devices able to convert chemical energy to electricity through controlled oxidation-reduction (REDOX) processes. They behave as open systems which produce electricity as long as they are fed with fuel and oxidant. Commonly fuels, hydrogen (H_2) and methanol (CH_3OH) are compounds with small molecules and high density of chemical energy, considered energetic vectors. Basically, a fuel cell consists in two electrodes, anode and cathode, separated by an electrolyte. The cathode is always the positive electrode and the anode is the negative one as the electrons flow from negative to positive electrode. During operation the fuel is continuously fed to the anode while oxidant (oxygen or air) is continuously fed to the cathode.

Fuel cell systems are used as power sources for mobile, automotive and stationary applications including back-up power systems [3]. They can be single energy conversion devices (the low temperature operating fuel cells) or can operate in co-generation electrical/thermal energies (the high temperature fuel cells).

The advantages of fuel cells are the high efficiency and the decrease or even absence of harmful emission. Wang et al. [3] give a value up to 60% of efficiency in electrical energy conversion and for systems in co-generation of electrical and thermal energies the global efficiency reach 80%, both applications yielding less than 10% pollution. Particularly, the electrochemical reactions from inside the hydrogen fuel cells yield just water and heat.

Hydrogen can be produced in many ways: reforming of fossil fuels, hydrolysis of hydride type compounds or water electrolysis. It can be stored in special tank, under pressure, absorbed in hydride or metal organic framework, included into the structure of organic liquid compounds, or compactly stored in cryogenic form. The oxidant is gaseous oxygen extract from air or even purified air.

An economy based on fuel cell should be the economy of the future, free of greenhouse emissions and independent of fossil fuel. There are still many challenges related of materials and processes optimization for reaching the targets of operating life 5000 hours for automotive and 40,000 hours for stationary application [4] in terms of cost reduction.

2. Classification

According to the type of electrolyte and fuel which determine the electrodes reactions there are several kinds of fuel cells which differs by the electrolyte and others behaviors: Alkaline Fuel Cells (AFC), Proton Exchange Membrane Fuel Cells (PEMFC), Alkaline Fuel Cell (AFC), Direct Methanol Fuel Cells (DMFC), Microbial Fuel Cell, Phosphoric Acid Fuel Cells (PAFC), Solid Oxide Fuel Cells (SOFC), Molten carbonate Fuel Cells (MCFC).

The **Proton Exchange Membrane Fuel Cells (PEMFC)** use protons (H^+) as available mobile ions and a hydrated proton exchange membrane with the role of electrolyte. PEMFCs operate at temperatures between 30 and 100°C, with platinum electro-catalyst and carbon based electrodes.

The half reactions at each electrode are:

- oxidation of hydrogen to anode:

$$H_2 \rightarrow 2H^+ + 2e^- \tag{1}$$

- reduction of oxygen to cathode:

$$\tfrac{1}{2}O_2 + 2e^- \rightarrow O^{2-} \tag{2}$$

- The overall reaction is:

$$H_2 + \tfrac{1}{2}O_2 \leftrightarrow H_2O + Q \tag{3}$$

where Q is thermal effect—heat—produced in reaction.

In **Alkaline Fuel Cells (AFC)** the charge carriers are hydroxyl ions (HO^-) and the electrolyte is alkaline solution, usually 30–45 wt% KOH in water. Particularly, electrolyte solution is immobilized in a solid matrix, typically asbestos. The operating temperature range varies between 65 and 220°C and with platinum or transition metal catalysts to electrodes [5].

At the anode, hydrogen is oxidized in reaction with mobile hydroxyl ions, releasing electrons and producing water:

$$2H_2 + 4HO^- \rightarrow 4H_2O + 4e^- \quad (4)$$

Resulted electrons pass to cathode through an external electrical circuit and generate electrical current. The reaction at the cathode consists of oxygen reduction with the electrons coming from the anode, in the presence of water, forming HO, which keep going the charge transport:

$$O_2 + 4e^- + 2H_2O \rightarrow 4HO^- \quad (5)$$

The overall reaction is:

$$2H_2 + O_2 \rightarrow 2H_2O \quad (6)$$

Even that the overall reaction is the same as that in the PEMFC the half reactions to each electrode differ.

Direct Methanol Fuel Cells (DMFC) is a special kind of polymer electrolyte fuel cell which is fed direct with methanol. Both Acid and Alkaline Fuel Cells could be feed direct with methanol, but there differ considering the mobile ions and the electrode reactions.

In an acid fuel cell a mixture of methanol and water is supplied at anode where methanol oxidation takes place [5, 6]:

$$CH_3OH + H_2O \rightarrow 6H^+ + 6e^- + CO_2 \quad (7)$$

The half electrochemical reaction at the cathode is oxygen reduction:

$$3/2 O_2 + 6e^- \rightarrow 3O^{2-} \quad (8)$$

In the Alkaline Fuel Cell, methanol oxidation occurs to anode in the presence of hydroxyl ion:

$$CH_3OH + 6HO^- \rightarrow CO_2 + 5H_2O + 6e^- \quad (9)$$

To cathode, the oxygen reduction occurs in the presence of water and results hydroxyl:

$$3/2 O_2 + 3H_2O + 6e^- \rightarrow 6HO^- \quad (10)$$

For both fuel cells, the overall reaction is:

$$2CH_3OH + 3O_2 \rightarrow 2CO_2 + 4H_2O \quad (11)$$

Microbial Fuel Cells (MFCs) are bio-electrochemical systems that can use microorganisms to produce electricity from a wide variety of organic compounds [7], being considered a promising source of sustainable energy, and thus they are considered a fast growing [8, 9] area of fuel cells research. Since the abiotic fuel cells are using platinum as catalyst (expensive), while MFCs are using microorganisms as biocatalysts, MFCs may be potentially more advantageously than abiotic fuel cells [10]. In a MFC, the transfer of electrons to anode surface occurs firstly via transformation of organic substrates from anodic chamber usually to CO_2 and generating electrons, eventually discharged on surface of the anode, and also protons. The excess of protons migrate to the cathode via a separator (i.e., Nafion) [11], while oxidizing species like oxygen are accepting electrons at the cathode surface, thus generating an electrical current through an external circuit. The electrochemical processes within a typical MFC can be described by the following generic chemical equations:

Anode:

$$C_6H_{12}O_6 + 6H_2O \rightarrow 6CO_2 + 24H^+ + 24e^- \quad (12)$$

Cathode:

$$4H^+ + O_2 + 4e^- \rightarrow 2H_2O \ (E = 0.816V) \quad (13)$$

Nevertheless, since the voltage generated by a MFC is relatively small (in most cases between 0.5 and 0.8 V), multiple MFC systems can be stacked in series to increase the overall voltage and power [12], or in parallel to increase the overall current intensity. Nevertheless, this technical approach is challenging due to voltage reversal [13, 14] and other problems (e.g., energy losses, fluctuation of power outputs, etc.) associated with the non-linear nature of biological systems [15] (i.e., variability in exponential growth of microorganisms, etc.). Therefore, further research is needed in the field of MFCs in order to make them reliable tools in generating sustainable energy.

Phosphoric Acid Fuel Cell (PAFC) is acid type fuel cell with ionic conduction provided by protons (H^+) transport. The electrolyte is concentrated (100%) phosphoric acid, immobilized, in the liquid phase, into a silicon carbide (SiC) matrix. Phosphoric acid has poor ionic conductivity at low temperature under 100°C but express the optimum conductivity between 150 and 220°C [5]. At this temperature, platinum electro-catalyst is tolerant at CO poisoning. Both anode and cathode catalysts are Pt/C and Pt alloys on carbon.

Reactions occurring at the electrodes are the same that occurs in PEMFC (Eqs. (1)–(3)).

Major applications of PAFC are in stationary power plants and on-sitecogeneration power plants [6].

In **Molten Carbonate Fuel Cells (MCFC)**, the electrolyte is a binary mixture of molten carbonate salts, such as lithium and potassium/sodium carbonate. These salts, highly conductive in the molten state at 600–700°C are included into a ceramic matrix of double oxide $LiAlO_3$. The mobile ion is carbonate $(CO_3)^{2-}$, produced from carbon dioxide fed to cathode. The electrodes half reactions are:

Cathode reaction:

$$2CO_2 + O_2 + 4e^- \leftrightarrow 2CO_3^{2-} \quad (14)$$

Anode:

$$H_2 + CO_3^{2-} \leftrightarrow CO_2 + H_2O + 4e^- \quad (15)$$

Overall reaction:

$$2H_2 + 2CO_3^{2-} \leftrightarrow 2H_2O + CO_{2(recycle)} \quad (16)$$

At this high temperature the Ni (anode) and NiO (cathode) are suitable catalysts to promote the electrochemical reactions, and the noble metal catalysts are not required. The MCFCs can use common hydrocarbon fuels as sources able to provide hydrogen and CO_2 by internal reforming. Severe corrosive electrolytes and high operating temperatures require special materials as nickel or high-grade stainless steel for cell hardware [5, 6].

Solid Oxide Fuel Cells (SOFC) is total solid devices operating in the high temperature range of 600°C -1000°C. However, in the last few years it is a noticeable effort to decrease the operating temperatures. A metal catalyst is not needed because the reaction kinetics are improved using the high operating temperatures which constitute an advantage of these fuel cells technology. Compared to other types of fuel cell, SOFC are relatively resistant to small quantities of sulfur in the fuel therefore can be used with coal gas.

The ion conduction is accomplished by the oxygen ions O^{2-}, which pass through an electrolyte composed from zirconia (ZrO_3) doped with 8–10% mole ratio of yttria (Y_2O_3).

The major components of an individual SOFC cell include the electrolyte, the cathode, and the anode and the following reactions take place:Anode reaction:

$$H_2 + O^{2-} \rightarrow H_2O + 2e^- \tag{17}$$

or

$$CO + O^{2-} \rightarrow CO_2 + 2e^- \tag{18}$$

Cathode reaction:

$$½O_2 + 2e^- \rightarrow O^{2-} \tag{19}$$

Overall reaction:

$$H_2 + ½O_2 \leftrightarrow 2H_2O \tag{20}$$

The electrolyte, such as zirconium oxide stabilized with yttrium oxide, determines the operating temperature of the fuel cell and is used to prevent the two electrodes to come into electronic contact by blocking the electrons. It also allows the flow of charged ions from one electrode to the other to maintain the overall charge balance.

The operating principle of such a cell is illustrated in **Figure 1**.

At anode, electrochemical oxidation of fuel gas such hydrogen H_2 or carbon monoxide CO takes place (Eqs. (17) and (18)).The anode scatters the hydrogen gas over its whole surface and conducts the electrons that are freed from hydrogen molecule, to be used as power in the external circuit.

The oxygen, usually carried by air, is reduced by reaction taking place the cathode, described in equation from above (Eq. (19)). The cathode distributes the oxygen fed to it onto its surface and conducts the electrons back from the external

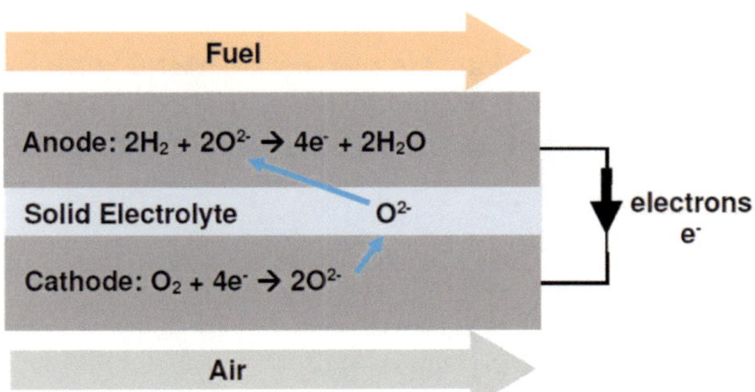

Figure 1.
Working principle of a solid oxide fuel cell (SOFC).

circuit where they can recombine with oxygen ions, passed across the electrolyte, and hydrogen to form water. The overall reaction (Eq. (20)) generates water.

Anode materials are based on metal ceramic mixture, named cermet, typically Ni-ZrO_2, Co-ZrO_2 and cathode materials is a p-type semiconductor usually $LaMnO_3$ doped with strontium. Both electrodes have highly porous structure for rapid transport of gaseous reactants or products.

The choice of these materials is a compromise between a number of factors necessary to meet some requirements: chemical, phase and morphological stability, chemical compatibility and with other components, dense electrolyte, etc.

For the electrolyte, in addition to the yttria stabilized zirconia mentioned above, have been considered compositions such as gadolinium doped ceria and scandium doped zirconia [16]. These compositions are more conductive than yttria stabilized zirconia, permitting a further reduction of the operating temperature by 50–100°C.

Currently, most cathodes are based on doped lanthanum manganites such as strontium doped lanthanum manganite or other materials with the perovskite structure [17–21]. Most used for the anode material is a cermet of nickel and yttria stabilized zirconia [22]. This material is sensitive to contaminants like sulfur and requires desulfurization of the anode feed. Some other materials such as copper—cerium oxide anodes are being developed.

Planar and tubular are the different geometries of solid oxide fuel cells [23–26]. The air and hydrogen flow though the unit of flat stacks via channels built in to the anode and cathode in the planar design. The tubular design has the advantage of much easier to seal air from the fuel. In this geometry the flow of air or fuel is through the inside of the tube and the other gas is passed along the outside of the tube. From the performance perspective the planar design is at this time better than the performance of the tubular design.

SOFCs are extremely useful in wide ranging applications such as stationary fuel cell applications and for transportation systems.

3. Fuel cells: thermodynamic and electrochemistry

Equation of overall reaction into Proton Exchange Membrane Fuel Cell (PEMFC) is:

$$H_2 + \frac{1}{2}O_2 \leftrightarrow H_2O \tag{21}$$

According to first principle of thermodynamic, enthalpy of reaction (ΔG) is the sum of Gibbs free energy (ΔG) and a term depending on entropy (ΔS) (Eq. (22)):

$$\Delta H = \Delta G + T\Delta S \tag{22}$$

In terms of energies, the fraction able to generate useful work for a system evolving at constant pressure and temperature is variation of Gibbs free energy. In an operating fuel cell, the only work produced is electrical, described by (Eq. (23)):

$$\Delta G = -W = -q \cdot E, \tag{23}$$

where

q is electrical charge, [q] = C (Coulomb)

E is electrical potential [E] = V (Volt)

For 1 mole of hydrogen, the charge transferred in reaction is the product between the number of transferred electrons (n) and the molar charge of electrons (F = 95,485 Coulombs/mol), so:

$$\Delta G = -nFE \tag{24}$$

where F, named Faraday's constant represents the specific charge for 1 mole of electrons, F = 96.485 C/mol;

The value of the maximum cell potential, obtained at constant, standard temperature and pressure is:

$$E^{\circ} = -\Delta G^{\circ}/nF \tag{25}$$

Gibbs free energy could be calculated from (Eq. (22)) taking into accounts the tabulated values of standard enthalpies and entropies of formation (**Table 1**) [1, 2]:

$$\Delta G = \Delta H - T\Delta S = \sum h_f^{\circ} - T\sum S_f^{\circ} \tag{26}$$

For water:

$$\Delta H_{H_2O}^{\circ} = h_f(H_2O) - h_f(H_2) - h_f(O_2) = -286.2\ kJ/mol \tag{27}$$

$$\Delta S_{H_2O}^{\circ} = s_f(H_2O) - s_f(H_2) - {}^1\!/_2 s_f(O_2) = -0.163\ kJ/mol \cdot K \tag{28}$$

Hence, the standard Gibbs free energy for liquid water formation is:

$$\Delta G_{H_2O_{(l)}}^{\circ} = \Delta H_{H_2O}^{\circ} - T\Delta S_{H_2O}^{\circ} = -237.42\ (kJ)/mol \tag{29}$$

Taking into account the expression of cell potential (Eq. (25)) and the value of $\Delta G_{H_2O_{(l)}}^{\circ}$ calculated above (Eq. (29)) we obtain the theoretical value of potential of PEMFC:

$$E_{H_2O(l)}^{\circ} = -\frac{\Delta G}{nF} = \frac{237.42kJ/mol}{2 \times 96\ 485^{A\cdot s}\!/_{mol}} = 1.23V \tag{30}$$

This is the theoretical value of PEMFC potential if the product is liquid water. If the water vapor results from electrochemical process, the value of theoretical potential is changed corresponding to values of standard enthalpy and entropy: $E_{H_2O_{(g)}} = 1.18V$ [1].

The values of PEMFC potential are determined for reaction at equilibrium, in standard conditions: 1 mole hydrogen consumption, constant temperature, 25°C and pressure, 1 atmosphere (760 Torr). At equilibrium, the net current is equal to 0, although the reaction proceeds in both directions simultaneously.

The influence of reactants concentration on the cell potential is revealed by Nernst equation:

$$E = E^{\circ} - {}^{RT}\!/_{2F} \ln \frac{c_{H_2O}}{c_{H_2} \cdot (c_{O_2})^{1/2}} \tag{31}$$

Reactants	h_f (kJ/mol)	S_f (kJ/mol·K)
H_2	0	0.1306
O_2	0	0.2051
$H_2O(l)$	−286.02	0.0699
$H_2O(g)$	−241.98	0.1888

Table 1.
The values of standard enthalpy and entropy of formation for reactants and products into PEM fuel cell reaction [2].

where

- E^o is standard potential;
- R is the universal gas constant: R = 8.314 J/(K·mol);
- T is the absolute temperature (K);
- 2 is number of electrons transferred in reaction;
- F is Faraday's constant;
- cH_2, cO_2, cH_2O are concentrations of reactants and products of electrochemical reaction.

In the fuel cell, chemical energy of fuel is converted in electricity through electrochemical reactions. The intensity of electric current generated through an electrochemical process is proportional with the charge transfer and reactants consumption, according to Faraday's law:

$$i = nFj \tag{32}$$

where

- nF is the charge transferred in reaction for 1 mole of reactant consumption (C/mol), and
- j is the flow of reactant per unit area ($mol/s{\cdot}cm^2$).

Since electrochemical process within fuel cell is split in two electrode reactions, oxidation on the anode and reduction on the cathode, each reaction generates an electron transfer, from reactant to anode and from cathode to reactant. The net current generated into operating fuel cell is a sum of these two currents, from cathode (j_c) and to anode (j_a):

$$i = nFj_c - nFj_a \tag{33}$$

The intensity of anode current is negative because the current sense is opposite with the electrons flow.

The reactant flow is proportional to concentration of reactant species on the electrode. For PEMFCs, this could be expressed by the further equations:

$$j_c = k_{ox} \cdot (c_{O_2})^{½} \text{ and } j_a = k_{red} \cdot c_{H_2} \tag{34}$$

The current in the operating PEMFC is:

$$i = nFk_{ox} \cdot (c_{O_2})^{½} - nFk_{red} \cdot c_{H_2} \tag{35}$$

At equilibrium, the net current is equal to 0, but the reaction proceeds in both directions simultaneously with the same rate. Based on Arrhenius equation and the Transition State Theory, the rate constant in the electrochemical process depends on the potential of fuel cells, which rich the open circuit voltage for zero current.

$$k_{ox} = Ae^{\frac{-\alpha_{ox} F E_{eq}}{RT}} \tag{36}$$

$$k_{red} = Ae^{\frac{-\alpha_{red} F E_{eq}}{RT}} \tag{37}$$

where α_{ox} and α_{red} are the so called transfer coefficients.

The equilibrium flow of reactants gives so called the exchange current density (i_O):

$$i_0 = nFk_{o,ox}\left(c^0_{o_2}\right)^{1/2} e^{\frac{-\alpha_{H_2} F E_{eq}}{RT}} = nFk_{0,red}c^0_{H_2}e^{\frac{-\alpha_{O_2} F E_{eq}}{RT}} \tag{38}$$

The express of current density in the PEMFC could be written function of the exchange current density and the potential of the cell:

$$i = i_o\left\{e^{\left[\frac{-\alpha_{H_2} F(E-E_{eq})}{RT}\right]} - e^{\left[\frac{\alpha_{O_2} F(E-E_{eq})}{RT}\right]}\right\} \tag{39}$$

This equation, known as Butler-Volmer equation gives a relation between current density and the potential at the electrode of the PEMFC. Note that the reversible or equilibrium potential at the fuel cell anode is 0 V by definition, and the reversible potential at the fuel cell cathode is 1.229 V (at 25°C and atmospheric pressure) and it should vary with temperature and pressure [2].

4. Materials and components of proton exchange membrane fuel cell

Proton Exchange Membrane Fuel Cells (PEMFC), developed initial in USA by General Electric Company, was used in the U.S. Space Program at the beginning of 1960. The first electrolyte membrane was a polymer based on sulfonic acid substituted polystyrene. Few years later, in 1967, a new proton exchange membranes with better strength and chemical resistance came to improve the PEMFC [27]. This membrane was **Nafion**, from DuPont—a perfluorosulfonic acid membrane, based on polytetrafluoroethylene (PTFE) backbone, grafted with short ethylene-propylene chains terminated with sulfonic acid groups [28]. Until now, Nafion remains the standard proton exchange membrane.

Most of the PEMFCs work in the temperature range of 50–80°C generating between 0.35–0.7 W/cm^2 power densities in hydrogen-air systems at pressure levels varying from ambient pressure to 2.5 bar [29].The power density can be multiplied by coupling more cells in a stack. In this way it is possible to attain the volumetric power density of 1 kW/l, target value for automotive applications.

In the study of the cost and expected future performance of PEMFCs for vehicles, Whiston et al. [30] presents the best cost life times and performances for current fuel cells stacks: costs between 40–500 $/kW with power densities ranged from 0.5 to 4 kW/L and the lifetime varying between 1200 and 12,000 hours.

The architecture of Proton Exchange Membrane Fuel Cells (PEMFC) (**Figure 2**) shows a flat multi-layered structure with a polymer electrolyte membrane core and several specific layers ordered in pairs on both sides of the membrane. The two parts of the PEMFC placed in the mirror on both sides of the polyelectrolyte membrane have a generic name anode and cathode side. The anode side of the fuel cell contains the negative electrode, where fuel (H_2) is oxidized (hydrogen oxidation reaction HOR), while the cathode side is the half cell with the positive electrode, where the oxygen reduction reaction (ORR) occurs.

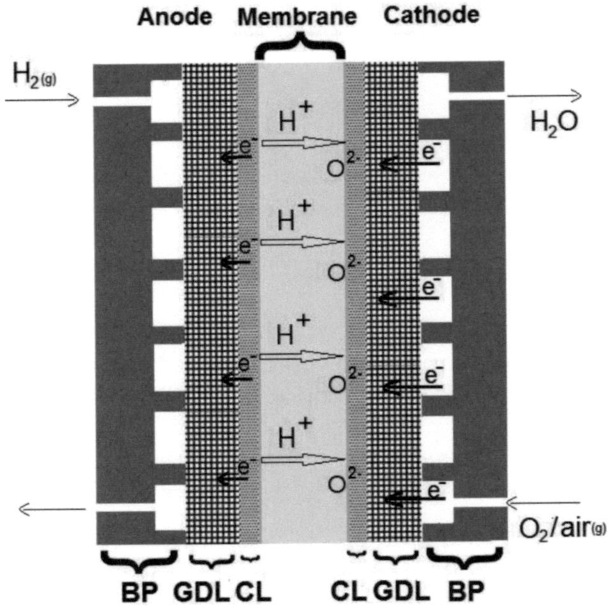

Figure 2.
Schematic presentation of PEMFC.

The role of each component of the PEMFC could be understood starting from the functioning principle of the device and its overall reaction: chemical energy of hydrogen molecule is converted into electricity through electrochemical processes. Using oxygen as oxidant into the REDOX process the only reaction product are water and heat.

Both reactants enter the PEMFC on their side: hydrogen (H_2) by the anode inlet and oxygen (O_2) by the cathode. The gases flows through the channels engraved on the Bipolar Plates (BP). From the flow channels, the reactants pass the Gas Diffusion Layers (GDL) toward the reaction sites in Catalyst Layers (CL). To the anode, hydrogen is oxidized yielding protons (H^+) and electrons. The protons cross the Proton Exchange Membrane (PEM) and the electrons pass through an external circuit, yielding electrical current. On the cathode side catalyst layer, oxygen is reduced to O^{2-}ions by the electrons from external circuit. Oxygen ions react with protons resulting water.

Hydrogen oxidation reaction (HOR): $$H_2 \rightarrow 2H^+ + 2e^- \quad (40)$$

Oxygen reduction reaction (ORR): $$2H^+ + \frac{1}{2}O_2 + 2e^- \rightarrow H_2O + Q \quad (41)$$

Bipolar Plate (BP) connects two adjacent cells in the stack. BP consists in a graphite plate with flow channels engraved on both faces of the plate; one face contains the flow field from cathode side of one cell and the other contains the anode flow field of adjacent cell. Sometimes, the BP is composed by two flow plates from different materials. Flow field provides a pattern for even distribution of reactant gases by in plane direction of the device (**Figure 3**). At the same time it collects and removes the reactants and products into the flow channels, which has less than 1 mm depth and width separated between them by ribs of 1 mm width [31]. Furthermore, the conductive behavior of graphite plates enable electronic conduction and polarization, so that a face of BP has negative charge on the faces toward cathode side of a cell and positive charge to the other face toward the anode side of adjacent cell. This makes possible the serial connection of many cells result a cells stack with additive power character. Also the BP gives the stack solidity.

The Bipolar Plates requirements are [2, 4]:

- impermeable for gases, in order to minimize the gas leakage to 0%

  [$<2 \cdot 10^{-6}$ $cm^3/(cm^2 \cdot s)$];

- high electrical conductivity;

The resistance of the flow plate, including the flow channels should be less than 20 $m\Omega cm^2$;

- good thermal conductivity (>20 W/mK);
- compressive strength (>2 MPa);
- minimal thickness;
- high chemical resistance.

The structure of BP integrates the two flow field, one of each face of the plate, and also the cooling system. The geometry of distribution channels network considerable contributes to the fuel cell performance. Moreover, the material properties such as density and strength are important.

Bipolar plates operates in harsh chemical and electrochemical environment: highly oxidizing environment at the cathode under oxygen/air flux and 1 V potential versus the normal hydrogen electrode, or strong acid solution at the anode, due to the ions present in anode compartment H^+, $SO^-{}_3$, F^-. Therefore, the BP should be resistant to oxidation, to acid and to hydrogen embrittlement.

Flow plates represent a major part in weight from total PEMFC and contribute to 80–90% to the weight and volume of entire stack, determining the power density of a PEMFC stack [29].

Gas Diffusion Layers (GDL) are important components of PEMFC, inserted between the bipolar plate and the catalytic layer on both side of the membrane, in order to ensure an uniform distribution of reactant gases, a properly management

Figure 3.
Bipolar plate pictures.

of water, both, who enters with humidified gases and that produced to cathode by electrochemical reaction.

GDL is composed from carbon fibers orderly arranged in carbon cloth (**Figure 4(a)**) or randomly in carbon paper (**Figure 4(b)**) forming a macro-porous layer with pore size between 10 and 30 μm and thickness between 200 and 400 μm [29]. The carbon fibers are produced mainly by electro-spinning of polyacrylonitryle (PAN) solution. The fibers network is treated with a hydrophobic agent in order to impede water retention. Usually the hydrophobic agent is polytetrafluoroethylene (PTFE) used in proportion of 5–30 wt% from total mass of GDL preserving a large void volume, typically 75–85% [28, 32].

On the inner face of the cathode GDL is deposited a thin, micro-porous layer (MPL), with porosity between 150 and 500 nm, allowing to enhance reactants transport and water removal. MPL consist in a thin layer of carbon black powder mixed with PTFE, with thickness around 30 μm [31], on GDL. MPL control water diffusion to cathode allowing an appropriate amount of water to be held at the interface MPL - catalyst layer to keep the membrane hydrated. MPL is applied also on the GDL from anode side in order to maintain catalyst active layer dried. The difference in permeability between the two layers (GDL and MPL) seems to be responsible to this water management behavior.

Main functions of GDL are:

- transport of reactant gases with uniform distribution toward catalytic sites;
- transport of electrons to (cathode) and from (anode) catalytic sites;
- removing the heat of reaction;
- rigid support for membrane and catalyst layers;
- water management.

Gas transport through GDL could be obstructing by liquid water, therefore the water removal is so important. A good transport rate and uniformity in gases distribution depend on both in plane and through plane gas permeability of GDL. In absence of flooding the oxygen transport is mostly diffusion driven and the permeability of material is sufficient for transport of gaseous reactants. Because the GDL is much thicker than MPL, the permeability of GDL is determinant in reactant gases transport.

(a) (b)

Figure 4.
Picture of the fragments from carbon cloth sample (a) and carbon paper (b).

Transport of electrons and heat occurs through carbon fibers which form GDL. In plane transport of electrons and heat are higher than through plane values. This anisotropy in GDL together different thickness of GDL and MPL are important factors for uniform in plan distribution of heat and potential.

Wet-proofing material increases the contact resistance. On the other hands, due to difference in thickness between GDL and MPL, the latter has a minor effect on the through-plane conductivity.

At the cathode GDL—gas flow channel interface, the interfacial resistance for reactant transport will be significantly increased due to the presence of liquid water. This happens mainly for oxygen transport through cathode GDL, where excess water could be yielded in cathode reaction.

Catalyst layer (CL) is one of the most important components of PEMFC assembly. While the bipolar plates and gas diffusion layers are responsible only for optimal mass, charge and heat transfer, the two catalyst layers together with the membrane, entitled Membrane Electrode Assembly (MEA) constitute the real heart of device. Therefore, catalysts are the center of the entire electrochemical process, playing an important role in accelerating electrochemical cell reactions: oxidation of the hydrogen gas and reducing the oxygen gas to water. The catalyst layer is often named electrode.

In order to achieve this function, the catalyst layer must provide three different pathways: one for reactant gases to catalytic centers, another for ions (H^+) transport from active site to membrane (to anode) and from membrane to active site (to cathode), and the third path is for electrons. The catalytic layers must have appropriate porosity and hydrophobicity for gas transport and water management. Moreover, ion and electron conductors together with the catalyst are mixed in composed material which form catalyst layer. There are three phases and three boundaries in the CL: between the gas phase reactant, the ion conductors and the electro catalytically active sites with electron conduction capacity. The catalyst particles must be in contact directly with the electron conductors therefore, the electro conductive carbon particles are used as support for catalyst nanoparticles. Hydrophobic substances, generally PTFE, serve as a catalyst binder and maintain the hydrophobicity in the catalyst layer, moreover polymer electrolyte, usually Nafion give the way for ionic transport.

With typically thickness between 5 and 100 μm and a porosity of 40–70%, the electrode consists of a metallic catalyst deposited on a carbon substrate, usually Pt/C. The catalytic particles with particle size of 1–10 nm are bonded in a compact layer by hydrophobic binder, usually PTFE, and an ionomer, usually, Nafion, provide pathway for proton transport. Also, the binder confers mechanical stability to catalyst [33].

The carbon-based materials used as catalyst substrates because of their high conductivity include carbon powder, graphite, and active carbon. In the last 10 years, new form of carbon based materials were synthesized and tested as carbon substrate in catalyst layer: nanotubes and graphene. The carbon substrate must have also larger porosity with average pores diameters in the range of mesoporous materials (2–50 nm).

From the initial design of PEMFCs in the 1960th, when the amounts of platinum at the electrode were 35 and 28 mg/cm^2, this value decreased up to day at 0.5–0.1 mg/cm^2 [33]. For decreasing the amount of platinum, while maintain the desired performance of the catalyst material, several properties of the catalyst layer should be optimized including the level of adsorption of the reactants, the hydrophobicity level, and rate of transfer for ions and electrons. In addition, the catalyst durability is an important design constraint.

Because oxygen reduction reaction (ORR) is slower than the hydrogen oxidation reaction (HOR), the cathode catalyst layer requires high catalyst content in order to maintain a tolerable response speed. The catalyst loading depends on thickness of the catalyst layer.

CL can be prepared by several methods, including casting, painting and brushing, injection molding, impregnation, spraying, thin layer deposition and electro deposition.

The first step in producing the electrode is the preparation of the catalytic ink. This must contains the main components of catalyst layer: metallic catalyst deposited on carbonic support, usually Pt/C, dispersed in appropriate solvents mixture, the hydrophobic binder (PTFE) and ionomer (Nafion), both in form of aqueous suspension. The ratio between catalyst particles and binder has major importance in optimization of electrode properties. B. Millington et al. prepare catalyst ink using Nafion as binder in the weight ratio of 2:1 supported catalyst to Nafion [34].

The catalyst ink is then spread on thin layer either on the membrane or on the gas diffusion layer. In the first case, the membrane with the two catalyst layers on both faces forms Catalyst Coated Membrane (CCM), in the second case the GDL with the catalyst layer is named Gas Diffusion Electrode (GDE). By putting together the CCM and the GDL or the GDE with the membrane obtain the Membrane Electrode Assembly (MEA).

However, platinum is an expensive metal catalyst and is easily poisoned. Platinum alloys are used to reduce the content of Pt catalysts without significantly decreasing the catalytic performance. Examples of Pt alloy catalysts that include other metals are Pt-Ru, Pt-Co, Pt-Pd, Pt-Ru-Co, Pt-Co-Cr and combinations of Pt with Fe, Co, Cu and Ni [33].

Proton Exchange Membrane or Polymer Electrolyte Membrane **(PEM)** is somehow a symmetry center of PEMFC and one of the main components of the PEMFC. PEM play the role of electrolytic bridge: it provide a conductive path for the ionic species (H^+, H_3O^+, etc.) and prevent unwanted phenomena: mixing of gaseous reactants and passing the electrons through membrane.

The membrane must fulfill several requirements in order to accomplish the operating functions and to maintain the integrity and stability in the PEMFC aggressive operating environment, including (i) chemical and electrochemical stability under acid and oxidizing environment; (ii) dimensional stability to water-uptake (in the range from liquid water to water vapors) while preserving the mechanical strength and durability; (iii) low gas permeability through membrane; (iv) high proton conductivity in the operating conditions [2, 4]; and (v) low cost and easy to recycle.

But the most important requirement is the highest ion conductivity/low resistance to ion transport. In general, the fuel cell performance is strongly dependent on the proton conductivity through membrane, the **value of proton conductivity for Nafion membranes being around 0.1 S/cm** [35]. The most important factor governing the efficiency of the fuel cell is the concentration of proton donor moieties, such as sulfonic or phosphonic groups. Other important factors that affect the proton conductivity of almost all PEM (except the acid-base complexes—based on polybenzimidazoles and inorganic acids) are the degree of hydration and the ionic channels morphology [36].

Distribution of the ionic channels inside the amorphous polymer matrix results from the polymer microphase separation into hydrophilic and hydrophobic domains. It is very important to control the physical balance between the hydrophilic and hydrophobic domains, because the first domain is responsible for proton, water and/or methanol transport, while the second one confers stability and mechanical strength against serious swelling. The rate of proton transport into PEM is directly influenced by the water content of the membrane.

Several parameters can characterize the protons transport capacity of PEM materials: Ion Exchange Capacity (IEC), Sulfonation Degree (SD) or Equivalent

Weight (EW), and the hydration degree (λ). IEC represents the milliequivalents of sulfonic acid groups per gram of dried polymer and the sulfonation degree (SD) is the average number of sulfonic acid groups corresponding to a monomer unit. IEC and SD were determined by titration method and by calculation, function of IEC, respectively. Also, EW is defined as mass of dried polymers (g) per number of sulfonic acid groups. The hydration degree of electrolyte polymer is often described as the number of water molecules per fixed ionic group (HSO^-_3) [37].

The value of proton conductivity of PEM is determined through potentiostatic measurement of PEM resistivity, either ex situ only for polymer electrolyte film, or in situ, in the MEA or in the PEMFC.

The state of the art proton exchange membrane still remains Nafion membrane from DuPont. Nafion is a perfluorosulfonic polymer, based on polytetrafluoroethylene backbone grafted randomly with short chain. The grafted chain is composed by 2–3 izopropylene oxide units followed by one ethylene oxide unit, capped on sulfonyl acid group (**Figure 5**).

There are several type of commercial Nafion varying by equivalent weight and thickness (**Table 2**).

Many other materials, polymer, polymer blends or polymer composites were tested for proton exchange membrane. Polymers used as proton exchange membrane can be classified in fluorinated and non-fluorinated backbone. Between non fluorinated materials, most of them are high performance sulfonated polyarylene: polystyrene (**Figure 6(a)**) and block polyethylene-block polystyrene, poly (phenylene oxide) (**Figure 6(b)**), poly(ether ketone) (**Figure 6(c)**), poly(ether sulfone) (**Figure 6(d)**), polyimide (**Figure 6(e)**) [39] and all. A lot of variation from this generic structure for example: polyether-ether ketone ketone, polyarylene substituted with different substituent was tested for improving the PEM properties.

For sulfonated polyarylene PEMs the transport of protons depends on the hydration degree, in other world depend on the amount of liquid water in the membrane. These PEMFCs work properly at temperature under 100°C.

However, working at high temperatures, above 120°C, improve catalytic activity and thermal management in PEMFC. An attempt to improve water retention inside the PEM uses the hybrid membranes, with inorganic fillers included inside the polymeric matrix. Moreover, fillers with hydrophilic behavior: silicates, titanium dioxide, zirconium dioxide, heteropolyacids (phosphotungstic, phosphomolibdenic) and carbon nanotube [39], also improve the ion conductivity, the thermal and mechanical properties.

Composite organic–inorganic PEM can be prepared through different techniques, involving direct mixing of inorganic fillers into polymer solution or melt, followed by recasting of the composite film or in situ techniques. Homogeneous distribution of the filler inside the polymer matrix is very important in order to obtain an isotropic distribution of material properties. In this respect, choosing the solvents and mixing procedure could enhance the degree of dispersion of the

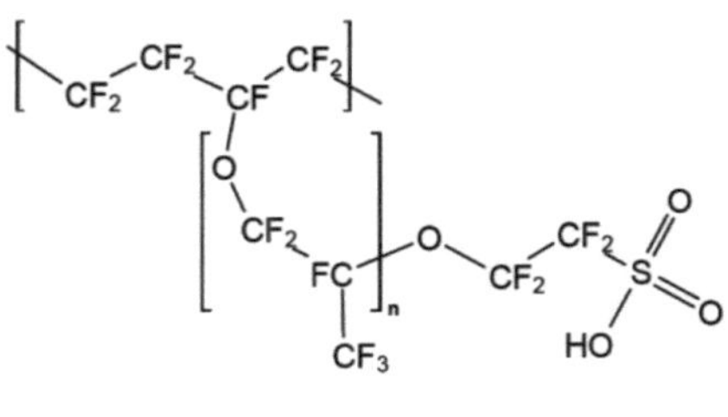

Figure 5.
Structure of perfluorosulfonic ionomer Nafion.

Commercial Nafion membrane	Equivalent weight g/(-SO_3H)	Thickness (μm)
Nafion 112	1100	50.8
Nafion 117	1100	183
Nafion 211	1100	25.4
Nafion 212	1100	50.8

Table 2.
Characteristics of commercial Nafion membranes [27, 35, 38].

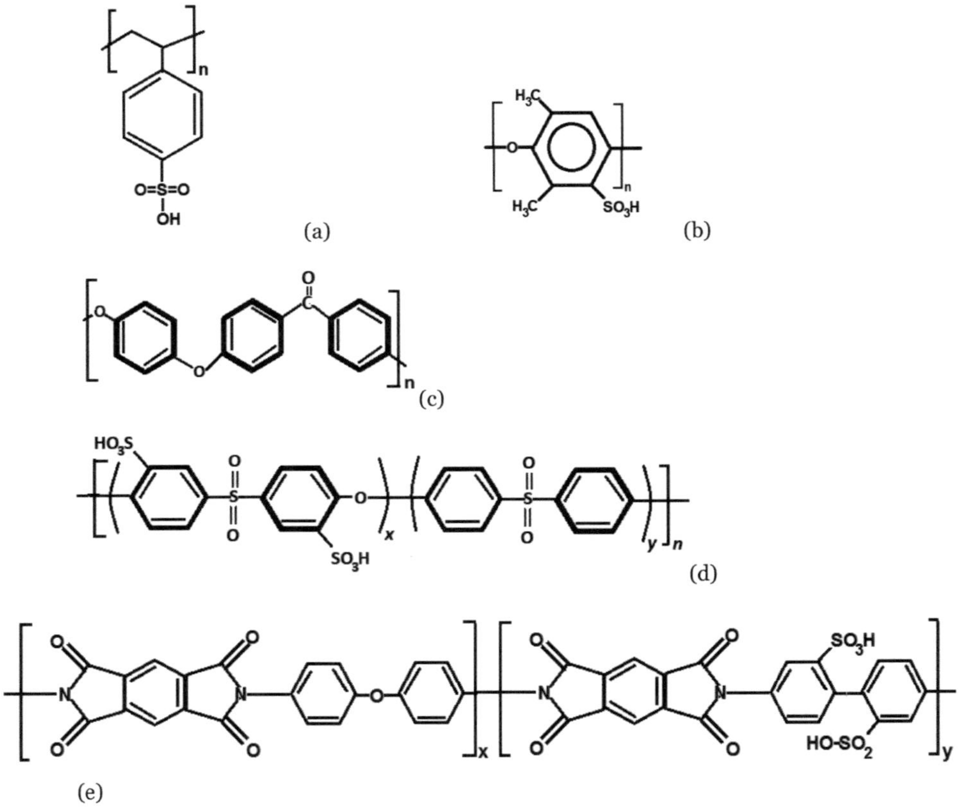

Figure 6.
Structure of several representative sulfonated polyarylene tested for PEM: (a) sulfonated polystyrene, (b) sulfonated polyphenylene oxide (dimethyl substituted), (c) polyetherketone, (d) sulfonated polysulfone, (e) sulfonated polyamide.

nanoparticles in the polymer matrix. Better dispersion depends on chemical compatibility between inorganic filler and polymer, too.

The in situ techniques for composite PEM preparation involve either the inclusion of the nanofiller into the polymer matrix during polymerization from mixed solution of filler—monomer, either the synthesis of the inorganic filler inside the polymer matrix by modified sol-gel technique [36].

Another kind of proton exchange membrane for high temperature PEMFC are acid-base membrane, which has a polybenzimidazole base structure (**Figure** 7) doped with phosphoric acid or heteropolyacids. These membranes expressed proton conductivity at temperature above 100°C even with low humidification or in the dry state.

DOI: http://dx.doi.org/10.5772/intechopen.93032

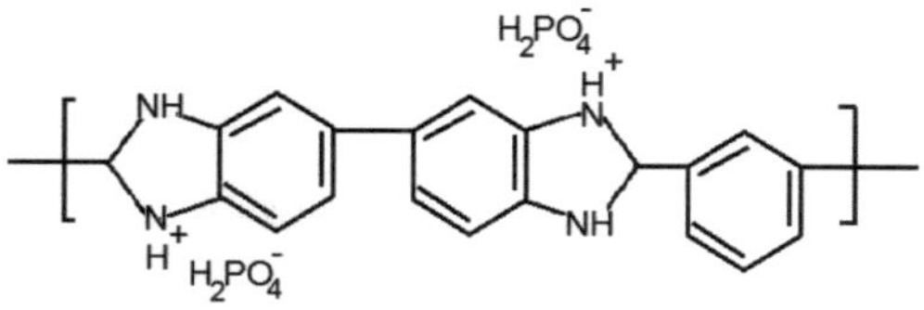

Figure 7.
Chemical structure of polybenzimidazole repetitive unit equilibrated with phosphoric acid molecules.

Amphoteric heterocycles containing nitrogen atoms such as pyrazole, imidazole, triazole, and tetrazole has both proton donors and acceptors behaviors acting as proton carriers even in dry environment [40].

Design of PEM materials is a problem of major importance for tailoring and fit the materials properties with functional requirements. The hydrolytic acid and oxidant environment from PEMFC, variable operational temperature and moistening act as stressors that materials have to deal with.

5. Conclusions

Technologies that produce electrical energy from clean, renewable fuels, with high efficiency and without harmful emissions could resolve the stringent problems of current atmospheric pollutions. In this regard, fuel cells are devices able to convert chemical energy into electricity through efficient electrochemical processes, so much time they are fed with suitable reactants: fuel and oxidant.

PEMFCs are type of fuel cells suitable for mobile and automotive applications due to low operating temperature, sustained operation at a high current density, low weight, compactness, the potential for low cost and volume, long stack life, fast start-up and suitability for discontinuous operation [41]. Compared to plug-in battery electric vehicles, PEMFCs are more clean power sources for automotive. Several cars producers develop Fuel Cells powered vehicles: Toyota Mirai, Honda Clarity and Hyundai ix35 PEMFC [42].

Despite the apparent simplicity of the PEMFCs functional theory, there are some obstacles in commercialization of PEMFCs related to production cost and durability. High cost of platinum catalyst or the Nafion membranes, the instability given by dissolution of platinum and corrosion of carbon support are just a few problems which must be solved.

The cost and performance targets for PEMFC proposed are $30/kW and 3 kW/L respectively [30]. Also, the ultimate durability target for vehicles is 8000 hours. The durability was defined as the time until the stack's rated power reduces to a value that is 10% less than its beginning-of-life rated power under the drive-cycle durability protocol [30].

The priorities for research and development (R&D) reflected by the funding allocated for their development was in order: catalysts and electrodes, performance and durability, membranes and electrolytes, and testing and technical assessment.

There are several challenges to wide-spread commercialization of the technology, including a number of technical barriers to overcome: technical and commercial. Technical challenges are considered increasing durability by reducing degradation, optimize electrode design, improve impurity tolerance at the anode, the water management.

Several aspects could be taking into account in order to reduce the cost of devices for commercial applications: reducing material cost especially the use of platinum, increasing power density, reducing system complexity, and improving durability.

Author details

Irina Petreanu*, Mirela Dragan and Silviu Laurentiu Badea
National Research and Development Institute for Cryogenic and Isotopic Technologies - ICSI, Ramnicu Valcea, Romania

*Address all correspondence to: irina.petreanu@icsi.ro

References

[1] Larminie J, Dicks A. Fuel Cell Systems Explained. 2nd ed. Chichester: Wiley&Sons; 2003. p. 433. DOI: 10.1002/9781118878330

[2] Barbir F. PEM Fuel Cells: Theory and Practice. 2nd ed. San Diego, California: Elsevier Academic Press; 2005. p. 456. DOI: 10.1016/B978-0-12-078142-3.X5000–9

[3] Wang Y, Chen K, Mishler J, Cho SC, Adroher XC. A review of polymer electrolyte membrane fuel cells: Technology, applications, and needs on fundamental research. Applied Energy. 2011;**88**:981-1007. DOI: 10.1016/j.apenergy.2010.09.030

[4] Yu Y, Li H, Wang H, Yuan X-Z, Wang G, Pan M. A review on performance degradation of proton exchange membrane fuel cells during startup and shutdownprocesses: Causes, consequences, and mitigation strategies. Journal of Power Sources. 2012;**205**:10-23. DOI: 10.1016/j.jpowsour.2012.01.059

[5] Carrette L, Friedrich KA, Stimming U. Fuel cells—Fundamentals and applications, 1. Fuel Cells, No. 2001;**1**: 5-39. DOI: 10.1002/1615-6854(200105)1:1<5::AID-FUCE5>3.0.CO;2-G

[6] U.S. Department of Energy. Fuel Cell Handbook. 7th ed. West Virginia, USA: EG&G Technical Services, Inc; 2004. p. 427. Available from: https://netl.doe.gov/sites/default/files/netl-file/FCHandbook7.pdf

[7] Logan BE, Wallack MJ, Kim K-Y, He W, Feng Y, Saikaly PE. Assessment of microbial fuel cell configurations and power densities. Environmental Science & Technology Letters. 2015;**2**:206-214. DOI: 10.1021/acs.estlett.5b00180

[8] Franks AE, Nevin KP. Microbial fuel cells, a current review. Energies. 2010;**3**: 899-919. DOI: 10.3390/en3050899

[9] Santoro C, Arbizzani C, Erable B, Ieropoulos I. Microbial fuel cells: From fundamentals to applications. A review. Journal of Power Sources. 2017;**356**: 225-244. DOI: 10.3390/en3050899

[10] Lovley DR. Bug juice: Harvesting electricity with microorganisms. Nature Reviews Microbiology. 2006;**4**:497-508. DOI: 10.1038/nrmicro1442

[11] Logan BE, Hamelers B, Rozendal R, Schröder U, Keller J, Freguia S, et al. Microbial fuel cells: Methodology and technology. Environmental Science & Technology. 2006;**40**:5181-5192. DOI: 10.1021/es0605016

[12] Rahimnejad M, Ghoreyshi AA, Najafpour GD, Younesi H, Shakeri M. A novel microbial fuel cell stack for continuous production of clean energy. International Journal of Hydrogen Energy. 2012;**37**:5992-6000. DOI: 10.1016/j.ijhydene.2011.12.154

[13] An J, Kim B, Chang IS, Lee H-S. Shift of voltage reversal in stacked microbial fuel cells. Journal of Power Sources. 2015;**278**:534-539. DOI: 10.1016/j.jpowsour.2014.12.112

[14] Oh SE, Logan BE. Voltage reversal during microbial fuel cell stack operation. Journal of Power Sources. 2007;**167**:11-17. DOI: 10.1016/j.jpowsour.2007.02.016

[15] Lobo FL, Wang X, Ren ZJ. Energy harvesting influences electrochemical performance of microbial fuel cells. Journal of Power Sources. 2017;**356**: 356-364. DOI: 10.1016/j.jpowsour.2017.03.067

[16] Han M, Tang X, Yin H, Peng S. Fabrication, microstructure and properties of a YSZ electrolyte for SOFCs. Journal of Power Sources. 2007; **165**(2):757-763. DOI: 10.1016/j.jpowsour.2006.11.054

[17] Harboe S, Sohn YJ, Guillon O, Menzler NH. Investigation of LSM-8YSZ cathode within an all ceramic SOFC. Part I: Chemical interactions. Journal of the European Ceramic Society. 2020;**40**(10):3608-3617. DOI: 10.1016/j.jeurceramsoc.2020.02.004

[18] Dragan M, Enache S, Varlam M, Petrov K. Perovskite-based materials for energy applications [online first]. In: Perovskite Materials, Devices and Integration. Rijeka: IntechOpen Limited; 2020. DOI: 10.5772/intechopen.91271

[19] Enache S, Dragan M, Soare A, Petrov K, Varlam M. Environmentally friendly methods for high quality lanthanum cobaltite perovskite catalyst synthesis. Progress of Cryogenics and Isotopes Separation. 2019;**22**(1):39

[20] Dragan M, Enache S, Varlam M, Petrov K. Perovskite-type material Lanthanum Cobaltite LaCoO3: aspects of processing route toward practical applications. In: Cobalt Compounds and Applications. Rijeka: IntechOpen Limited; 2019. DOI: 10.5772/intechopen.86260

[21] Enache S, Dragan M, Varlam M, Petrov K. Electronic percolation threshold of self-standing Ag-LaCoO3 porous electrodes for practical applications. Materials. 2019;**12**(15):2359. DOI: 10.3390/ma12152359

[22] Buccher IMA, Singh A, Hill JM. Anode- versus electrolyte-supported Ni-YSZ/YSZ/Pt SOFCs: Effect of cell design on OCV, performance and carbon formation for the direct utilization of dry methane. Journal of Power Sources. 2011;**196**(3):968-976. DOI: 10.1016/j.jpowsour.2010.08.073

[23] Hussain J, Ali R, Akhtar MN, Jaffery MH, Shakir I, Raza R. Modeling and simulation of planar SOFC to study the electrochemical properties. Current Applied Physics. 2020;**20**(5):660-672. DOI: 10.1016/j.cap.2020.02.018

[24] Nguyen XV, Chang CT, Jung GB, Chan SH, Yeh CC, Yu JW, et al. Improvement on the design and fabrication of planar SOFCs with anode–supported cells based on modified button cells. Renewable Energy. 2018;**129**(B):806-813. DOI: 10.1016/j.renene.2017.03.070

[25] Soydan AM, Yildiz Ö, Durğun A, Akduman OY, Ata A. Production, performance and cost analysis of anode-supported NiO-YSZ micro-tubular SOFCs. International Journal of Hydrogen Energy. 2019;**44**(57):30339-30347. DOI: 10.1016/j.ijhydene.2019.09.156

[26] Kong W, Zhang W, Huang H, Zhang Y, Wu JXY. Analysis of micro-tubular SOFC stability under ambient and operating temperatures. Journal of Materials Science and Technology. 2018;**34**(8):1436-1440. DOI: 10.1016/j.jmst.2017.12.009

[27] Rao V, Kluy N, Wenbo J, Stimming U. Proton-conducting membranes for fuel cells. In: Handbook of Membrane Separations Chemical, Pharmaceutical, Food, and Biotechnological Applications. 2nd ed. Boca Raton: CRC Press; 2015. pp. 567-614. DOI: 10.1201/b18319

[28] Maiyalagan T, Pasupathi S. Components for PEM fuel cells: An overview. Materials Science Forum. 2010;**657**:143-189. DOI: 10.4028/www.scientific.net/msf.657.143

[29] Bruijn F., Makkus R, Mallant R., Janssen G. F. Materials for state-of-the-ArtPEM fuel cells, and theirsuitability for operation above 100°C. T.S. Zhao, K.-D Kreuer, Trung Van Nguyen. Advances in Fuel Cells. Vol. 1. Amsterdam, The Netherlands: Elsevier; 2007. p. 235–336. 10.1016/S1752-301X(13)60001-0

[30] Whiston M, Azevedo I, Litster S, Whitefoot K, Samaras C, Whitacre J. Expert assessments of the cost and expected futureperformance of proton exchange membrane fuel cells for vehicles. PNAS. 2019;**116**(11): 4899-4904; first published February 25, 2019. DOI: 10.1073/pnas.1804221116

[31] Kulikovsky AA. Fuel cells basics. In: Analytical Modelling of Fuel Cells. 1st ed. Amsterdam: Elsevier; 2010. pp. 1-38. DOI: 10.1016/C2009-0-30566-8

[32] Latorrata S, Stampino PG, Cristiani C, Dotellivier G. Performance evaluation and durability Enhancementof FEP-based gas diffusion media for PEMFuel cells. Energies. 2017; **10**:2063. DOI: 10.3390/en10122063

[33] Majlan EH, Rohendi D, Daud WRW, Husaini T, Haque MA. Electrode for proton exchange membrane fuel cells: A review. Renewable and Sustainable Energy Reviews. 2018;**89**:117-134. DOI: 10.1016/j.rser.2018.03.007

[34] Millington B, Whipple V, Pollet BG. A novel method for preparing proton exchange membrane fuel cell electrodes by the ultrasonic-spray technique. Journal of Power Sources. 2011;**196**: 8500-8508

[35] Product Bulletin P12. Available from: https://www.fuelcellstore.com/spec-sheets/chemours-nafion-115-117-1110-spec-sheet.pdf

[36] Petreanu I, Marinoiu A, Sisu C, Varlam M, Fierascu R, Stanescu P, et al. Synthesis and testing of a composite membrane based on sulfonated polyphenylene oxide and silica compounds as proton exchange membrane for PEM fuel cells. Materials Research Bulletin. 2017;**96**:136-142

[37] Petreanu I, Ebrasu D, Sisu C, Varlam M. Thermal analysis of sulfonated polymers tested as polymerelectrolyte membrane for PEM fuel cells. Journal of Thermal Analysis and Calorimetry. 2012;**110**:335-339. DOI: 10.1007/s10973-012-2442-z

[38] Jones D. Perfluorosulfonic Acid Membranes for Fuel Cell and Electrolyser Applications. Available from: https://www.sigmaaldrich.com/technical-documents/articles/materials-science/perfluorosulfonic-acid-membranes.html

[39] Mishra AK, Bose S, Kuila T, Kim NH, Lee JH. Silicate-based polymer-nanocomposite membranes for polymer electrolyte membrane fuel cells. Progress in Polymer Science. 2012;**37**: 842-869. DOI: 10.1016/j.progpolymsci.2011.11.002

[40] Du M, Yang L, Luo X, Wang K, Chang G. Novel phosphoric acid (PA)-poly(ether ketone sulfone) with flexible benzotriazole side chains for high-temperature proton exchange membranes. Polymer Journal. 2019;**51**: 69-75. DOI: 10.1038/s41428-018-0118-7

[41] Jourdani M, Mounir H, Marjani A. Latest trends and challenges In proton exchange membrane fuel cell (PEMFC). The Open Fuels & Energy Science Journal. 2017;**10**:96-105. DOI: 10.2174/1876973X01710010096

[42] Liu W, Peng Z, Kim B, Gao B, Pei Y. Development of a PEMFC dynamic model and the application to the analysis of fuel cell vehicle performance. IOP Conference Series: Materials Science and Engineering. 2019;**628**: 012006. DOI: 10.1088/1757-899X/628/1/012006